NATIONAL ENERGY PLANNING AND MANAGEMENT IN DEVELOPING COUNTRIES

ISPRA — COURSES

ON ENERGY SYSTEMS AND TECHNOLOGY

A series devoted to the publication of courses and educational seminars given at the Joint Research Centre, Ispra Establishment, as part of its education and training program. Published for the Commission of the European Communities, Directorate-General Information Market and Innovation.

Already published:

ENERGY STORAGE AND TRANSPORTATION

PROSPECTS FOR NEW TECHNOLOGIES

Edited by G. BEGHI

1981, x + 497 pp. ISBN 90-277-1166-6

THERMAL ENERGY STORAGE

Edited by G. BEGHI

1982, vi + 505 pp. ISBN 90-277-1428-2

PERFORMANCE OF SOLAR ENERGY CONVERTERS: THERMAL COLLECTORS AND PHOTOVOLTAIC CELLS

Edited by G. BEGHI

1983, viii + 527 pp. ISBN 90-277-1545-9

Additional volumes in preparation

The publisher will accept continuation orders for this series which may be cancelled at any time and which provide for automatic billing and shipping of each title in the series upon publication. Please write for details.

NATIONAL ENERGY PLANNING AND MANAGEMENT IN DEVELOPING COUNTRIES

Lectures delivered at a Management Training Seminar held at the Joint Research Centre, Ispra, Italy, May 3-14, 1982

Edited by

H. NEU and D. BAIN

Systems Analysis Division,
Joint Research Centre, Ispra, Italy

D. REIDEL PUBLISHING COMPANY

A MEMBER OF THE KLUWER ACADEMIC PUBLISHERS GROUP

DORDRECHT / BOSTON / LANCASTER

Library of Congress Cataloging in Publication Data

Management Training Seminar (1982 : Joint Research Centre, Ispra, Italy)
National energy planning and management in developing countries.

(Ispra courses on energy systems and technology and information science)
1. Underdeveloped areas–Energy policy–Congresses. 2. Underdeveloped areas–Power resources–Congresses. 3. Underdeveloped areas–Energy consumption–Planning–Congresses. 4. Underdeveloped areas–Energy consumption–Management–Congresses. I. Neu, H., 1920- . II. Bain, D., 1944- . III. Title. IV. Series.
HD9502.A2M343 1982 333.79'09172-4 83-4407
ISBN 90-277-1589-0

CIP

Commission of the European Communities Joint Research Centre Ispra (Varese), Italy

Publication arrangements by
Commission of the European Communities
Directorate-General Information Market and Innovation, Luxembourg

EUR 8498 EN

Published by D. Reidel Publishing Company
P.O. Box 17, 3300 AA Dordrecht, Holland

Sold and distributed in the U.S.A. and Canada
by Kluwer Boston Inc.,
190 Old Derby Street, Hingham, MA 02043, U.S.A.

In all other countries, sold and distributed
by Kluwer Academic Publishers Group,
P.O. Box 322, 3300 AH Dordrecht, Holland

Printed in The Netherlands

TABLE OF CONTENTS

EDITORS' INTRODUCTION

While much has been written on different aspects of energy and development, there exists little in the way of material which attempts to integrate all the various elements necessary for effective energy planning and management. This book results from an attempt to remedy that deficiency.

The contributors represent a variety of disciplines and approaches, with a particular effort having been made to bring together experts from the different national traditions in the European Community member countries. While there are important contributions from the developing countries themselves, the emphasis is on the European experience of energy planning and management and the lessons to be learned from this for countries where the energy system is still dominated by traditional fuels or where the transition to a fully developed energy infrastructure is still in its early stages.

There can be no question of a simple transfer of experience since the national energy systems of the industrialized countries evolved in a more-or-less ad hoc fashion, with the suppliers of different fuels making their investment decisions in isolation from each other and with little in the way of co-ordinated national energy strategy. Only the belated recognition of the vulnerability of such non-systems to events such as the OPEC oil price shocks persuaded the developed countries to make a considerable intellectual investment in studying the dynamics of energy systems and in creating energy management techniques and planning tools.

From the point of view of the developing countries, the hope must be that this new energy planning and management research is even more beneficial to them, starting as they do from a much earlier stage in energy system development, than to the countries for which it was undertaken. This is not to say that the energy problems of developing countries can be solved easily - particularly since they are inseparable from the range of other severe and well known crises confronting presently the developing world - but at least they can, in principle, be tackled more systematically and with greater understanding than was possible in Western Europe during the process of industrialization.

H. Neu and D. Bain (eds.), National Energy Planning and Management in Developing Countries, vii–xi.

This book does not pretend to be an encyclopedia or handbook of energy planning and management. It is rather a compendium of different pieces of information which may be of assistance to Third World energy policy-makers. Information, in this sense, comprises not simply data or "scientific knowledge" but also the informal and contextual knowledge essential to the practical operation of energy systems. If the material presented here constitutes a valued information source in this braod sense, it will have fulfilled the editors' hopes for it.

The contributions published here were presented at a seminar on "Energy Planning and Management in Developing Countries", held at the Commission of the European Communities, Joint Research Centre, Ispra Establishment, in May, 1982. The seminar was one module in a programme aimed at (a) facilitating a North-South joint learning experience and (b) providing a management training exercise (see Annex I for a fuller description).

While the structure of the book is necessarily loose, the different contributions fall naturally into two parts. In Part I the range and dimensions of the energy problems confronting developing countries are analysed from different standpoints, while, in Part II, the emphasis is on tools, techniques and institutional arrangements as aids to energy policy-makers. Given the variety of academic and practical backgrounds of Third World energy officials - comprising engineers, economists and administrators - the level of detail and of explanation may seem at times to verge on the elementary. Yet what may be a statement of the obvious to an economist (e.g. the need to take full account of opportunity costs) may well be a new concept for an electrical engineer, and *vice versa*. We would expect most readers to encouter a certain amount of familiar material but we would urge them to pass this over in favour of other concepts which are new to them, although regarded as commonplace by other readers.

The following summary will give readers a brief indication of those contributions which are likely to interest them most.

In Part I, Ramesh *Bhatia* gives an economist's view of the dimensions of the problems facing developing countries as a result of oil price increases. He draws special attention to the way in which the energy cost problem interacts with other aspects of development to create a particularly intractable situation.

Gerald Foley concentrates on the fuel-wood crisis, emphasizing how its impact differs according to varying levels of subsistence and according also to varying degrees of commercialization in local economies. Drawing on his experience of devising "low energy strategies" he indicates, through detailed analysis of consumption patterns, technologically feasible solutions to the rural energy problem.

Ariane Van Buren takes up Foley's argument that commercialization of the traditional fuel sector is an essential starting-point for resolving the rural energy crisis and builds this into a case for making the transition to cash transactions the focus of analysis in energy and development studies. In so doing she proposes not only a new analytical perspective but also a strategy designed to improve both institutional efficiency and user control.

Jyoti Parikh also concentrates on the rural energy sector and explores the implications not only of commercialization but also of structural changes in demand. A large part of her contribution is devoted to a detailed economic and technical evaluation of the minimum conditions necessary for the introduction of various energy technologies, both conventional and renewable.

The next two contributions deal with the very different problems posed for energy planners and managers by the phenomenon of rapid urbanization. Bertrand Chateau draws on his substantial experience of analysing energy demand patterns in the urban sectors of industrialized countries to suggest guidelines for assessing the continuing impact of urbanization on Third World energy needs. In his contribution Oliviero Bernardini underlines the fact that urbanization with its creation of new needs and a new life-style, can often make nonsense of statistics for energy consumption per head of population or energy consumption per unit of production since the aggregated figures conceal radically different demand for commercial energy. Despite this, however, he demonstrates that the urban and rural energy crises are ultimately interdependent since only strategies designed to help the rural areas to retain an expanding population can solve the crisis of exponential growth of demand for commercial energy in the cities.

The last two papers in Part I examine energy prospects from the supply side. Both present detailed challenges to the conventional wisdom of only a few years ago. Paul Basile, for his part, demonstrates that fossil fuels are abundantly available for the foreseeable future, at least on a global basis. The case for a re-examination of the merits of conventional fuels is also made, indirectly, by Markus Fritz who employs extensive survey material to refute the idea that new and renewable energy sources are particularly suited to the needs of developing countries, at least until such time as the various technologies concerned have matured.

In Part II the non-specialist will find, in Helmut Neu's contribution, a necessary introduction to the consistent use of energy statistics, as developed by the Statistical Office of the European Communities. Analytical aids such as energy chains, national energy balances and energy flow diagrams are discussed in some detail.

Gerald Leach concentrates on the detailed analysis of energy end-use, site by site, which forms a necessary adjunct to the highly aggregated information contained in national energy balances. He provides numerous examples of how such a "bottom-up" approach can suggest technical improvements or fuel substitutions not apparent from a "top-down" view of energy demand and supply systems.

In the first of his two papers Pierre Vernet describes a method adopted by the Latin American Energy Organization (OLADE) for creating and using energy balances as a planning tool. Vernet's second paper deals with the problems involved in gathering accurate energy information in developing countries and is based on his personal experience in Ecuador.

The collection, systemization and harmonization of such energy data as does exist is a task which has attracted much recent attention from international agencies. One example of the resulting activity is the EEDB data bank developed by the Vienna-based International Atomic Energy Agency (IAEA), This data-bank, which covers both economic and energy information for a large number of countries (including developing countries) is described by Jean-Pierre Charpentier in the first of his two contributions.

Charpentier's second contribution consists of a review of econometric approaches to energy planning together with brief descriptions and evaluations of several of the energy models (not solely econometric models) currently available. With policy-makers in developing countries under increasing pressure to use various energy models Dominique Finon's paper serves a consumer protection function. Himself an experienced model-builder, Finon demolishes many of the exaggerated claims and expectations associated with energy modelling, yet offers constructive advice on what models can, in fact, achieve.

One of the most widely-used modelling approaches used in Europe is that incorporated in the MEDEE series of energy demand models. Bruno Lapillonne, one of the creators of the MEDEE approach, gives an account of its origins and development, together with an assessment of how it may be adapted to assist energy planners in developing countries. With its capacity to create scenarios based on various hypotheses regarding future energy end-use patterns, the MEDEE approach is particularly favoured for long-term energy planning.

For the shorter term, however, and particularly as regards short-term energy demand management, conventional economic analysis remains the key tool. Ramesh Bhatia and Mohan Munasinghe both employ such analysis - the former from the standpoint of political economy, the latter adopting a more econometric approach - to suggest energy management strategies based on price-regulation, taxation, subsidies and controls. In his second paper Munasinghe also builds on standard economic theory in his exposition of project analysis techniques applicable to energy policy-formation, this time from the supply perspective.

In the final paper Sergio Garribba examines the problem of actual implementation of plans, management strategies and other measures. The institutional framework for effective energy planning and management is at the heart of the energy problem (and may, in some cases, be a large part of the problems in developing countries). Garribba's analysis of the situation in Kenya and his detailed proposals for institutional frameworks elsewhere provide a starting point for considering what organizational changes, if any, are required in any particular country.

Acknowledgements

This book, together with the programme on which it is based, was made possible by the support and financial assistance of the European Commission's General Directorate for Development and the General Directorate for Energy. Particular thanks must go to Mr. **Bernard Henry**, Manager of the Education and Training Programme at the Joint Research Centre's Ispra Establishment, and his staff in the Ispra Courses Secretariat, all of whom contributed greatly to the success of the programme.

In planning the structure and the content of the seminar we benefitted from the advice of colleagues in the Joint Research Centre and elsewhere in the European Commission and of experts from a large number of outside institutions. The list of individuals is too long to print in full but we feel that it is only fair to make special mention of the dynamic, stimulating and highly professional role played by **Gerald Foley** and **Ariane van Buren**, from the International Institute for Environment and Development (IIED) in London, not only in coordinating the workshops but in all aspects of the programme conceptualization.

Our special thanks go also to Mrs. **Dorit Schlittenhardt**, our divisional secretary, without whose high organizational abilities and endless patience this book would never have seen the light of day.

Above all we are indebted to the contributors, not only for the content of their papers, but also for their readiness to revise texts and submit them in the form of camera-ready copy. We are especially appreciative of the efforts in this respect of those contributors - the majority - for whom English is a foreign language.

Ispra, December 1982 HN, DB

ENERGY AND DEVELOPMENT: DEMAND, CONSTRAINTS AND POLICY OPTIONS

Ramesh Bhatia

Institute of Economic Growth, University of Delhi, India.

The interrelations between energy and development can be discussed within the following three dimensions:

(i) Changes in energy consumption as determined by the rate and pattern of development;

(ii) Impact of higher energy costs or non-availability on growth, employment, income distribution and regional development;

(iii) Implications of alternative scenarios of meeting demand for energy for "basic needs" and development.

ENERGY CONSUMPTION AND DEVELOPMENT

Demand for commercial energy grew rapidly in the developing countries during the 1960s and, although more slowly, during the 1970s. Annual growth rates varied considerably between countries depending upon a number of demographic and economic factors. Using time-series data for a number of countries, it has been shown that demand for commercial energy depends to a large extent upon real GDP but there are other important influences besides income. These include (a) changes in the structure of the economy since different sectors have different levels and patterns of energy requirements, (b) degree of urbanization because this affects household demand for energy and (c) availability of different sources of primary energy and forms of secondary energy. The empirical results show that income elasticity of energy demand in developing countries is substantially higher than unity, in the neighbourhood of 1.3

H. Neu and D. Bain (eds.), National Energy Planning and Management in Developing Countries, 1–10.

on average. This means that a 1% increase in GDP is associated with a 1.3% increase in consumption of energy, on average. In the case of oil, the income-elasticity coefficient is higher, around 1.8, on account of use of oil in transport, agriculture and power generation.

Thus, based on past data, development seems to result in growth of consumption of commercial energy, particularly oil, which is higher than the growth of GDP. However, it may be noted that past data reflect choices of development strategies which were based on a low price of oil and need not be followed in the future when the basic assumptions regarding cost of oil have changed. Although energy aspects may not determine the development strategy, energy considerations may be highly relevant to specific choices connected with urbanization, location of industries and industrial policy. Examples may be found in industrial policy concerning mix of products, in technology policy concerning modes of production, and in regional policy concerning the location of economic activity. Energy issues are also major factors in decisions about level and pattern of urbanization and about systems of transportation.

It may also be necessary to raise a question regarding life styles in developing countries. It is widely recognized that these developing countries need not follow the Western life styles which are associated with high energy intensities of food production and distribution, personal transport and high consumption. However, most of the developing countries are following so-called Western life styles without being aware of the consequences. It may be necessary to reappraise the philosophical basis of all economic activity and review the fundamental assumption of Western societies that human welfare depends mainly on consumption of goods and services. High energy costs (and shortages) may be used as a catalyst to sponsor such a process of development and life styles where the true happiness of human beings is the ultimate goal of all human endeavour.

Then, there is the question of providing energy to relatively poorer sections of the society living in rural areas of the Third World. A large number of countries depend on so-called non-commercial sources of energy to the extent of 70-90% of total energy requirements. Large populations in rural areas in these countries primarily depend on fuelwood, charcoal, agricultural residues and dungcakes for meeting their requirements of cooking and heating; and, human and animal power for crop production, irrigation, agroprocessing and rural transportation. All these energy resources are used at very low levels of efficiency resulting in what can be termed as "avoidable wastages". The need for additional food, as well as the energy required for cooking it, has led to deforestation in many areas.

Although requirements of timber and construction of new projects have added to the process of deforestation in many regions, use of fuelwood and charcoal for cooking has also made its contribution. Besides the environmental consequences, sheer social costs of collection and transportation of fuelwood by poor people should provide enough incentives for a programme of improved efficiency in use of traditional fuels. In addition to these, programmes of making commerical fuels (e.g. soft coke for cooking, kerosene for lighting), available at a reasonable price would be required for meeting the "basic needs" of these populations.

ENERGY AS A CONSTRAINT TO DEVELOPMENT

The non-availability and high cost of energy may act as a constraint to development both directly and indirectly. This may show up in adjustments that the country may have to make in the short-run as well as in the impact of higher cost of oil imports on other development imports resulting in a slow-down of growth in the medium and the long term. As has been noted by the World Bank reports, the cost of net oil imports of oil-importing developing countries (OIDCs) has risen almost ten-fold in the past decade, i.e. from 5.4 billion (1980) US dollars in 1970 to 49.3 billion US dollars in 1980. Cost of net oil imports in 1980 was almost 18 billion US dollars higher than in 1975 although the quantum of oil imports was very similar in the two years. Thus, although the OIDCs attempted to restrict growth of oil imports, their import bills increased significantly (50-60%) on account of increases in real price of oil in 1979-81. In 1980, their net fuel import bill amounted to 5.3% of GDP (74 billion dollars) compared with 2.8% of GDP in 1978. A large part of these increases in the deficits has been financed by drawing from their reserves and by short-term borrowings which have adverse impact on development in the medium and long term.

According to the World Development Report 1981, middle-income, semi-industrialized countries relied mainly on external financing and structural adjustments (export promotion and import substitution) to meet the external shocks. Almost one-third of the balance-of-payments accommodation to the changing environment occurred through slower growth. Primary producing countries resorted to substantial external financing to reduce the adverse impact on growth. For populous South Asia, workers' remittances and aid accounted for the adjustment process and adverse effects on growth could be avoided. In the case of least developed countries, relying on a few crops for their export earnings, adverse export price effects were significant. Most of their adjustments were through additional real external financing.

Although adverse impact on growth has been reduced through external financing in the short-run, this would have resulted in increases in the external debt of OIDCs. Total external debt as well as external debt service has substantially increased in almost all the South and South East Asian countries over 1972-78. The impact of these borrowings will be felt when debt servicing becomes a substantial part of export earnings, thus reducing the import of raw materials and machinery.

In a detailed study of the effects of the rise in oil prices on four South Asian countries, it has been pointed out that the rise in oil prices resulted in substantial loss of real incomes in India, Sri Lanka, Bangladesh and Pakistan. The estimates of loss ranged between 1.5 - 2.5% of GDP for India, 2.5 - 6% for Sri Lanka, 2.5 - 8% for Pakistan and 4 - 13% for Bangladesh. It was also found that this loss of income was two to three times the average net overseas official loans received by Sri Lanka and India while for Pakistan the loss and loans were of the same order of magnitude.

There are not many empirical studies on the impact of higher oil prices on employment, income distribution and regional development. The available studies show that the factors that contributed to a small fall in employment could be traced back, at least partly, to the difficult balance of payments situation arising from higher oil prices. While employment in major industries declined, it increased in energy industries competing with oil, namely coal, gas and electricity. Similarly, no assessments have been made on the impact of higher prices on income distribution and regional development. For the four South Asian countries studied, the impact of the price rises was probably broadly progressive, at least in urban areas, on account of changes in prices of gasoline, kerosene, diesel oil and electricity. However, to the extent kerosene prices were raised, it would have adversely affected the welfare of poorer sections in rural areas for whom kerosene accounts for a major portion of out-of-pocket cash expenses. Higher transport costs could encourage development of local industries thus reducing concentration of economic activity. At the same time, higher costs of transporting raw materials and final products may go against location of industries in backward areas. There is a need for empirical investigations in this important area.

Perhaps, it is too early to make precise estimates of the medium and long term impacts of high oil prices on OIDCs and situation would also vary considerably from one country to the other. However, one conclusion seems to be unavoidable and that is that the OIDCs should make substantial efforts towards reducing their dependence on oil imports if they wish to minimize the adverse growth impact of higher oil prices. This is discussed at length in the next section.

POLICY OPTIONS

The policy options available to the OIDCs may be grouped under four heads:

- (i) Improving the efficiency of use of both commercial and non-commercial sources of energy;
- (ii) Substituting oil by other forms of energy and/or other factors of production such as labour and capital;
- (iii) Increasing the production of commercial energy;
- (iv) Promoting the use of new and renewable sources of energy both in urban and rural areas.

Efficient use of energy

Estimates by the World Bank and many individual studies show that there is considerable scope for increasing the efficiency of use of energy in developing countries. Potential fuel savings are estimated to range between 20 - 25% in industrial units, trucks/buses, irrigation pumps, kerosene stoves, etc. Some of these savings could be achieved through what is known as implementation of good "housekeeping" measures. However, the experience in some countries, notably India, shows that implementation of energy conservation programmes requires an integrated approach which is generally lacking in developing countries. The success of energy conservation measures, either through higher prices or through legislation, depends on the actions of a large number of consumers: households, entrepreneurs, managers, transporters, traders and government officials. These consumers respond differently to different signals and react in a variety of ways, usually unknown. The consumer is to be made aware of the existence and financial viability of a variety of measures for fuel conservation. He has to be convinced that these measures are, in fact, effective and these technologies would work. Even when he is convinced of the technical and financial viability of the measure involved, he has to be provided with an incentive to adopt that measure along with the provision for financial assistance. All these ser-

services would require a well-coordinated organizational set-up able to provide the necessary extension service, financial inputs and after-sale service with respect to new technologies.

Available experience in the promotion of improved stoves in the countryside also underlines the inherent difficulties in finding solutions which are in conformity with life styles, cooking habits, food consumption patterns, allocation of time of household members to various activities, etc. Even a moderate success requires intensive development and extension work which is difficult to duplicate on as a large scale. In some cases, improving the efficiency of use of energy may conflict with other objectives such as promotion of small scale industries, employment and development of backward regions. For example, if energy-efficient diesel engines, kerosene stoves and lamps were to be subsidized by the government, it would result in the closure of a large number of small scale units spread all over the country. Such trade-offs will have to be properly evaluated before any energy conservation measure can be pushed by the government.

Inter-fuel and inter-factor substitution

Some of the adverse impacts of higher oil prices could be reduced if oil could be substituted by other forms of energy (coal, oil, gas, renewables) or other factors of production such as labour and capital. In fact, if labour could be substituted for energy on a significant scale, directly or indirectly, this would be a beneficial aspect of the oil price hikes. However, the evidence in this regard is rather mixed, although there are areas where labour could be a good substitute for energy. However, there is urgent need to study interfuel and inter-factor substitution possibilities at a disaggregated level, say, for a sub-industry or a manufacturing unit. This becomes crucial because in many developing countries, the existing market prices for capital, labour, foreign exchange and energy products may not necessarily reflect their "true opportunity costs" owing to market imperfections and/or administrative controls (minimum wages, subsidized capital or ceilings on interest rates, import quotas, etc.). Given these distortions, it may be necessary to know that changes in relative prices of different fuels as well as relative price of energy to other factors of production do not encourage substitution which is not desirable from the viewpoint of the society.

One other factor which reduces the effectiveness of any policies of conservation or interfuel substitution is that energy costs form a small proportion of the total costs in the case of many industrial units, trucks, buses and railways. Moreover, the suppliers of these products/services are in a position to pass on the increases in fuel costs to the consumers and, hence, have no incentives to allocate scarce resources of management and skilled labour to energy conservation. Besides, user choices among alternative energy sources may be determined by factors other than relative prices and the important considerations may be auxiliary costs of handling and storage, tax/subsidies on investment, security of supply, uniformity of quality and the effect of fuel-mix on the quality of processed materials. Thus, like improving efficiency of use of energy, interfuel and interfactor substitution has also inherent difficulties due to rigidities in the developing countries.

Increasing the production of commercial energy

The available estimates show that in the OIDCs, commercial primary energy production would increase from 7.8 mbdoe in 1980 to 15.2 mbdoe in 1990. The average annual percentage growth during 1980-90 is 6.9% for OIDCs compared with 3.9% for the world as a whole. Production of oil and gas together is estimated to increase from 3.5 mbdoe in 1980 to 6.1 mbdoe in 1990. The capital requirements for additional oil production are of the order of 40 billion dollars (in 1980 dollars) which are large in relation to expenditures in recent years. The economic rate of return on domestic oil production, given present and projected international oil prices, is likely to be reasonably high. Hence, OIDCs would find it economic to increase domestic production subject to financial, technical and managerial constraints. However, in view of the limited oil resources available, it may not be in their interest to raise production levels substantially during the next decade. A step-up in domestic production may solve the energy crisis in the short run but may result in fast depletion of exhaustible reserves. In such cases, production increases may be constrained by additions to reserves through new discoveries of oil and gas fields.

Besides domestic considerations, the OIDCs may find it difficult to raise financial resources for investments in oil and gas sectors. Of the total external loan disbursements of 7.3 billion dollars to OIDCs for commercial energy development in 1978, oil accounted for only 1.42 billion dollars, the bulk going to the electric power sector. Even in the case of private loan disbursements, the share of oil was 22.5% (1.23 billion dollars out of 5.45). The projections of investment requirements in the commercial energy sector in OIDCs (annual averages during

1986-90) are: 40 billion dollars in electric power; 4.7 billion dollars in oil; 2.7 billion dollars in gas and 1.5 billion dollars in coal. These annual investments are almost double the estimated investments in 1980. These are high levels of investment and existing institutional arrangements may not be in a position to provide all the external finances required. For example, the estimates of World Bank lending programmes show that over a 5-year period of 1981-85, the desirable lending programme for oil and gas would be 8 billion dollars as against the total project cost of 24 billion dollars. With such high investment requirements for the proposed programme, it is difficult to envisage that OIDCs would be in a position to find resources for additional production of commercial energy such as coal, oil and electricity. Besides, "Oil Proliferation", i.e. the strategy of investing large amount of capital in exploring for new petroleum resources in non-OPEC developing countries, is likely to face serious political, manpower and financial constraints.

Role of new and renewable energy sources

New and renewable energy sources can be grouped into three categories: (i) solar, wind and hydro power; (ii) liquid and gaseous fuels from biomass, and (iii) fuelwood plantations. The role of these energy sources in the energy transition over the next 15 - 20 years will depend on the extent to which solutions to various technical, economic, social and organisational problems can be found. Although each of these sources could provide effective substitutes for oil in agriculture, industry, household and transport sectors, the experience in the past few years in developing countries does not seem to be very encouraging. A detailed review of this experience in several countries would be outside the scope of this paper. However, a brief overview would suggest that the impact of new and renewable energy sources in substitution for oil has been rather marginal in many developing countries, with the possible exception of China. The following factors could be listed as possible reasons for this slow growth of renewable energy sources:

(i) High cost technologies

Some of the available technologies such as photovoltaic cells, windmills, micro-hydro stations, geothermal energy, solar water heaters, etc. are either too expensive in terms of capital costs or are too specific to particular sites. The capital costs of photovoltaic systems are relatively much higher than alternative conventional systems. In a comparison of energy alternatives for irrigation pumping in India, it was found that the costs of photovoltaic cells would have to be reduced to 30 cents per peak watt (as compared to a 1980 price of 7 - 8 dollar per peak watt) to make them competitive with a system using biogas and diesel. Some studies have indicated that the potential for wind in Africa, India, Indonesia is not significant because of high costs or low wind speeds and wind energy can be considered only a location-specific resource for mechanical drive and electrical energy. In the case of micro-hydro units, the judgements are that the development of this source is likely to be held back by the scarcity of managerial and engineering talent.

(ii) Technologies requiring high organizational input

The utilization of the potential of some of the renewable energy sources requires a highly organized set-up which takes care of the technical, economic and social aspects of development and use of these resources. Biogas systems, solar cookers and fuelwood plantation (or social forestry) schemes fall in this category. Available studies show that the success of these schemes would depend on the managerial and organizational framework in which they are set.

(iii) Technologies not studied in depth

There are some technologies which have either not been adequately developed so far or have not been studied and analysed in depth. "Producer gas" from crop residues on a small scale for irrigation pumping belongs to the first category, while ethanol from sugarcane/cassava belongs to the second. There is a need for R+D and field-testing work, which is required for "producer gas" as well as in-depth analysis of impacts of large-scale fuel alcohol(ethanol) programmes on the balance of payment, crop substitution, food prices and employment.

Although most of these renewable sources would be available, it is very unlikely that they would have any substantial impact on oil imports in OIDCs. This is not an argument against encouraging the use of renewables, but a realistic assessment of the role these resources can play in reducing foreign exchange deficits in the medium term.

CONCLUSIONS

The above discussion shows that there are no easy options available to the oil-importing developing countries in the medium term, say, 15 to 20 years. Efforts would have to be made on all fronts to avoid the adverse effects on balance of payments, growth and employment. These would include negotiating with OPEC for a concessional price for LDCs, rescheduling of their external debt, and obtaining additional funds for exploring and producing more commercial energy. On the domestic front, it would require creating organizations which provide an integrated package of services required for conservation and interfuel substitution; and, a concerted strategy towards development and diffusion of new and renewable sources of energy. A comprehensive social benefit-cost analysis of alternative technologies/programmes may be used in ranking the available options. The urgency for taking appropriate decisions is common for all oil-importing developing countries even though the actual policy-mix may vary from country to country.

RURAL ENERGY PLANNING IN DEVELOPING COUNTRIES: A NEW FRAMEWORK FOR ANALYSIS

Gerald Foley
EARTHSCAN, London W1P0DR, Energy Information Programme; Senior Visiting Research Fellow, Beijer Institute of the Royal Swedish Academy of Sciences, Stockholm.

ABSTRACT

A schematic structure for the rural fuel economy is presented. The static and dynamic characteristics of each sector are described. The analysis is then applied to a number of practical examples to help elucidate the reasons for past programme failures and suggesty methods of approaching the design of future rural energy programmes.

There is, at the moment, a reasonable agreement on the broad outlines of the rapidly worsening energy problems of the rural areas in the developing world. The worst aspect is the depletion of fuelwood, with the attendant effects of land-erosion, environmental degradation, and increased time spent collecting fuel. Other effects are the use of dung and vegetable residues as fuel, instead of returning them to the land. Global energy problems also penetrate to the rural areas in the form of rising prices and a frequent unavailability of diesel oil, kerosene and petrol.
Many efforts have been made to deal with these problems, but little, in practice has been achieved. Few governments in the poorer developing countries have national energy policies; fewer still have policies which are realistic and effective; and even fewer have any energy policies at all for their rural areas. Such initiatives as have taken place have, therefore, lacked a policy framework and tended to be sporadic and uncoordinated. Most have also relied heavily on financial and

H. Neu and D. Bain (eds.), National Energy Planning and Management in Developing Countries, 11–39.

technical assistance from abroad, and hence been subject to changing western views on development policy and its proper aims and focus.
Awareness of the failure of conventional development programmes to reach the poorest among the rural populations of the developing world has been growing rapidly over the past decade and a half. With the notable exception of Korea /1/, few rural electrification programmes, for example, have made any substantial impact on rural energy needs. At the same time, the full dimensions of the energy plight of the world's poorest rural dwellers are becoming better known. A notable contribution to alerting world opinion to the seriousness of the fuelwood problem was made by Eckholm in 1975 /2/, and a considerable amount of work has taken place since. The UN Conference on New and Renewable Sources of Energy devoted a great part of its attention to the possible role of small-scale rural energy technologies in developing countries; and major efforts to promote such technologies are being considered by many international aid agencies.

The typical favoured project, in the recent past, has thus been small-scale and decentralised; preferably based upon renewable energy sources; and designed to be carried forward relying on indigenous skills and resources. Almost invariably such a project is a failure. The catalogue is dismal and almost endless. All over the developing world there are broken windmills, empty biogas pits, withered community tree plantations, unused solar cookers, rusting charcoal kilns, diesel engines without spare parts, and derelict small-scale hydro installations. This is not to say there have been no successes; but certainly the failures have outnumbered them. Even where technical and social success have been demonstrated, the rate at which these seem able to disseminate themselves is minute in comparison with the rate at which rural energy problems appear to be worsening /3/.

In many cases the causes of failure can be identified. They include, far too often, unreliable or inappropriate technology; cost problems; lack of local technical skills; unawareness of local customs or indifference to them; administrative corruption; and, frequently, a complete lack of interest, or even active hostility on the part of the people supposed to benefit from the project.
Measures can be taken to try to overcome most of these, but they still do not offer any assurance of success. It is the thesis of this paper that, in addition to the above, the causes of failure in the basic analytic approach to rural energy planning

and project design.
There is, despite the adverse economic climate of the world, still a considerable amount of goodwill, and significant amounts of money available for deployment in energy programmes in the rural developing world. The problem is to find a way out of the present conceptual impasse so that programmes can be devised which have a better chance of success than those of the past. This requires an examination of the structure of the rural fuel economy in the poorest developing countries. These can be taken to include all those with an annual per capita income of less than $ 500, as defined in the World Bank's World Development Report for 1980.

SCHEMATIC STRUCTURE OF THE RURAL FUEL ECONOMY

Figure 1 shows a simple, schematic division of the rural fuel economy. Reality, of course, blurs the edges of any abstraction such as this. Nevertheless, it offers a useful method of distinguishing a number of very different mechanisms, and how they interact with each other and the world about them. The primary distinction is between the subsistence economy and the cash economy. It is doubtful if any fully subsistence economy still exists on a significant scale anywhere in the world. But this is not so in the case of energy. In the majority of the poor developing countries the greater part of the rural fuel supply lies completely outside the realm of the cash economy. Fuel is gathered as a free good from common land, or is acquired as a waste product from agricultural operations. Frequently, fuel collection is the responsibility of women who are, themselves, excluded from cash transactions, thus introducing a further partition between the fuel and the cash economies. The energy used in farm activities is generally supplied by the family, using its own muscle-power, or relying upon that of domestic animals. Except for the rare event of the purchase of working animals, cash, again, does not enter into any of the energy transactions which occur.

In the cash economy there is a wide variety of possible commercial fuels. The list includes firewood, charcoal, kerosene, bottled gas, coal briquettes and electricity. They are employed for cooking, heating, lighting and other uses. Motive power is provided by diesel, petrol, electricity; sometimes by windmills, or waterwheels. Electricity generation is usually by fuel-oil, diesel, or hydro-power; but it may also rely on coal, or even nuclear. Human labour is used on a large scale, often without cash payments, but in general is seen to have a cash value or opportunity cost. None of this, however, excludes the possibility of using free fuels - any more than it is excluded in Scandinavia or North America where people often cut and gather their own fuelwood.

Within each of these two broad divisions, further distinctions can be made. In the subsistence economy the division is between domestic consumption and productive consumption, or as it has been described elsewhere - energy for survival and energy for development /4/. In the cash economy there is a threefold division, between energy for domestic use, energy for services, and energy for production. The proportion consumed in services increases with national income. For illustrative purposes (and perhaps ill-advisedly) a tentative indication of per capita GNP figures is shown in parallel with the gradation from the lowest levels of subsistence upwards into the diversified commercial fuel economy.

Subsistence Domestic Consumption

Whenever it is readily available, wood appears to be the preferred fuel in this sector. Only when wood is scarce are other materials used. These include dung and straw, both of which have many other possible and preferable uses; vegetable waste materials; and even leaves, twigs, and tree-roots. Wood is usually foraged from common lands and forests, often at considerable distances from the home. Cooking is generally done on an open, three-stone fire, either indoors or outdoors depending on the climate and local custom.

A major characteristic of the subsistence domestic sector is that it is, by its nature, impervious to commercial substitutes for its traditional fuels. As long as the provision of domestic fuel is seen as a non-commercial activity by those who engage in it, it will be disconnected from any transactions requiring cash outlays, be these for cooking stoves, solar cookers, or appliances using new fuels such as biogas.
Such devices may, however, be used and approved by individuals when they are given them. Much valuable work has been done on understanding the constraints imposed by traditional cultural patterns upon innovation in energy use, and many of the mistakes of earlier efforts will not be repeated. But the prospects for self-sustaining dissemination of an improvement in methods of energy consumption are what count if a substantial impact on consumption is to be made, and these are not related solely to whether people approve of new devices. As long as

domestic fuel provision remains a non-cash activity it is, by definition, incapable of diffusing throughout itself anything which requires a cash outlay. The process of careful adaptation of devices to suit local conditions and cooking methods, on which so much stress is now rightly laid, is only part of the battle.

Another notable feature of this sector is that, as long as wood is readily available, the quantities used tend to be extremely high. Many surveys of wood use have now been conducted, some showing a consumption of over 2 tonnes per head per year /5/. An annual figure of about 0.8 tonnes per head per year can be taken as a reasonable average. This is about 12GJ per head per year, and by the standards of industrial society is astonishingly high. For a family of six, it is an annual consumption of 72GJ; this can be compared with the 5.9GJ used by the average UK family for cooking /6/. In other words the average wood-using family in a poor developing country may be using more than twelve times as much energy as an average UK family. The figure for total family consumption, including space-heating and all appliances, in the UK is, in fact 79GJ, roughly what is used for cooking alone in the rural Third World.

The stage at which woodfuel has a preponderant position in the rural domestic fuel sector could be described as typically African, in that it is characteristic of most of Africa south of the heavily droughtstricken countries of the Sahel, but it also includes countries such as Nepal, Thailand and Peru. In general, however, the depletion of fuelwood has progressed much further in Asia and Latin America, and the subsistence sector domestic sector fuel mix has a smaller proportion of dung, a greater use of dung and vegetable residues, and a lower total energy use.

Subsistence Productive Consumpion

Subsistence production is, essentially, the non-cash agricultural work carried out by families for their own sustenance. Strictly speaking, this is all that should be included under the heading of subsistence, but, in practice, the pure subsistence existence is rare. Money is obtained from work elsewhere, or families engage in some cash-cropping or handicrafts, in order to be able to purchase essential commerical goods. Energy, however, remains non-commercial, and depends virtually entirely on human and animal muscle power. In

Africa, south of the Sahel, the use of draught domestic animals is much less common than it is in Asia and Latin America.

In contrast with domestic energy consumption, the quantities of energy used in subsistence production are tiny. The work output of a human being in an eight-hour day is unlikely to be greater than 1/4 kWh (0.9MJ), the amount contained in 2.5 cl of gasoline, or about 60 grams of wood. A bullock has a work output of perhaps 4 kWh in a day, less than that in a half litre of gasoline.
There are two important characteristics here: the extremely low totals of energy used in a typical family farm, and the low level of peak power inputs. Even with two bullocks in harness together, the maximum power at the disposal of the subsistence farmer is just about 1 kW /4/. Traditional agricultural practices have evolved in accordance with these low levels of available energy and power, and the sequence of activities along the productive chain is related to both. The sequence of land preparation, planting, tending, harvesting, processing, transport and selling or consuming must remain, at all points, compatible with the available energy and power. In fact, so well adjusted are some traditional agricultural systems to the available energy supplies that they cannot even absorb the small amount of energy available in the form of unemployed members of the family. In Nigeria, for example, the marginal productivity of labour in the subsistence farming sector is "generally reckoned to be zero, or even negative" /7/.

A comparison with the energy provided by mechanical means is instructive. Consider a 15 kW tractor, operating for an eight hour day. Its total work-output is 120 kWh, equivalent to the output of nearly 500 people or a herd of 30 bullocks. In addition, its peak power output renders it capable of a multitude of tasks quite outside the capabilities of a subsistence farmer. Introduction of such a machine into one part of the annual sequence of activities could well change its productivity completely, but without a concomitant increase in capabilities at all other critical stages the gains would be unrealisable. They might, in fact, make matters worse than they had been in the first place.
The subsistence productive sector is thus a tightly closed system, with small delicately balanced energy flows and power levels. It is extremely difficult to increase its energy consumption, or transfer resources of any kind into it. There are numerous examples of attempts at mechanisation creating supply bottlenecks, displacing existing and proven methods of production, increasing vulnerability to market forces, and

ultimately leaving farmers worse off than they had been before. On the other hand, it is difficult to see how substantial improvements in subsistence production can be made without considerable increase in its energy consumption. This is in total contrast with the domestic sector, where the direction of progress lies in a reduction in energy consumption.

Cash Domestic Consumption

There is little need to spell out the characteristics of this sector. It ranges from the poor family just emerging from subsistence and having to divert a proportion of its income to purchase fuel, upwards in income to those with consumption patterns little, if at all, different from their counterparts in the industrialised countries.

As the level of expenditure on fuel increases, energy problems, in general, diminish. The crude open fire on the ground gives way to stoves and other domestic appliances. Energy use becomes more efficient, and the quantities consumed decrease. The range of end-uses widens. If a particular fuel goes up in price, or supplies become unreliable, there are alternatives which can be purchased. Whatever the annoyances of power cuts and unreliable distribution systems, few people with middle-class incomes in the developing countries suffer consistently from a basic shortage of domestic energy supplies - any more than they do in the industrial countries.

Cash Productive and Services Consumption

Again there is little need for description, not because these sectors are unimportant, but because they are familiar. Virtually all stages of agricultural production become more effective and reliable with the judicious use of energy-consuming devices. Analyses of energy use in agriculture show a steady increase in crop output as the energy used in farming increases. For example, corn production in the USA in 1970, using intensive mechanised methods, was over five times as productive, per hectare, as subsistence corn production using purely manual labour in Nigeria. The Nigerian production used 30 times as much manual labour as that in the US, but only a sixtieth as much energy per hectare in total (counting the energy used in the manufacture of fertilisers, as well as that used directly in fuel, and labour) /8/. This, in a simplified manner, could be said to incorporate the whole energy difference between subsistence and commercial production.

Progression from subsistence can be measured by the degree to which supplementary energy uses are assimilated into the agricultural system.
In addition, and as agricultural production increases, numerous associated and additional energy uses come into being. They include those connected directly with agriculture such as crop processing, transporting, drying, packing, marketing. Derived industries such as brick-making, workshops, forges, implement-making, vehicle repair, and fuel distribution become necessary. Productive activity begins to merge into a variety of service functions, the scope and complexity of which are closely related to the level of economic activity. The role of the services sector within the developing rural economy can be both abused and misunderstood. The evils of rapacious middle-men or parasitic and obstructive bureaucracies are not to be condoned; but these are not reasons for trying to eliminate the functions of distribution, marketing and administration. Contrary to the somewhat puritanical view which is sometimes held in development studies - and, indeed, among politicians of the new conservatism in the industrial countries - the service functions are not a drain on economic development, but an integral part of it. A balance between sectors of the economy is obviously required at each stage of development, but it is noteworthy that Sweden and the US, two of the most prosperous countries in the world, derive 63% of their GNP from the service sector /9/.

As energy consumption becomes more sophisticated and end-uses diversify the proportion supplied by human and animal power - which dominate energy use in the subsistence sector - diminishes. In the highly developed rural economy, the quantities of energy supplied by muscle-power are negligible compared with those provided by commercial sources. At some stage, as energy use increases and broadens in application, electricity, whether from a local plant or a connection to a grid, becomes an economic possibility, and a social and political necessity.

ENERGY TRANSITIONS AND THE INTERACTION BETEEEN SECTORS

The above description deals with what might be called the static characteristics of the different sectors of the rural fuel economy, viewing it as a fixed economic hierarchy. Figure 1 illustrates this, with its schematic gradation from the economy of wood and muscle-power though to that of fully commercialised

Figure 1

Figure 1

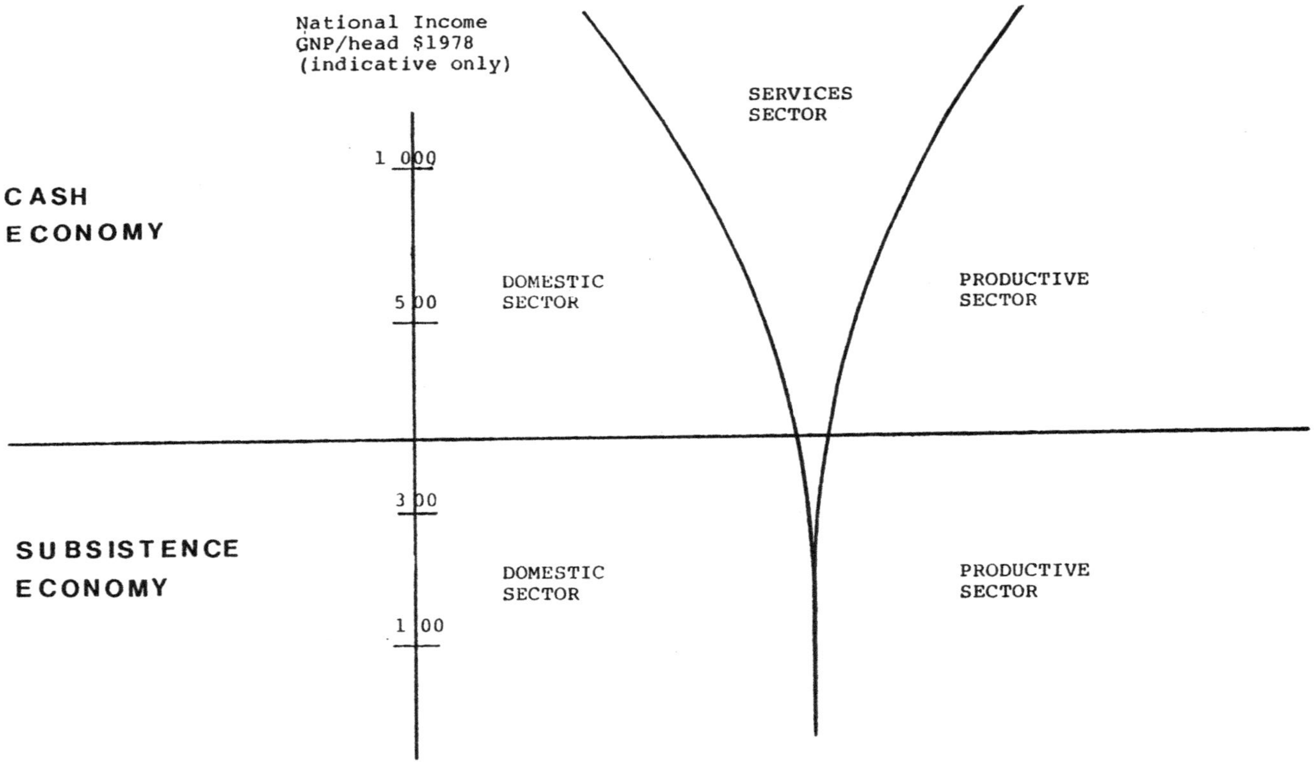

Figure 1. Schematic division between Energy-using Sectors in the Rural Fuel Economy in Developing Countries.

diversity. Reality, of course, will always be more complicated, and will be a function of local conditions rather than statistical abstractions such as the national average per capita income. Indeed, the full spectrum of development, from the poorest level of subsistence upwards, can just as easily manifest itself inside a single rural community as within a country or a comparison between countries.

None of these sectors, however, is static within itself. Change occurs as circumstances improve or worsen. Sectors interact upon each other. It is these processes of change with which intervention is concerned, facilitating them, or opposing them. Each must be understood if the intervention is to be effective, and have the results expected of it. It is therefore worth, again, examining each of the sectors in turn.

Subsistence Domestic Sector

The imperviousness of the subsistence fuel system to commercial intervention has been noted. It therefore follows that when scarcity begins to be felt, consumers will follow not the logic of a cash economy, but that of their own circumstances. Women will walk further for wood, or collect lower quality pieces; there will be less cooking; children will help more, or men may get additional wives; there will be a move towards the use of inferior fuels, or materials for which there are other uses, such as dung and straw. There will not be any investments in improvements in efficiency or commerial substitutes but rather a succession of enforced changes determined within the constraints of a subsistence economy.

Another distinguishing feature of the subsistence fuel economy is that once the rate of consumption of free-good fuels exceeds the rate of natural replenishment, the progress towards depletion is unstoppable. This has been described as "the tragedy of the commons" /10/. It happens because, when people are consuming a free-good which is collected from common lands, it is to everyone's individual advantage to continue consuming even when the resource is disappearing. On the other hand, it is to no individual's advantage to refrain from consuming, or to undertake replenishment for the common good. The classic historic example is that of overgrazing of common land.

In the past there have been societies which have lived more or less in equilibrium with their environment - some of the huntergathers, for example. What they consumed as common goods from their environment was naturally replenished. But such societies are now rare, and in most places where fuelwood is available as a free-good the rate of consumption is above that of natural replacement.

Breaking out of this inexorable progression towards complete depletion of resources requires fundamental social and economic changes. The common resources have to be brought under individual, or communal control; and investments have to be made in replenishment measures which will make up the difference between consumption and the rate of unaided natural regeneration.

Few subsistence societies now have the capacity to make such changes. Contrary to what is sometimes still romantically believed, subsistence societies are rarely stable and self-correcting in their patterns of energy consumption - whatever they may have been in the past. For the most part they contain in themselves a fatal self-destructive dynamic which comes into action once they have passed the threshold of the natural replacement rate of their wood resources.

This auto-destructive process within the subsistence economy can take one of two routes. In the first, the destruction is literal: the society cannot meet its needs for cooking fuel, cannot find adequate substitutes, and has to move, or perish. This is said to have happened to some of the early Mediterranean societies, and appears to be an imminent possibility in some of the Sahelian countries at the moment.

The alternative is that some economic progress begins to occur within the subsistence society. Then, typically, a transition from the free-good fuel economy begins to occur. Fuelwood comes on sale, charcoal may come into use; if economic progress continues, the other commercial fuels, kerosene, bottled gas, coal, or electricity become available. This is precisely the transition which has occurred in the industrial countries and the more prosperous developing countries, and is still under way in many of the others.

This transition is the key to the resolution of the subsistence energy problem. It opens the way to investments in improving the efficiency of energy use, thus beginning to bring fuelwood consumption down to manageable levels. It brings in the use of substitutes, further relieving the pressure on free good resources, and perhaps leaving them available for those who remain poor.

Consider, for example, the case of dung, which is so widely used in many areas. When it is dry it has an energy content similar to that of wood; but it can never replace the quantities of wood which are typically used for cooking in open fires in a subsistence economy. Reliable data on the availability of dung from domestic animals in the rural areas of the developing world are, for understandable reasons, rather sparse. Nevertheless some estimates can be made. The dung output of a mature European beef cow is almost exactly 1 tonne dry weight per year /11/. The typical subsistence cow or buffalo is simply scavenging its food from marginal lands and, in consequence, is badly undernourished. Some Indian figures for the early 1970s give s shortfall in dry fodder of 55%, and green fodder of 78% /12/. Dung yields would consequently be much lower than those for the well-nourished European animal. Taking an average yield of 50% lower, and allowing a 50% utilisation (apart from that not collected, insects and microbiological action in a warm climate rapidly consume dung after it has been deposited), the available dung per domestic animal is unlikely to be more than about two thirds of a kilogram per day. The annual yield per animal comes to about 3.75GJ. The number of cattle required to meet the energy needs of a woodburning subsistence society comes to around 20 per family, an impossible number for almost all areas of subsistence farming. If dung is to be an effective substitute for fuelwood it means that general energy use must become much more efficient than in the solely wood-burning subsistence economy, or that substantial numbers of consumers switch to other fuels, leaving the dung to be foraged and used by the very poorest.

The sheer quantity of wood consumed in subsistence consumption, in fact, makes an exclusively fuelwood solution virtually impossible for most areas. A fuelwood plantation, with care, reasonable rainfall, and protection from animals and fire, can be expected to yield around 10 tonnes per hectare per year. The family of six, with an annual fuelwood consumption of around 5 tonnes, would therefore need about half a hectare dedicated solely to its fuel needs. If the trees were growing 2 metres apart, it would mean up to 750 trees per family. Given the minute size of peasant land-holdings, in most places this is quite inconceivable.

The transition to the cash fuel economy need not, however, be complete. The open fire is frequently the only source of light available to a family. It is therefore kept burning more brightly, and for longer, than it is required for cooking. Introduction of kerosene in small quantities, and simply for lighting, can permit a considerable reduction in fuelwood consumption immediately, and open the way to the introduction of efficient closed stoves and cookers.
This energy transition - it must be noted, however - does not originate within the subsistence domestic sector itself, since domestic energy consumption is economically non-productive.

Thus, there has to be a transfer of economic resources into the domestic energy sector, which will often require a breaking of the barriers between men's and women's work, effectively giving a cash value to women's time. It has been noted for the case of Senegal that, in addition to technical programmes to improve the efficiency of energy use in the rural areas, people will need to be educated into accepting that fuel, which was formerly a free good, will have to be paid for in the future /13/.

Scarcity may play a part in stimulating an energy transition, but it cannot be regarded, or relied upon, as a primary driving force. Economic advance is the crucial factor. This is amply borne out by the fact that when people become richer they abandon the use of the open fire on the ground, and move to the use of cooking stoves and a variety of fuels, irrespective of whether trees are plentiful or not. Increasing efficiency in fuel use, and diversification of the energy supply, are natural concomitants of rising economic welfare.

Table 1 gives a list of countries, mostly with preponderantly rural populations, together with their per capita incomes, and the proportion of national fuel needs supplied by wood. The comparison is crude, and probably raises more questions than it answers. Nevertheless, it lends support to the proposition that there is an almost inevitable move away from firewood as incomes increase.

Subsistence Productive Sector

Although there are numerous, and obvious, problems in this sector it does not possess the intrinsic auto-destructive tendencies manifested in the subsistence domestic sector. In many parts of the world, traditional farming could well carry on its centuries-old methods far into the future. If this were adequate to meet the needs of the rural poor there would be a case for leaving well alone, and avoiding the risks of disruptive intervention. But subsistence agriculture is not keeping pace with the rise in rural populations, let alone generating the surpluses required for improving living standards; in some places its output is actually falling. If an accelerating slide into widespread famine is to be avoided, there will have to be sustained and considerable increases in the yields from subsistence farming in most of the developing countries.

Energy shortage, in itself, is however rarely the decisive factor in preventing economic development in this sector. The size

Table 1: Fuelwood Consumption and per capita Income

	Wood as % of total energy consumption	Quantity wood per head tonnes/year	Proportion of population in rural areas %	GNP per head $ 1978
Angola	74	1.0	79	300
Benin	86	0.6 - 2.3	86	230
Burundi	89	0.2	98	140
Cameroon	82	0.96	65	460
Cent. African Empire	91	0.8	59	250
Chad	94	0.8	82	140
Chile	13.3	–	29	1410
El Salvador	37	–	59	660
Ethiopia	93	0.8	85	120
Ghana	74	1.0	64	390
Guinea	74	0.6	81	210
Honduras	45	–	64	480
India	34	–	78	180
Ivory Coast	46	0.7	62	840
Kenya	74	0.6 - 1.5	86	330
Liberia	53	0.7 - 1.5	67	460
Madagascar	80	0.6	82	250
Malawi	–	0.6	91	180
Malaysia	8	–	71	1090
Mali	97	2.1	80	120
Morocco	19	0.2	59	670
Mozambique	74	0.9	91	140
Nicaragua	25	–	47	840
Niger	87	0.5	87	220
Nigeria	82	1.0	80	560
Papua New Guinea	39	–	83	560
Rwanda	96	0.9	96	180
Senegal	63	0.5	75	340
Sierra Leone	76	0.8	75	210
Somalia	90	1.0	70	130
Sri Lanka	54	–	73	190
Sudan	81	1.2	75	320
Tanzania	94	1.4 - 2.3	88	230
Tunisia	42	0.2	48	950
Upper Volta	94	0.5	91	160
Zambia	35	0.8	62	480
Zimbabwe	28	0.8	77	480

Figure 2

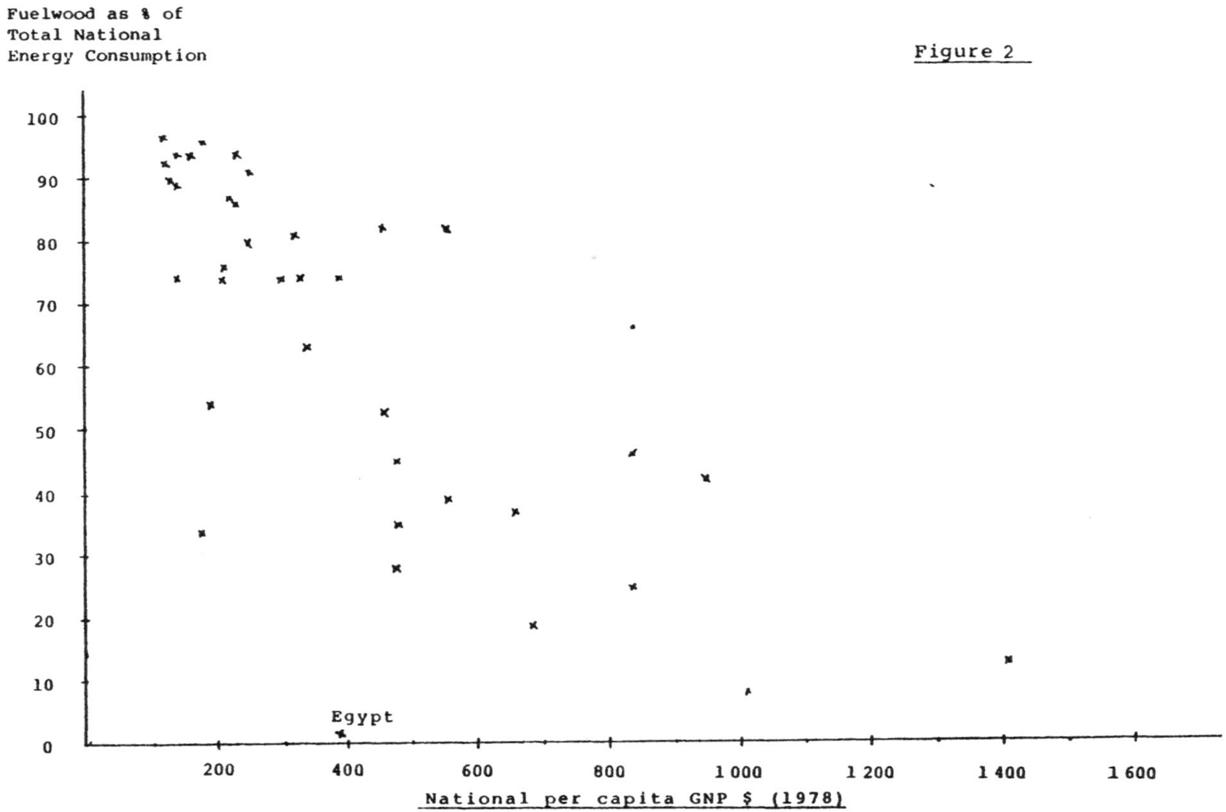

Figure 2. Percentage Fuelwood in National Energy Consumption and National per capita Income

Source; See Table 1

Figure 2

of land-holdings, for example, may be so small that investment in mechanisation, even if the capital were available to so, would not be justifiable. In Korea it is being found that farms in the 1.5 - 3.0 hectare size are proving too large for family working but too small for mechnisation, and farmers are either moving out of agriculture or increasing their acreage /1/. Where the problem, in practice, is a lack of effective demand for energy, or an absence of opportunities for economic deployment of energy-using equipment, widening the range of supply choices is not what is required. An attempt to provide new energy inputs, be these tractors or windmills, will not, in this case, bring any benefits - though it may well incur quite considerable costs.
The question of how to stimulate rural development is far from being resolved; but it is only within a total developmental context that energy can usefully be considered as an effective agent in promoting rural economic advance. The task is to identify those aspects of rural development in which a lack of energy, for precise and identifiable reasons, is inhibiting progress. Here, and rarely elsewhere, will it be found that economically justifiable, and hence potentially self-disseminating, energy initiatives are both necessary and possible.

Areas of greatest interest appear to be small-scale irrigation, cash crop processing, and, perhaps, crop drying. Where these contribute to increased cash crop production, or lessen the need to purchase supplementary food, they stimulate the process of transition into a more prosperous cash economy. They also make resources available for deployment into the subsistence domestic sector, thus promoting its energy transition. But, it must be repeated, energy itself will rarely be the crucial or first factor to be considered in endeavours to create the conditions necessary for an economic breakout from the trap of subsistence farming.

Cash Domestic Sector

As long as cash remains available, the energy problems of this sector remain under control. If, however, the flow of money, which depends upon the prosperity of the services and productive sectors, begins to slacken, or if the price of fuel begins to rise rapidly, then the system can begin to slip downwards. At the lower end of the economic scale, on the boundaries of the subsistence sector, where the transition has just barely been achieved, economic problems can force people back into scavenging their fuel again - with all the problems this brings.

Under such circumstances, subsidising kerosene, or perhaps other fuels, may be the only realistic choice open to governments. The problems of subsidies are notorious; and each country will have its own particular difficulties. One, which sometimes causes worry, particularly among those who view, with justifiable apprehension, the depletion of world oil resources, can be dismissed fairly readily. Supplying the world's poor with kerosene for cooking would do little to worsen the long-term global problem of dependence upon fossil fuels. There are about 1,800 million people in the rural areas of the world's poorest developing countries. This is about 300 million families. If each were to have a kerosene stove and consume as much energy for cooking as does the average UK family, the total additional oil consumption from this massive shift would be only 40 million tonnes of oil per year. This is just 1.3% of the total oil production in the world. It represents a 14% saving in the energy consumption of the US car fleet - something that is well on the way to happening. It is a fifth of the reduction which took place in OPEC oil production in the year 1979-80. It is the amount of oil which would be released by building six 2,000 MW baseload electricity-generating stations, using coal or nuclear power, to replace the same capacity of existing oil-fired plant.

Cash Productive and Services Sectors

In this sector, too, thereis a huge gradation, beginning at the margin of subsistence, and ranging upwards into prosperous and diversified farming communities. At the lower end, the maintenance of productive capacity depends critically on general economic conditions, rather than on the price of fuels; in particular, it depends on the prices obtained for agricultural commodities. In some cases these have fallen drastically in recent years; cocoa, sugar and jute are notable examples. When farmers are unable to cover the costs of cash crop production from the income they derive from sales, they have little choice but to revert to subsistence, or abandon farming. This has been noted, for instance, in Jamaica where some once-prosperous farmers have "fallen into a state of subsistence, and used their savings to leave farming altogether, while many have fled the country for menial jobs abroad" /14/.

Although energy is undoubtedly an element in the difficulties now met by farmers, care is needed in defining the precise energy-problems which they face. If a rise in the price of fuel is the only disruptive factor, it can be dealt with by substituting

a cheaper, but equally effective, fuel source; or, and this is generally easier, the efficiency of energy-use, and production in general can be increased to counter the increased fuel costs. If, however, the increased fuel costs are but one aspect of a general combination of adverse circumstances, attacking energy in isolation will not necessarily improve conditions for individual farmers.
There may, for example, be a general deterioration across the whole services sector as well, with difficulties in transport and marketing, shortages of spare parts, difficulties in repair and maintenance, and breakdown in rural administration. Providing substitute energy systems at the farm, while the distribution system is in ruins helps no one.

At the upper economic end of these sectors, rural energy consumption patterns shade into those established in the industrialised countries and face the same long-term uncertainties. Here, technical and cultural linkages are strong and whatever future the industrial world devises for itself - be this based upon a sophisticated use of renewables, or an equally sophisticated use of coal and nuclear, or indeed a combination of these - it will be one to which the developing countries, too, will ultimately aspire. The complex problems of transition from present energy patterns within the industrial countries were, in fact, discussed in considerable depth in the Beijer Institute's Nobel Symposium on the "European Transition from Oil" held in Stockholm in 1980 /15/. The long-term strategic objective in the upward energy transition in the developing countries should be one which enables this trajectory of development to intersect that on which the enlightened industrial countries are themselves now proceeding.

IDENTIFYING PROBLEMS, ASSESSING SOLUTIONS AND DECIDING POLICIES

It cannot be over-emphasized that the structures and sectoral dynamics being discussed here are schematic, and not an attempt to portray reality in any particular instance. What is being presented is a theoretical proposition about how reality may be examined. The test of such a theoretical approach is whether it is of any practical assistance in dealing with actual conditions. In the case of rural energy planning, three major areas of concern can be distinguished. They are the identification and definition of problems; the assessment of possible solutions; and the formulation of general policies.

It is also necessary to point out that the exercise here is an attempt to be analytic, rather than normative. The basis of effective action is an ability to assess what is possible, and to predict the consequences of different actions. But the actual decisions made in particular circumstances will depend upon both the developmental goals, which will vary between cases, and the resources available at the time. The subsistence way of life is doomed; but what replaces it is far from determined. The position was eloquently described by Julius Nyerere in 1973 when he said: "The moment the first enamel pot or factory-woven cloth is imported into a self-sufficient communal society, the economic and social structure of that society receives its death blow. Afterwards it is merely a question of time and whether the members of that community will be participants or victims of the new economic order." /16/. The question being exqmined here is how energy fits within the complex of change which circumstances now inevitably impose on the rural communities of the developing world.

Identifying Problems

By most standards, the depletion of fuelwood is the gravest energy problem facing the developing world. As part of the work carried out by the Forestry Division of FAO in preparation for the UN Conference on New and Renewable Sources of Energy, a systematic global survey /3/ of the fuelwood supply-position in all the major developing areas of the world (except China) was carried out. Areas of acute scarcity, in which people do not have enough fuelwood to meet their present needs, were identified; so also were the deficit areas, in which, although needs can be met at present, the rate of consumption is greater than that of replenishment. Trends were then projected forward to see how the position will have changed by the year 2000.

The present zones of acute scarcity include much of the Sahel, and the savanna areas of east and southern tropical Africa; the high lands of South and Central America; the dry tropical forests of the eastern regions of Brazil; and the high lands to the north of the Indian sub-continent. Zones presently in deficit extend widely beyond all these areas to include virtually the whole of the developing world's dry tropical areas. When the trends are projected forward twenty years many of the present deficit areas have moved into acute scarcity, and apart from the bountifully endowed areas of tropical rain-forest, fuelwood consumption everywhere exceeds the rate at which it

is being replaced.
It this is taken as defining the developing world's fuelwood problem - and used as a basis on which wood-replenishment strategies might be devised - it is, of course, thoroughly misleading. A projection forward of the early European fuelwood problem would have revealed an equally unmanageable future. If the problem is defined as one of maintaining fuelwood supplies then it is quite insoluble. Provision of the sheer quantities of wood presently being consumed is inconceivable, no matter what the efforts made to promote large-scale or small-scale fuelwood plantations.
This is neither to denigrate the work of FAO nor to deny the problem of supplying cooking fuel, but to point to the need to see it within an appropriate reference framework. Fuelwood, in fact, is only a problem when the rural economic condition is one of static or worsening poverty. An upward economic movement resolves the fuelwood problem as it did in Europe, and as it has done in Korea during the past decade and a half, by creating diversity and improved efficiency in fuel use. The Korean forestry achievements since the mid-1960s are indeed remarkable, but during that time the use of fuelwood has been in steady decline. In 1966 it supplied 55% of the total energy used in the Residential and Commercial Sectors, but this had declined to 19% in 1979; moreover, the total consumption of fuelwood had fallen by 31% /1/.

In no sense is this to be construed as saying that tree-planting on a large scale will not be required. In the case of Korea, over 2,000,000 hectares of forest have been planted, for example, during the past twenty years. In any likely and acceptable future, fuelwood will play an indispensable role in supplying domestic fuel in most of the countries now heavily dependent upon it. But the fuelwood problem must not be defined in isolation. It is not one of ensuring that trees are planted at a rate which matches projected rates of consumption, but rather of ensuring that an energy transition, which will reduce the level of fuelwood consumption to manageable proportions, takes place.

Assessing Solutions

Here again, it is a question of selecting a number of examples and examining them in the light of this analysis to see if it, in fact, helps to identify more precisely how they might best be deployed. The selection is relatively arbitrary, but spans a range from simple local technologies to the high technology of

photo voltaics.
Cooking stoves have many advantages over the open three-stone fire. They are far more economical; they are more comfortable to use; and they reduce the hazards of fire and the ill-effects of smoke. Women who use open fires are well aware of their disadvantages; in a survey in Fiji, where most of the cooking in the rural areas is done over an open fire, 83% of the women interviewed wanted better cooking facilities /17/. Efforts have been made to introduce cheap cooking stoves in many areas, but reports generally reveal that the rate of dissemination is slow. Personal observation in three countries suggests that the possibility of diffusion is closely related to the degree of commercialisation of fuel which has occurred, which is, in turn, broadly related to general levels of income. In Guatemala a high rate of success appears to be being achieved by the Lorena stoves. It seems significant that, in Guatemala, the national per capita income is $ 910 per year, that fuel is widely traded, and electricity is available in many areas. If nothing else, the link between the open fire and lighting is easily broken; but it does also appear as though the use of a stove is connected with a more broadly-based upgrading of living standards and the quality of dwellings.

In Senegal, where there are considerable efforts being put into stove dissemination, relying upon an adaptation of the Lorena stove, called the Ban ak Souf, the prospects appear less promising. There, national per capita income is $ 310 per year, rural sales of fuelwood are rare, and there is little use of commercial fuels outside the towns. The transition to a commercial fuel economy, in other words, has not greatly advanced. In Tanzania, where the per capita GNP is considerably lower, at $ 230, the prospects for immediate promotion of stoves appear to be extremely slight. These three countries, thus appear to represent different points along the energy transition from subsistence; each offers different, and to some extent predictable, opportunities for successful intervention in the domestic subsistence sector.

Another example is biogas. This has been very successful in China, less so in India, and of negligible impact elsewhere. Many of the reasons for a failed biogas programme can be identified by simply examining the contextual requirements for a successful introduction of this technology. But underlying many of the initiatives which have been taken, or proposed, has been the assumption that, somehow, biogas can be used to ease the burden on woodfuel resources and take up some of the demand for cooking fuel. In fact, biogas is a commercial fuel requiring quite substantial investments, and far beyond the reach of the typical subsistence family relying on a wood-fuelled open fire. Its place is therefore within the lower reaches of the cash economy, where it needs to be evaluated against its commercial competitors. In China, the system of communal effort, and individual contributions by means of work-points makes the economic analysis difficult; but it is noticeable that biogas is used, and promoted, as a substitute for electricity /18/. It has also been observed in Bangladesh that biogas plants are most readily adopted by people who already have electricity /19/. This is not an argument for abandoning biogas developments, but for recognising where they are most likely to succeed.

It is also becoming clear that under some circumstances biogas can have severely detrimental effects upon the very poor. The difficulty of introducing resources into the subsistence sector has been noted; the contrary is not so. It is all too easy to draw resources from that sector, and biogas is a pertinent example. When fuelwood is scarce, dung can supply the fuel needs of the very poorest, provided those who are better off have alternatives. When the more affluent start to use biogas, they may economise on wood, or electricity, but they may well be doing so by consuming the dung on which the poorest depend.

Turning to another example, solar cookers are among the most notorious failures among the energy initiatives made in developing countries. There is no need to recite the many reasons why, at a technical and social level, they do not provide an adequate substitute for traditional methods of cooking. Nevertheless, it is still true that it is possible to use concentrating solar collectors to heat water, or cook, in a way which would be competitive, for example, with electricity. The efforts to introduce solar cookers as a means of saving fuelwood in the subsistence sector were bound to fail. Yet there could be a role for them somewhere in the cash productive or in the services sector. All that is required is an economic demand for day-time boiling water in relatively small quantities. Medical services come readily to mind.

Solar voltaic water-pumps have been tried in a number of areas. The machines have worked reasonably well, though there have been some technical problems - surprisingly enough, even in the photo voltaic cells themselves /20/. No one, however, had advanced a serious case for how the undoubted benefits of these machines can be made available to the millions of villages which could use them in the developing world.

Solar pumps have peak outputs, in the versions usually tried, in the range 0.25 - 0.5 kW. This is within the range of power-outputs of draught animals and human beings; moreover, the daily output of a solar pump is similar to that of a draught animal, or the work expended by the people of a village in collecting water. Solar pumps thus fit naturally into the pattern of subsistence existence, and it is easy both to demonstrate, and realise, their potential benefits in reducing human drudgery. They are, indeed, attractive.

On the other hand, a solar pump installation is likely to cost anything between $ 5,000 and $ 30,000. There is little hope of a subsistence community raising this amount of capital, or if it did, having a solar pump as its first priority. While solar pumps will remain perfectly acceptable as donations, and may indeed be welcomed enthusiastically, they have little hope of self-sustaining economic take-off within the sector towards which they are usually aimed. The place of solar pumps, if there is one, is in the upper reaches of the cash economy.

Solar voltaic systems for small battery charging, however, could have a very large future. The transistor radio is an item of high priority for many millions of poor people; and even for those who frown upon luxuries for the poor, it has self-evident utility as an element in education and extension services. Yet, the cost and trouble of dry-cell replacements quickly renders many transitors totally useless, and the investment in them a waste. Small cheap solar voltaic systems (around $ 15) already exist, however, which can either recharge a standard rechargeable radio battery, or from which a radio can be run directly. Here, there is an evident opportunity for introducing photo voltaics into the rural services sector in the majority of developing countries. There is every possibility of the system becoming economically self-sustaining, and leading eventually to local manufacture of everything but the solar cells.

Further up the scale of cost, a brief-case sized unit containing solar cells and a 10 ampere-hour rechargeable battery system is being marketed by at least one company /21/. Lights, fans, and other small electric appliances, such as are now run, as a matter of routine, from a car battery, can be used. The cost is about $ 800, far beyond the reach of the individual family, but becoming interesting for communal facilities.

Neither of these, of course, directly meets the needs of the very poor. Neither do most technical innovations. But if the objective is to use photo voltaics to enhance rural living, meet some energy needs, stimulate some measure of local economic activity, and give developing countries a toehold in a rapidly advancing technology, it is applications such as these in the cash economy which offer the most promise.

Deciding Policies

The theme of this paper has been that it is essential to locate energy initiatives at their appropriate level in the rural fuel economy if they are to have a possibility of becoming integrated, and self-sustaining measures for easing rural energy problems. This is not a matter of creating a hierarchy of actions, or making mutually exclusive choices. The objective should, rather, be a range of effective and complementary measures.

In recent years, efforts to introduce small-scale energy technologies in developing countries have resulted in many failures. Understandable tensions have arisen between the traditional engineering approach to energy provision and these more tentative attempts to introduce often unproven techniques into imperfectly understood rural systems. A well-run hydro-electric project will produce a dam and a power station; a forestry project usually resuls in a large number of trees; but a rural stove programme, with the best will in the world, may result in no more than a few score heaps of cracked mud. To the conservative of mind the balance of risk is clear.

But to exclude the small-scale in favour of the possibly more reliable large-scale is wrong. It is to miss the point. Both are required. Without the large-scale infra-structural projects, national economic development is impossible. The fact that many mistakes were made with such projects in the past was not a reason to abandon them, but to get them right. Large-scale development, however, relies upon heavy government intervention and usually a large foreign subvention. By its nature it is concentrated, capital intensive, and standardised; it cannot, and need not, be otherwise. These characteristics are strengths, not weaknesses, and make possible the massive investment programmes of such agencies as the World Bank.
The price, however, is that the large concentrated capital projects cannot reach the dispersed rural poor who exist largely outside the money economy. Meeting their needs is not antithetical to making provision for the basic infrastructural needs of the country, but a logical and necessary concomitant.

Returning to the schematic illustration of Fig. 1, the same point can be made by saying that large projects tend to meet the needs of the higher levels of the productive, services and domestic sectors of the economy. The task of the small-scale project is to help to create a bridge between the subsistence and the cash economy. Unless this happens, the effect of investment in large-scale projects is to widen the gulf between the minority and the majority in most developing countries - precisely the reverse of what is normally intended.

The fundamental policy question for the rural sector is therefore concerned with creating a momentum of development. True, there will often be a need for emergency measures, for donations to prevent things getting worse, or for policies which protect those unable to fend for themselves. But the underlying policy objective must be the promotion of a transition from the subsistence to the cash economy.

Commercialisation of fuel is rightly feared for the hardship it can bring, and the extra burdens it imposes on people whose financial resources are already meagre, yet ultimately it is the key to the resolution of the rural energy problem, and particularly that of fuelwood. It is quite obvious that over the next decades, and probably for longer, wood will remain the primary source of fuel for very large numbers of people. The task is to create the conditions under which the necessary tree-planting and economies in fuel use which will make this possible, can take place. This means that fuelwood must be seen within a dynamic of development; it must be used as an element in the energy transition from subsistence.
This cannot happen unless fuelwood becomes integrated with the self-propelling economic processes of rural living. Donations and government programmes will be necessary to provide technical knowledge and infrastructural facilities, but ultimately replanting must happen with the enthusiasm and effort of local people behind it. Encouraging the production of fuelwood as a cash crop for sale in the urban areas and later in the rural areas themselves would appear to be an essential element in any strategy with a hope of having a significant impact on the fuelwood problem.

Charcoal-making provides another opportunity. Partially burning wood to make charcoal is a process which is frequently castigated for its large energy losses - they are comparable with those incurred in the production of electricity - but the case against it is not quite as clearcut as might at first sight appear. The criticism on energy grounds tends to miss the point that charcoal is an extremely attractive fuel /22/, readily saleable in cities and with a market far wider than fuelwood. It can even become an important export. Making charcoal drives off energy-containing volatiles; but it adds a considerable amount of value to the original raw material. Thus charcoal-making, provided it is linked with wood replenishment measures, could become a major source of rural benefit, rather than the evil it is now often seen to be. The way is then opened to further technical improvements in charcoal-making, and eventually the use of metal kilns and retorts to capture the tars and volatiles which are wasted in traditional methods.

Breaking the link between using the open fire for cooking and as a source of lighting has been noted as an essential element in the promotion of efficiency in fuel use. The connection between dwelling quality and economy of fuel use is also important. Better housing is one of the highest priorities for huge numbers of people, yet simple masonry techniques, or blockmaking equipment, are beyond their reach. An improved stove built in one of the poor sisal-pole and mud huts in which many people have to live in East Africa is an improbability; but given the possibility of building a block swelling, many people would be willing to incorporate a stove within it. Linking stove-building with programmes for dwelling improvements would certainly increase the likelihood of success.

Thus, it is when the subsistence economy begins to emerge into the cash economy that the opportunities for effective action begin to increase. The use of the term, cash economy, in this context obviously does not imply any substantial degree of wealth, or people who are in no need of assistance in their struggle to live. Even in the most prosperous developing countries, rural incomes tend to be far below what would be the minimum acceptable standards in an industrial economy. It simply means that when fuel has entered the realm of cash transactions some of the constraints imposed by the nature of the subsistence fuel economy are lifted and there are greater opportunities for introducing economy and diversity into the energy supply.

The task becomes one of finding an appropriate niche in the rural fuel economy for some of the wide range of technical devices and techniques now available or under development. Some examples have been discussed and many more could have been chosen. Although there have been many failures in the past a great deal has been learned from them. It is now fairly clear what will not work; but the picture is not all gloomy. With proven technologies, the range of applications and contexts in which they will work is also becoming clearer.

In some cases local conditions will demand specially adapted, or what used to be called "appropriate" technologies; but probably less so than was assumed in the past. The criteria of economy, durability, safety in use and proven performance which are applied to consumer goods in the industrial countries are very much what is required in the developing world also. The test of whether something is attractive and economical to consumers in the industrial world, as opposed to the assumed needs of a poor person in a developing country, is by no means infallible, but it has probably been too little used in the past. It is imperative that poverty and lack of alternatives are not seen as excuses for inventing second-rate technologies which would not meet the reliability standards of industrial society. The margin between survival and dying is so fine for a large proportion of subsistence dwellers that they are no more willing to risk the use of untested technologies than, one hopes, would be a builder of passenger aircraft.

Yet, and despite the fact that the general possibilities for effective aid become clearer, and the problems and potential of different technologies are better recognised, the identification of practical working projects is still difficult. This is particularly so in the subsistence domestic sector. It is here the greatest problems are to be found, yet where the opportunities for successful action are most restricted. This means that, for the immediate future, progress is likely to be slow. Mistakes will be made, as everywhere else in development.

The proportion of total aid flows that can be absorbed in dealing with rural energy problems is also likely to be considerably smaller than many concerned people would like. It will certainly be less than that which can be spent on large capital projects. A considerable proportion will have to be spent on simply finding out how the rural economy works, and what it is possible to achieve within the constraints imposed by it. But unless these efforts are made, and properly supported to the maximum that can usefully be spent, the more spectacular, and politically rewarding, large capital projects lose most of their point and purpose.

Technical initiatives, moreover, must not just fit in an appropriate developmental context; they must also be part of a continuum of development along which people may travel. The gradation of subsistence, commercialisation, diversification and economy of fuel-use given here is not simply a means of identifying the level in the economy at which a particular energy initiative might succeed. It is also a trajectory along which subsistence energy use itself might be expected to evolve. If this can be achieved, it will provide for the developing countries an energy future at least as secure as that which the industrial countries are trying to prepare for themselves.

In the end, however, such long-range speculation is not the primary purpose of this paper. It is not attempting grand designs or proposing comprehensive solutions. Neither is it, in any way, minimising the energy problems, particularly those relating to fuelwood, facing the developing world; they are almost stupefying in their magnitude and urgency. It has, instead, aimed simply at finding a way of identifying, somewhat more surely than in the past, some of those small, practical and immediate actions which may be taken, and which are capable of self-sustaining dissemination among the people they are intended to benefit.

ACKNOWLEDGEMENTS

This is a substantially revised version of a talk first delivered at the Beijer Institute on 9th December 1981. I am particularly grateful to Prof. Gordon Goodman and Dr. Lars Kristoferson for offering me the hospitality of the Beijer Institute on that occasion.
Many friends and colleagues have contributed to the work that went into the preparation of the original paper, and into this revision. I would, however, like to make a special record of my debt to Ariane Van Buren for all she contributed to the ideas expressed here during the time of our collaboration in IIED.

REFERENCES

1. *Kim, Y.H., Energy Strategies for Subsistence Agriculture: the Korean Experience, paper presented at El Colegio de Mexico Symp., June 1981.
2. Eckholm, E. P., The Other Energy Crisis: Firewood Worldwatch, paper No. 1, Worldwatch Institute, 1975.
3. FAO, Map of the Fuelwood Situation in the Developing Countries, 1981.
4. *Stassen, E. M. and Van Swaaij, W. P. M., Energy Technologies for Rural Development; Processes and Applica-

tions, paper presented at El Colegio de Mexico Symp., June, 1981.
5. Leach, G. et al., A Low Energy Strategy for the United Kindgom, Science Reviews, 1979.
6. Barnard, G. W. and Hall, D. O., Biomass for Developing Countries, report for Solar Energy Research Institute, 1981.
7. *Alolabi-Ojo, G. J., Energy Strategies for Subsistence Agriculture: The Nigerian Case, paper presented at El Colegio de Mexico Symp., June 1981.
8. Leach. G., Energy and Food Production, IIED, London, 1975.
9. World Bank, World Development Report 1980.
10. Hardin, G., The Tragedy of the Commons, Science Vol. 162, Dec. 1968.
11. Palz, W. and Chartier, P, eds., Energy from Biomass in Europe, Applied Science Publishers, London, 1980.
12. Crotty, R., Cattle Economics and Development, Commonwealth Agricultural Bureaux, London, 1980.
13. *Sene, El Hadji, Energy Strategies for Subsistence Agriculture: The Senegalese Experience, El Colegio de Mexico Symp., June 1981.
14. *Ventura, A., Energy Needs and Potential Supplies in Jamaican Subsistence Agriculture, El Colegio de Mexico Symp., June 1981.
15. Goodman et al., eds., The European Transition from Oil, Academic Press, London, 1981.
16. Coulson, A., ed., African Socialism in Practice, Spokesman, Nottingham, 1979.
17. Siwatibu, S., Rural Energy in Fiji: A Survey of Domestic Energy Use and Potential, Int. Dev. Res. Centre, Ottawa, 1981.
18. Van Buren, E.A., Biogas Training in China: A first Exchange with Developing Countries, United Nations Environment Programme, Nairobi, 1980.
19. Islam, Nurul, personal communication.
20. World Bank, Small-scale Solar-powered Irrigation Pumping Systems Phase 1, Project report, 1981.
21. Holecsol Systems, B.V. Eindhoven, Netherlands.
22. Foley, G., Van Buren, E.A., Substitutes for Fuelwood, Unasylva Vol. 32, No. 130, 1980.

*An edited edition of all these papers will be published in 1982 by Westview Press, Colorado, under the title Energy in the Transition from Subsistence Agriculture. Wionczek, M., Foley, G., Van Buren, E.A., eds.

RURAL ENERGY TRANSITIONS: THE COMMERCIALISATION OF WOOD AND OTHER SUBSISTENCE FUELS

E. A. van Buren
Senior Research Associate

International Institute for Environment and Development (IIED)

The ability to monitor whether transitions are actually occurring from one source of energy supply to another is crucial to energy planning. This paper addresses the rural areas where ignorance is pervasive. It begins with a framework for depicting energy and economic transitions. It then focuses on the commercialisation of woodfuel and organic residues. In a non-commercial form these are the primary source of fuel for most rural households. When commercialised, they provide essential rural income and supply fuels for industrial and urban demand. Studying woodfuel markets can help to monitor impending crisis, and to understand the conditions under which any counteracting solution will or will not be adopted. The depletion of fuelwood in Europe during the 16th Century is discussed for comparison.

Rural energy projects are a recent invention. Their proliferation has been accompanied by surprisingly little monitoring or evaluation. Energy is admittedly a new subject and has not yet accumulated a backlog of experience. The concern has been to get individual projects on the ground. There has been little systematic analysis of their utility to rural development. The energy problem has not yet been defined. There is no coherent framework for selecting projects or evaluating their significance or, indeed, assessing the very role of energy in develop-

H. Neu and D. Bain (eds.), National Energy Planning and Management in Developing Countries, 41–55.

ment.

Energy planners need more than just recommendations of what sould be done. They also need to know what changes have already taken place. Otherwise there is no basis for assuming that people will actually alter their habits in the future. Economic planners, in turn, need to know whether these shifts from one source of supply and pattern of consumption to another make a difference to economic development.

This paper describes a schematic framework for depicting these transitions. It assumes a progression from a peasant energy economy where fuel is used as an uncommercialised naturally-supplied resource, in a very limited number of forms; through to a developed energy economy where fuel sources and uses are diversified, multiplied and all regulated through market transactions. Between these two extremes, subsistence fuels become commercial and are substituted by more modern fuels. This paper focuses on that transition.

Many developing countries are now experiencing a double pressure upon their wood resources, due to an increasing subsistence usage and an attempt to substitute wood for imported industrial fuels. Pressure from both extremes may cause a convergence on the commercial markets for woodfuel. The operation of these markets is an area of critical ignorance.

THE FRAMEWORK

A quadrant diagram has been devised to represent the variety of ways energy is used in a developing country (Foley 1981). Two divisions, horizontal and vertical, refer respectively to the rural/urban dualism in underdeveloped economies and to the distinction between non-productive and productive activity. The four quadrants (1,2,3,4) parallel the progressive reliance of rural activity upon the monetary economy (Beijer Kenya Project 1982).

At first glance, to draw a division between subsistence and cash economies may seem unrealistic. Small farmers in developing countries are rarely without some cash remuneration, either for off-farm activity or for food production in surplus to that needed to feed the family. Cash is needed for goods which cannot be produced by the family, such as cooking oil, sugar, and kerosene for lighting. The degree of monetisation in any household is inversely proportional to the extent of that family's reliance upon its own production of food, tools, clothing, housing, and so forth (Guzman 1981).

	<u>Domestic Activity</u>	<u>Productive Activity</u>
<u>Monetary Energy Economy</u>	(3)	(4)
	--------COMMERCI	ALISATION-------
<u>Subsistence Energy Economy</u>	(1)	(2)

When we consider energy, however, the distinction is in fact realistic. In a peasant economy energy consumption is often entirely cashless. In most rural areas fuel is obtained only for subsistence; it is gathered from the open woodland by hand, usually by women and children, and its cost is measured only in labour-time. It is used primarily for cooking, but also for heating, rudimentary lighting, and protection at night. Productive activity, such as the cultivation and processing of crops to provide a basic food supply, is almost all manual. There is some draught animal power contribution where income and land tenure are sufficient for animal husbandry. Normally energy is only acquired in commercial forms for specific tasks which cannot be accomplished non-commercially, as for a more intense illumination by means of kerosene. When there is no money available, such energy uses are foregone.

The second division is made between domestic and productive uses of energy, because in the peasant economy only domestic activity uses fuels--as opposed to animate forms of energy. In production fuels do not appear, except in the tractorised cultivation which occurs only at a relatively advanced level of development or among the fewer wealthy farmers in an underdeveloped rural society. Most subsistence farmers have only muscular power to rely on, as they cannot afford fuel or the machines to use it. A long-bladed knife (panga or machete) is quite commonly the only implement owned.

Where muscles cannot supply the energy in the form needed, as in heat requirements, then a fuel will be used. Procuring fuel is labour-consuming and is done only when necessary. In the peasant economy there is little conversion of energy from one form to another, except the simplest conversion from organic matter into heat and muscular power into mechanical motion. Other than milling, water-lifting and charcoal-making, there is no upgrading, intensifying, or harnessing of energy to increase the quality or quantity of output.

At higher levels of income, where increasing proportions of all activities are attributed a monetary value, energy is used proportionately more in the form of fuel and less in the form of labour. As the value of human labour increases, draught animals and then fuels are progressively introduced and energy is upgraded. Thus, production as well as domestic activity becomes fuelled, and fuels are bought through a market rather than collected by hand. Fuel forms become denser and highly specific to a precise point of application. Economic development generally correlates with the productivity of energy use: the useful output derived from the primary energy input rises. Actual measurement of inputs and outputs may, of course, be problematic. Nevertheless the level of development can be closely correlated with the absolute quantities of per capita energy used in production.

The first quadrant of the schematic framework describes subsistence energy use for domestic purposes. This quadrant is dominated by the use of wood or organic residues for cooking. In fact this can account for up to 90-95% of all energy consumed in a peasant economy and a similar proportion of all the energy used in the most underdeveloped countries.

Per capita wood-use varies with climate, with methods of food preparation and with local customs. A figure of 1 cu.m., however, is quite common for annual average per capita consumption (Barnard and Hall 1981). This is about 800 kg, depending on the type of wood, or 13 kg per day for a family of six (equivalent to 40-50 kWh, or 50-70 GJ per year).

In the third quadrant in the diagram--the commercially fuelled domestic sector--the amount of energy used for cooking falls dramatically. In an average household in a northern industrialised country, as exemplified by the UK (Leach et al. 1979), cooking requires only one twelfth the energy (5.9 GJ) and all energy uses, including heating half the year and a wide range of electrical appliances, are accommodated within 79 GJ, a figure similar to that just for cooking in conditions of rural poverty.

An alternate calculation can been made, based not on the primary energy procured but on the final energy required at the point of application, or in 'end-use' (Stassen 1981). The food requirement of an average adult is about 10 000 KJ/day. The assumption is that 80 % of this would be in the form of grain and that rice is representative in needing 2 500 KJ/kg for cooking (rice itself contains about 14 300 KJ/kg, so that 0.56 kg would be an adequate daily ration) or 1 400 KJ/daily ration. If 75 % of the energy for meal preparation is spent in grain cooking, the total is roughly 1 870 KJ/day or 0.68 GJ/person/year. The efficiency of primary wood consumption (12 GJ/person/year) would therefore be approximately 6%. A compilation of figures cited for average efficiencies, in fact, puts most open-fire cooking at an efficiency of about 5-10% (Floor 1981). Whether analysed on the basis of the primary fuel requirement or starting with the heat needed at the point of application, the domestic energy consumption of the rural poor is clearly both extremely high and very inefficient.

By contrast, the energy used in a peasant mode of production, represented in the second quadrant, is very small. This is related to a basic cause of low productivity, namely the low or non-existent capacity for investment in planting materials, fertilisers, tools, land, or hired labour. Peasant production remains based upon land areas too small even to absorb the labour available within the family, at all but peak seasons. This has been classified as simple reproduction (Guzman 1981), striving mainly to satisfy the needs of consumption, using inefficient and rudimentary implements, and unable to attempt accumulation. Assuming that a family has at its disposal the labour equivalent to two adults (Caceres 1981) working 8 hours a day each, the energy delivered is 0.25 kWh per person (equivalent to the heat energy in just 2.5 teaspoons of diesel oil (Foley 1976), or 60 g of wood); the total is only 0.5 kWh per family per day.

If the farmer has a pair of bullocks, contributing 4 kWh per day each (the equivalent of roughly 3/4 gallon of diesel fuel used in a tractor at 25% efficiency), then the total energy harnessed in production might reach a maximum power output of 8.5 kWh per day (or about 32 kWh in tractor fuel). The figure for human labour alone represents only 10 percent of the delivered energy (and 0.6 percent of the primary energy) used for cooking and the basic chores of domestic survival. The figure which includes draught animal power, and which is optimistic for most rural conditions, is still only half of the energy used domestically.

These proportions, however, must be treated with caution. If the energy in muscle power compares roughly to the heat delivered to the pot, the analogue of the primary energy in firewood would have to be the food input to the labourer. For example, the primary energy required by a labourer in the form of food and energy to prepare it (10 000 KJ/day as food and 1 867 KJ/day as energy), 11 867 KJ/day are needed to produce a work output of 0.25-0.5 kWh or 900-1 800 KJ (at an efficiency of 7-15%). One might also compare the output of domestic energy use (10 000 KJ/person/day) to the output of productive energy use. But these calculations have little operational value. The point is rather that in subsistence all the energy required to feed and maintain a peasant household only manages, in a rudimentary agricultural system, to produce enough for survival. The productivity of the system is so low that there is scarcely any surplus and therefore certainly no accumulation of capital.

One characteristic of peasant production, therefore, is its low energy input. Only a fraction of the energy used in production in an industrialised economy (Leach 1975) is used here. Draught animal power brings this up to 1/3 the daily energy content of industrial production (ca. 90 kWh/day in UK). The skew in subsistence energy requirements becomes apparent when one juxtaposes the low energy use in production with the high energy use for cooking--roughly ten times greater than in an industrial economy.

While these figures are highly generalised, they give us some picture of the quantitative - and order of magnitude - differences in energy use at the extremes of economic development. Domestic energy consumption becomes far more efficient and variegated as an economy develops, while productive energy consumption (shown in the fourth quadrant) rises dramatically and assumes greater intensity and specificity with investment in machinery, in production lines, and in high-quality fuels. A commercial services sector is also created alongside the productive sector.

COMMERCIALISATION

Energy development is represented in a progression from one quadrant to another. This paper focuses on the transition across the horizontal line representing the commercialisation of energy use. This is the movement from non-monetary to monetary energy use, from quadrants 1 to 3, and from 2 to 4. Crossing the vertical division represents the transfer of resources between productive and domestic sectors. While this is clearly essential to the process of development, it is not within the present scope of this paper.

Commercialisation is the establishment of market exchange between producers and consumers. This has a multiple significance. The single most important aspect of the commercialisation of a previously non-commercial good is that it provides a new source of income. In India it has been estimated that 25% of the fuelwood in India is commercialised and that this employs 15 million people full-time in cutting, carrying, and selling in the market.

This is particularly important in rural areas where there are few ways of earning cash. Selling firewood or charcoal brings ready cash when crops fail or marketing channels delay. In most areas it is those who own the least land or none at all who sell woodfuel. This activity is usually associated with extreme poverty and points to the lack of other employment opportunities in these areas.

Secondly, commercialisation in the use of a natural resource is an incentive to accelerate extraction. If the trees and the land from which they are removed are unowned, there is no responsibility or incentive to replant, and depletion must result. If the trees and land are owned, replanting may occur and wood may become a cash crop. This will depend on the relative economics of producing food crops or a fuel crop. In any case, a polarisation develops between those who own land and those who do not.

The woodfuel trade can be organised in a variety of ways. The presence and proportion of middlemen interposed between producers and the point of sale will diminish the return that this activity brings to the rural areas. As cities grow and the producers and consumers become further apart, entrepreneurs who own or rent vehicles gain access to the markets, and such people are usually urban residents. The major part of the return from woodfuel marketing thus accrues to urban areas and constitutes a net drain upon rural resources. Table 1 shows this clearly for the woodfuel supply to the capital city of Nicaragua.

Table 1. Breakdown of fuelwood price within the commercial fuelwood supply to Managua (Nicaragua).

Land owner	Fuelwood cutter	Empresario	Market operator
7%	33%	40%	20%

Source: Park _et_ _al_., 1982

The commercialisation of woodfuel is caused by a combination of rural need for income and an effective demand for woodfuel in market centres. This occurs when people have cash to spend on fuel and it is easier to pay for it than gather it themselves. The consumers are either the rural elite, salaried earners, or commercial businesses; or the urban inhabitants far removed from the source of the fuel.

Diagram 2 sketches the forces driving commercialisation. Scarcity of woodfuel polarises the supply from those who have it, or access to it, for sale to those who do not. In the absence of disposable income, if there is any non-commercial alternative, it will be substituted for scarce wood. The move to inferior fuels - twigs, leaves, crop stalks, and dung - has been widely observed. Each of these have competing uses which then come under pressure. Ultimately these are only temporary solutions to woodfuel scarcity and they too eventually take on a price or are traded against favours and obligations to those who have them.

In Europe these forces drove woodfuel to depletion as early as the 16th Century (Nef 1977). By 1570, firewood-stealing and 'hedge-tearing' had become a common occurence among the poor in England (Dyer 1976). Then the wood-consuming industries were penalised and forced to switch to coal. Wood was a luxury fuel. Then even the rich overcame their aversion to coal. Finally, it was only those towns with access to peat or coal which survived and prospered.

The position in the Third World today is nowhere near this extreme. Yet the dilemma now is how to monitor the severity of the problem and find appropriate responses and management methods before the point of crisis is reached.

There is a great difference between scarcity as diagnosed by professional analysts, and scarcity as seen and felt by woodfuel users. The first is an absolute scarcity and is measured and understood abstractly. The second is very concrete. It is perceived in relation to a memory of the way things used to be, or in relation to the observable conditions of one's neighbours. Only this second scarcity will povoke action by the users.

Before woodfuel is commercialised, the question of scarcity remains elusive. The cost of fuelwood is measured only in the distance and time needed to

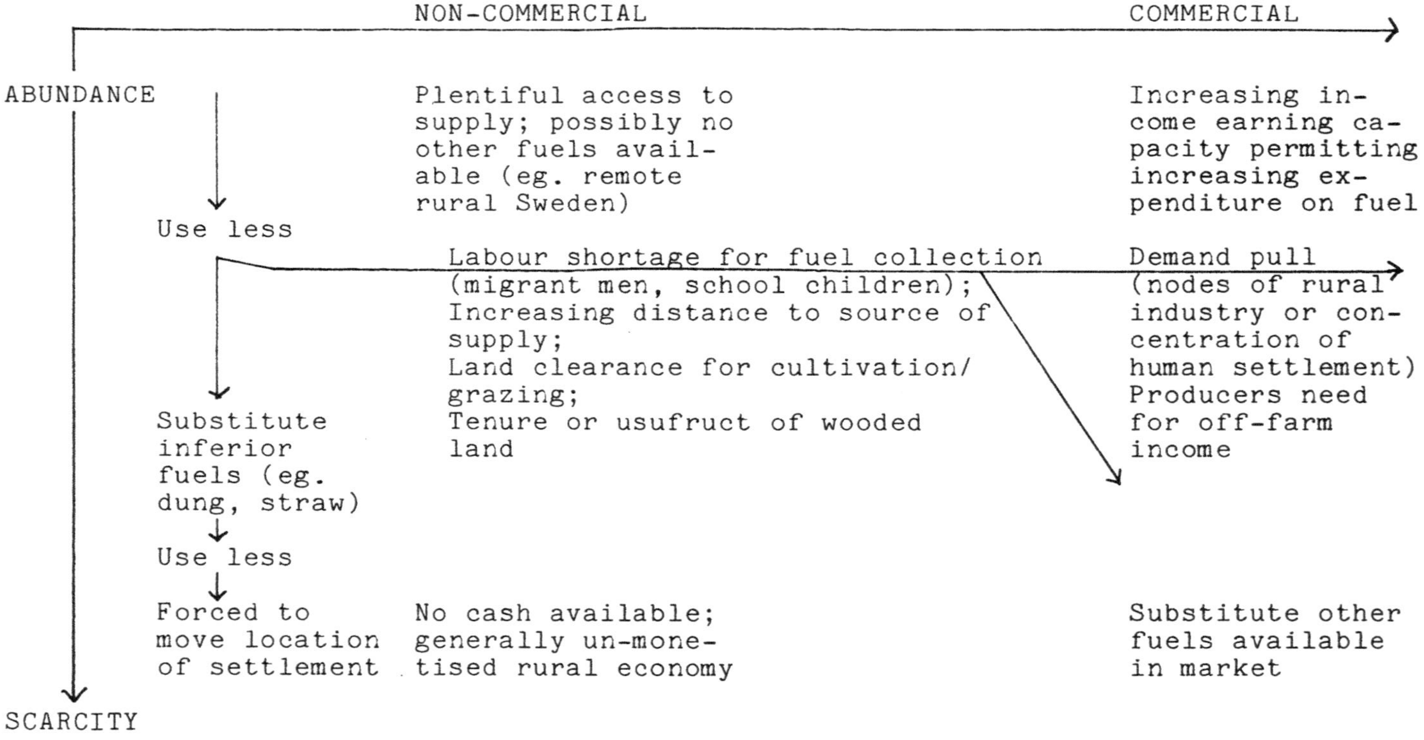

Diagram 2. <u>Schematic Description of Woodfuel Commercialisation</u>

gather it. Fuelwood gathering is generally the task of women; but women rarely own land, or control assets or information flows. The position is exacerbated by the lack of individual responsibility for common property resources. There are few mechanisms by which feedback from scarcity can be translated into effective ameliorative action. Depletion becomes inexorable. It was for this reason that the Fuelwood and Charcoal Technical Panel of the UN Energy Conference chose to concentrate on ways of coping with emergency in fuelwood scarcity.

Price is frequently cited as a measure of scarcity. Yet the prices of fuelwood are not necessarily rising faster than the prices of basic goods such as food; indexing prices to inflation, however, may not be possible where consumer retail price statistics do not exist.

In the course of the fuelwood crisis in Europe, the price of firewood was stable until 1540, then quadrupled by 1580; by 1620 it had risen ten-fold. While inflation was rapid for all goods (although moderate by today's standards), the price rises for firewood were apparently conspicuous (Dyer 1976).

If, through systematic study of woodfuel markets today, we can establish that the levels and rates of change of prices, volumes, and availability are worsening, then we may be in a position to tailor accordingly our proposals for solutions. The prices being paid for woodfuel may tell us at what price or total cost any other fuel can be substituted or investment made in appliances such as stoves to modify consumption.

Markets may also indicate the channels through which modifications can occur. Agroforestry solutions, for example, will need to be sited differently, according to whether they depend on market distribution of products or are intended to fit within patterns of subsistence consumption.

The commercialisation of an economic sector is a long-term transition that can take years to complete fully. This is equally true of energy. The question is then, how can we know whether and where it is taking place, or the rate at which and the way in which it is occurring?

MEASURES AND INDICATORS OF TRANSITION

In most places where the consumption of rural fuels has been studied, it has been observed that commercialisation is either not occurring or, indeed, has already taken place. Where wood is still uncommercialised, it is being procured or produced informally by the user, as and when needed. The studies that exist are mostly analyses of individual industries self-sufficient in wood, [tea estates (IIED 1980), tobacco-curing (Temu 1979), brick-burning (Mung'ala 1978),...] or anthropological descriptions of rural communities whose members collect wood from the common resource base around them (Deshler et al., 1979, Nkonoki 1980, Bajracharya 1981, Islam 1980, Brokensha 1978). Systematic comparisons between commercialised and uncommercialised areas, however, are rare and, at best, only qualitative; quantification is needed 1.

Everywhere, information on the transition is scarce - in developing countries today even more so than in Europe 400 years ago. This is partly because few records are kept for pre-industrial activities. Partly, it is because the problem has not been approached as an historical phenomenon. Consistent time-series data simply do not exist.

We therefore have to paint "before" and "after" pictures - a cross section of commercialisation by looking at the conditions evident in successive stages of the process - and to hypothesize the evolution in between. Specific parts of the model can then be spot-checked through field observation.

CONCLUSION

Most work in the energy field is going into designing solutions. No one knows whether these solutions will, in fact, be adopted. For this we need to have a better idea of what changes are, and have been, taking place. We need evidence of shifts that have actually occurred to know under what conditions future action will be taken. This should put into a realistic perspective the many energy technology and techniques now proposed. It may even give advance indication of whether they are likely to succeed.

NOTES

(1) The World Bank, for example, has held commercialisation as an implicit aim of its afforestation projects. Little analysis of woodfuel prices, however, has been done, other than in the appraisal of individual plantation projects (Spears 1982). A recent IBRD survey of the literature on energy pricing in developing countries scarcely touches on prices for woodfuel (Julius and Meta Systems 1981), leaving this as an uncharted area between the energy and forestry departments of development agencies.

REFERENCES

Arnold, John H., and deLucia, Russell, J., 1982 (September), "Rural Energy Surveys: The Thailand Experience" (draft paper).

Bajracharya, D., 1981, "Implications of Fuel and Food Needs for Reforestation: An Energy Study in a Hill Village Panchayat of Eastern Nepal", University of Sussex D.Phil Thesis.

Barnard, G., Hall, D., 1981, Biomass for Developing Countries, published by Pergamon Press (1982).

Brokensha, D., Riley, B., 1978, "Forest, Foraging, Fences and Fuel in a Marginal Area of Kenya", USAID Africa Bureau Firewood Workshop.

Caceres, R., 1981, "Estrategias Energeticas para la Agricultura de Subsistencia" to be published as Guatemalan National Case-Study in Wionczek, Foley and van Buren (1982).

Cipolla, Carlo M., 1980 (2nd Edition), "Before The Industrial Revolution: European Society and Economy, 1000-1700", Methuen & Co., Ltd., London.

Deshler, W., Fleuret, A., Fleuret, P., Fortmann, L., Reinig, P., 1979, "Ecological and Socio-Economic Profile of the Rural Poor in Tanzania", American Association for the Advancement of Science, Consortium of Affiliates for International Programmes.

Dyer, A.D., 1976, "Wood and Coal: A Change of Fuel", History Today, Vol.XXV, No.9 (Sept.) pp.598-607.

Fleuret, P., Riley, B., 197 , "Fuelwood Use at Lushato", University of California, Santa Barbara, Department of Anthropology.

Floor, W., 1981, "Comparison of cooking stove efficiencies" (mimeo).

Foley, G., 1976, The Energy Question, Penguin (2nd edition, 1981).

Foley, G., 1981, "Rural Energy Planning in Developing Countries: A New Framework for Analysis", Stockholm Beijer Institute Lecture.

Guzman, O., 1981, "Energia y Sector Agricola de Subsistencia" to be published as Mexican Country Case-Study in Wionczek, Foley and van Buren (1982).

Hardin, G., 1968, "The Tragedy of the Commons", Science, Vol.162, pp. 1,234-1,248.

Hyman, Eric L., 1982 (April), "Analysis of the Woodfuels Market: A Survey of Fuelwood Sellers and Charcoal Makers in the Province of Ilocos Norte, Philippines", East-West Centre, Hawaii.

IIED: International Institute for Environment and Development, 1980, Energy Policy in Tanzania, Commonwealth Fund for Technical Cooperation.

Islam, M. N., 1980, Study of the Problems and Prospects of Biogas Technology as a Mechanism for Rural Development: Study in a Pilot Area of Bangladesh (Village Resources Survey for the Assessment of Alternative Energy Technology), Dacca, Bangladesh University of Engineering and Technology/International Development Research Centre.

Leach, G., 1975, Energy and Food Production.

Leach G., Lewis, C., van Buren, A., Romig, F., Foley, G., 1979, A Low Energy Strategy for the United Kingdom, London, Science Reviews.

MacArthur, 1981, "Monitoring the Impact of Agricultural Projects on Low Income Groups: Principles and Problems", in Clayton, E., Petry, F., eds. Monitoring Systems for Agricultural and Rural Development Projects, Rome, FAO Economic and Social Development Paper 12 En.

Meta Systems Inc. and Instituto Tecnologico de Costa Rica Florestal Acesita S.A., 1982 (May) (revised), "An Examination of the Substitution of Woody Biomass Based Fuels for Oil in the Industrial Sectors of Costa Rica".

Mung'ala, P. M., 1978, "Estimation of Present and Likely Future Demand for Fuelwood and Charcoal in Machakos District, Kenya", University of Dar es Salaam (Forestry), MSc Thesis.

Nef, J.U., 1977 "An Early Energy Crisis and its Consequences", Scientific American, no.391 (November), pp.140-151.

Nef, J.U., 1932, "The Rise of the British Coal Industry", Vols.I and II, Routledge & Sons, Ltd., London.

Nkonoki, S., 1979, "Tanzania Rural Energy Consumption Survey", University of Dar es Salaam, Institute of Development Studies.

Nibbering, Waluyo, Aart Schrevel, and Palte Jan (ed.), 1980 (September), "Firewood Trading and Consumption in the Kali Konto Project Areas, East Java", Department of Geography of Developing Countries, University of Utrecht, the Netherlands.

Park, W., Newman, L.C., and Ford, K., 1982, Fuelwood Supply for Managua, Nicaragua: Sustainable Alternatives for the Las Maderas Fuelwood Supply Region, MITRE/CATIE, Virginia, USA.

Stassen, H. E. M., 1981, "Energy Technologies for Rural Development", in Wionczek, Foley, van Buren (1982).

Temu, A., 1979, Fuelwood Scarcity and Other Problems Associated with Tobacco Production in Tabora Region, Tanzania, University of Dar es Salaam, Division of Forestry, Record No.12.

Wionczek, M., Foley, G., van Buren, A., 1982, Energy in the Transition from Subsistence Agriculture, Boulder, Colorado, Westview Press.

PERSPECTIVES ON RURAL ENERGY: DEMAND, SUPPLY AND CONSTRAINTS

Jyoti K. Parikh

International Institute for Applied Systems Analysis,
A-2361 Laxenburg, Austria

INTRODUCTION

Rural energy planning has only recently received attention because the rural energy supply has consisted of gathered fuels, wood and waste, which, up till now, were supposed to be locally available, renewable resources. However, it is now recognized that this supply is not renewable if population pressure creates needs in excess of the capacity of the environment, in particular the land, to fulfill them. With rural development affecting more than 70% of the population in the developing countries and becoming an essential element for national development, its energy needs are gradually entering into the planning process. If rural development is intensified, rural economic activities will be different from what they have been in the past, and will call for a different energy consumption pattern. The present inefficient forms of traditional energy sources, supply of which is uncertain, would no longer suffice.

The present paper discussed the current energy consumption pattern, changes expected in this consumption pattern in the future and the possibilities of energy supply and constraints in exploiting them.

H. Neu and D. Bain (eds.), National Energy Planning and Management in Developing Countries, 57–76.

1. PRESENT ENERGY CONSUMPTION PATTERNS

Energy obtained directly or indirectly from agriculture often provides 90% of rural energy requirements. Considering fuelwood consumption from FAO forest statistics (1975), agricultural waste estimates by J. Parikh (1980) based on average coefficients of waste to cereal production, and commercial energy, i.e. coal, oil, gas and electricity consumption from the United Nations (1978), the shares of all these energy sources in the various world regions can be seen in Table 1. Africa, the Far East and Latin America obtain 67%, 50%, and 40% of their energy from non-commercial sources. Since the Far East has a much higher population, heavy deforestation has already taken place there, and the area under cultivation has expanded. Their total agricultural production is therefore higher. This explains why in the Far East, the share of energy from agricultural waste is higher. However, on a world level non-commercial energy does not contribute more than 7.7% of the total energy consumption. As percentages of the total world consumption, the developing countries consume 15.5% of the commercial energy and 21.7% of the total energy, including energy from non-commercial sources.

Figure 1 shows that some individual countries are considerably more dependent on non-commercial energy than the world regional averages indicate. The World Bank report (1975) on rural electrification indicates that only 4, 15 and 23% of the rural population of Africa, the Far East and Latin America was served with electricity in 1971.

2. STRUCTURAL CHANGES IN RURAL ENERGY DEMAND

Presently the household sector uses large fractions of rural energy and its supply comes from non-commercial fuels. However, significant structural changes should be expected to occur in the future as a result of the efforts for rural development. The aggregated projected energy demand is given in Table 2 and the changes expected in the structure of demand are qualitatively and quantitatively discussed below.

2.1 Household Sector

Because of population and economic growth, the energy required in households will approximately double and more commercial energy is expected to be used for lighting, comforts and simply because non-commercial fuels will have become scarce.

Table 1 - Commercial and Non-Commercial Energy Consumption for World Regions in mtoe, 1973 and 1978 (J. Parikh,1980)

Region	Commerc. Energy	Fuelwood Charcoal	Agricult. Waste	TOTAL
AFRICA				
1973	44.8	78.8	14.9	138.5
	(32.3)	(56.8)	(10.9)	(100.0)
1978	56.1	91.2	13.0	160.3
	(34.9)	(56.9)	(8.2)	(100.0)
FAR EAST				
1973	167.9	97.2	68.0	333.1
	(50.4)	(29.2)	(20.4)	(100.0)
1978	210.7	98.8	66.8	376.3
	(56.0)	(26.2)	(17.8)	(100.0)
LATIN AMERICA				
1973	206.7	66.6	21.1	294.4
	(70.2)	(22.6)	(7.2)	(100.0)
1978	262.5	58.9	14.3	335.7
	(78.2)	(17.5)	(4.3)	(100.0)

Table 2 - Projected demand for energy and rural energy for the three world regions in 2000, under the high rural development scenario

	Africa	Far East	Latin America
Population (millions)	763	2134	620
Rural population (millions)	475	1387	156
Energy			
Commercial energy cons. (mtce)	338	1094	1585
Non-comm. energy cons. (mtce)	179	336	127
Total energy cons. (mtce)	517	1430	1712
Total energy cons. per cap. (tce)	0.68	0.67	2.76
Comm. energy cons. per cap. (tce)	0.44	0.51	2.55
Comm. energy in total energy (%)	65.4	76.5	92.6
Rural energy			
Rural comm. energy (mtce)	117	165	206
Rural comm. energy cons. per cap. (tce)	0.25	0.12	1.32
Rural total energy cons.	296	501	333
Rural comm. energy in total rural energy (%)	40	33	62
Rural energy cons. per cap. (tce)	0.62	0.36	2.10

PERCENTAGES OF NON-COMMERCIAL ENERGY IN TOTAL ENERGY IN SELECTED DEVELOPING COUNTRIES (1978)

10–20	21–30	31–40	41–50	51–60	61–70	71–80	81–90	91–95
Chile Argentina Egypt	Malaysia	Colombia Brazil	Philippines Pakistan India Zimbabwe	Guatamala El Salvador Thailand	Sri Lanka	Bangladesh Mozambique Kenya Ghana	Zaire Sudan	Nepal Tanzania Ethiopia

Fig. 1 - Percentages of non-commercial energy in total energy in selected developing countries (1978).

2.2 Intensification of Agriculture

In 1972, 18% of world energy in agriculture was consumed by the developing countries. Their share will increase to 52% by the year 2000 (Parikh, 1980). This is logical even if not all the food required for the 4.6 billion poeple in the developing world is to be produced by them. The Far East, where 78% of arable land is already under cultivation and will have to support nearly 2 billion people from the same land, will alone require for its agriculture 18% of the total world energy in agriculture. Africa, the Far East and Latin America consumed in agriculture 0.9, 4.9 and 4.1% of the world energy in agriculture in 1972 and would require 2.3, 18.6 and 8.9% by the year 2000. Thus the average annual growth rate of energy in agriculture for the three regions, respectively, is 8, 10 and 8%.

2.3 Required Expansion of the Transport Sector

The interdependence of urban and rural areas could be expected to increase when the percentage of urban population is projected to go up from 23, 22 and 59% in 1973 to 38, 35 and 75% in 2000 for Africa, the Far East and Latin America, respectively.

The transport sector would have to be expanded to ensure the supply of at least the minimum of fuel in the rural areas and of food in the urban areas to meet basic needs. In addition to that, increased transport could also be viewed as a key to development. An improved transportation network expands the market for products, and the producer is no longer constrained by the demand in the local market. That brings a better price to the producer and the possibility to exploit economies of scale, which, in turn, provide an incentive to produce more. Other benefits of urban-rural transportation could be also felt in education, health and welfare in the rural areas.

Both passenger and freight transport would grow very rapidly, as there is a considerable suppressed demand in the transport sector; the lack of vehicles and roads and of communication in general has hindered the process of development.

2.4 Rural Industrial Sector

With the rise in non-agricultural population and efforts for rural development, a spurt in rural industrial activities should be expected. The scope for generating rural income from a wide variety of small- and medium-scale rural industries is quite significant. These industries could encompass manufacturing soap, bricks, pottery, charcoal, metal products, leather and agro-industrial products, all of which require con-

siderable inputs of energy (see Annex I).
There are a number of technical reports where a detailed analysis of individual technologies is given. Such descriptions will therefore not be given here. Instead, the focus will be on the issues of technology assessment and the difficulties of embedding new technologies in the rural environment.

3. PROGRAMS FOR SUSTAINABLE SUPPLY FROM RENEWABLE RESOURCES

Commercial energy is expected to penetrate even the rural areas of developing countries in the future, because:

- The limited supply of non-commerical energy cannot cope with rising demand;
- Commercial fuels are efficient, convenient to turn on and off, smokeless, and are generally preferable to present non-commercial energy forms. This preference will be of growing importance as income levels increase;
- The introduction of activities other than subsistence-level activities in rural areas, such as in the industrial and service sectors, as well as a rise in rural transportation, will generate a demand for commercial fuels in the absence of which the process of rural development will certainly be retarded;
- The time taken to collect non-commercial fuels, which is unaccounted for at present, will be accounted for in the future. Even at present some prefer to pay others who specialize in collection. When non-commercial fuels such as firewood, dung and waste are paid for, the user will increasingly shift to the preferred commercial energy as prosperity increases.

However, it should be pointed out that, while wood and dung used directly are non-commercial energy forms, schemes of wood plantations and biogas are considered commercial energy schemes as they require planning, management and investment, and consequently sustainable energy sources. While for many of the end-uses the non-commercial energy sources could be used directly, there would be the need for converting them into efficient fuels required for rural industrial development. The examples of use of dung, wood and crop-residues to obtain biogas, charcoal or alcohol are shown in Fig. 2.
What socio-techno-economic conditions need to be fulfilled for some of the schemes to be successful? Which countries could exploit which technologies efficiently? This is discussed below for four alternatives, biogas, wood, solar and alcohol technologies. Four schemes are considered below for meeting the

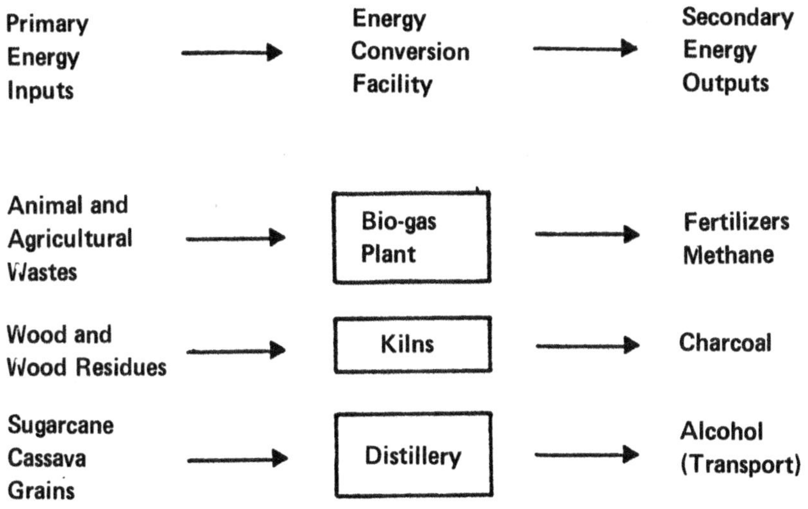

Fig. 2 - Energy from agriculture.

energy requirements of the developing countries. Their technical descriptions are not repeated here as they are given elsewhere in the literature. The options for electricity generation are discussed separately.

3.1 Biogas Generation

i) Ratio of cattle population

For biogas to be a major viable option for the entire village, 0.5 cattle per person would be required. Alternatively, the organic waste availability should be 1 to 2 kg/cap/day to meet cooking and lighting requirements entirely. Note that cattle dung contains more volatile matter than most other wastes. The countries having such possibilities are shown in Fig. 3.

ii) Temperature range

The desirable range is 15 to 40°C for an adequate yield of gas. In colder climates digesters have to be heated, making it more expensive.

iii) Capital cost of biogas

The cost of each family plant should not exceed $ 250[+)] for a 2 m^3 plant, if the free labour to maintain the plant is available. Alternatively, the total cost (capital, labour, maintenance for any plant- village- or family-size) should compete with cost of coal or kerosene imported to the villages. That cost then becomes 5 ¢/m^3 (assuming 14¢ per litre of kerosene). Of course, as kerosene prices rise, higher prices for biogas could be permitted.

iv) Agrarian society

If need exists for fertilizers as well as for fuels, the chances of acceptance are higher in an agrarian society compared to a nomadic or pastoral society; or on the other extreme, an industrial society or urban society. It also helps if a tradition for using dung exists, which is more prevalent in South East Asia compared to Africa or Latin America.

+) Throughout this paper 1974 prices are used.

							Columbia
							Chile
							Brazil
							Mexico
							Cuba
Sri Lanka	Khmer Rep.				Ecuador		Tanzania
Pakistan	Upper Volta	India			Bolivia		Sudan
Afghanistan	Mozambique	Burma	Nepal		Mali		Rhodesia
Tunisia	Morocco	Bangladesh	Guatemala		Kenya		Madagascar
Nigeria	Cameroon	Iraq	Uganda	Peru	Angola		Ethiopia
0.2	0.3	0.4	0.5	0.6	0.7	0.8	0.9

Only partial rural energy requirements for subsistence could be provided.

More than subsistence level rural requirements could be provided.

*The number of cattle per rural population. (If buffaloes and oxen are included, then some countries move to the other side, for example India).

J. Parikh (1980)

Fig. 3 - Number of cattle* per rural population in the developing countries

v) Village plants

The chances of success of a village biogas plant would be higher for villages with clustered dwellings rather than with dwellings scattered over large distances. Also if a village administration structure such as "Panchayat" in India exists the chances are better.

vi) Size of the village

Since it is possible to run biogas on human waste and a municipal sewage plant for an urban area is possible, there is no practical limit on the size of the village. In that case, however, the energy provided would not be sufficient for all the needs so supplementary fuels would be necessary.

If other fuel alternatives such as using coal dust, wood, or charcoal; or oil, gas and electricity exist, there may be less incentive for developing biogas resources. However, the number of developing countries with other alternatives appropriate for rural areas is rather small.

3.2 Wood Plantation Schemes

The important technical and other parameters defining success of the wood plantation schemes are summarized:

i) Yield per hectare

This is the most crucial parameter for making this alternative cost-effective as shown by the sensitivity analysis carried out. The acceptable range should be a 25 to 60 m^3 increment per hectare/year.

ii) Crop rotation cycle

The preferred range (see J. Parikh, 1980) for a crop rotation period (the years after which wood is ready for harvesting) should be 4 to 8 years. If it is 8 years or more, the yield should be very high (50 to 60 m^3/ha) and the land rent, low.

iii) Land rent

The rent of the land per year should not exceed $ 150/ha if the discount rate is 10% and wood is to compete with other fuels (kerosene at 14¢/litre).

iv) Labour-wage rate

The system suggested by Earl (1975) and J. Parikh (1980) is labour intensive and appropriate for villages in the developing countries for wages up to $ 2/day. Industrialized countries also talk about this system on a very large scale but with high investment in mechanization and low labour requirements.

v) Cost of harvested wood

Some flexibility in the afore-mentioned parameters is permitted as long as the cost per m^3 of wood is in the range of $ 4 to $ 10/m^3 in the final analysis. Then, the charcoal made can compete with fossil fuels transported to the rural areas, depending on the distance from the distributing urban areas.

vi) Land requirements

The land requirement per capita is 1/3 to 1/5 of a hectare, if the annual wood requirement is 1 m^3/person for an 8-year rotation cycle with a yield of 25 to 60 m^3/ha. A village of 500 persons, then, would require 70 to 100 hectares of land of energy for cooking and for industries such as pottery and brick making. This could be considered large for some of the developing countries with a high population density. However, some schemes could be initiated in the unproductive so-called forest areas, where very little wood grows at present.

3.3 Solar Applications

i) Solar radiation

The preferable annual average range is 250 to 500 cal/cm^2 per day. However, the frequency of cloudy days, regions with shadow (either due to mountains or to other causes), and nature of demand all have to be analysed together with the average figure.

ii) Solar thermal applications

Low temperature solar devices for heat requirements below 100°C, such as hot water, and space heating and cooling, are cost-effective today provided they are required for more than 3 months a year. High-temperature devices for temperature above 100°C and below 300°C, for which concentrating devices are preferable, would be cost-effective particularly if they are required for daily (non-seasonal) use. These include solar cookers, water distillators and solar pumps for drinking water.

iii) Load factors

It is unlikely that any solar devices would be cost-effective if they are required for only a short duration or for seasonal purposes (for a couple of months in a year or less), due to their high capital and land requirements. In all such cases, oil-based equipment (for example, heat pumps and diesel pumps) is likely to be cheaper even if oil prices are more than double the present prices (in real terms) because the oil required over a small period is not much and could be easily purchased from the interest on solar equipment capital requirements. The use of the solar pump for irrigation purposes with a 10 to 20% annual load factor has been discussed in detail elsewhere.

iv) Solar mechanical or electrical applications

Because of the current and projected prices of solar collectors or photovoltaics, their contributions probably will not be significant to the rural energy supply in the developing countries prior to the year 2000. The technology is not based on a self-help concept and the already small rural earnings would leave the rural areas in purchasing them.

v) Land constraints

Since solar energy could be harvested only to the extent of a few watts/m^2, the land required for a higher intensity use perhaps could prohibit its extensive use for highly populated developing countries.

3.4 Solar Cookers

It is worthwhile discussing a solar application in detail; the solar cooker is taken as an example.
Since a major part of energy use in the rural areas of many countries is for cooking (more than 70% for low-income countries, as shown in the analysis of energy for subsistence), solar cookers are the single important application which, if used by all, could alone relieve a major part of the burden on the forest and fossil-fuel resources in the developing countries.
Assuming that a family uses 2 kg of coal for one meal, 10.7 tons of coal could be saved during the 15-year lifetime of a solar cooker. Assuming a 10% discount rate and a delivered coal price of $ 15 per ton, the break-even cost of a delivered solar cooker is nearly $ 90.
The conservative estimate presented here considers only one cooked meal and no rise in real terms in the cost of fuel and availability at a subsidized rate, the present international selling price being $ 40 per ton.
The solar cooker could be made available today at this price. However, certain improvements are necessary:

a) the possibility of using the cooker within the home;
b) the possibility of controlling the heat;
c) easier installation and use in a limited space.

These improvements do not require a major scientific breakthrough; they can be developed if high priority is given to technological research. The designs should be tried out to suit the different traditions, social customs and eating habits of the people who are likely to use it.

3.5 Alcohol Programs

Recently, alcohol production from sugarcane, cassava and other crops has been considered by many governments. This can be blended with gasoline up to 20% to run cars (D. O. Hall, 1982). Brazil has already made considerable progress in this program. However, the following aspects need to be considered prior to putting investment in this area:

i) Requirements of gasoline

Up till now the use of alcohol obtained from such a distillation process has been limited to cars. The possibility of its use in diesel engines is perhaps remote. To what extent can developing countries invest in programs benefiting car-users, who are in the minority? What are the total requirements of gasoline and cannot the entire burden of importing gasoline be passed onto car-users? The developing countries of South East Asia and Africa have four cars per 1000 persons (Kahn and Hölzl, 1980). Even alcohol programs undertaken in the US can lead to significant repercussions elsewhere in the world (FAO, 1980).

ii) Land requirements

Unlike wood plantations, alcohol programs would require arable land and, of course, irrigation and fertilizers. Thus, they would be in competition with agriculture in most developing countries where the food shortage for the growing population is already a significant problem. Considering energy and agricultural "self sufficiency" using the average trade data of 1974-76 for energy and agriculture, major countries are classified in four parts, as shown in Fig. 4. The countries on the top left are possible candidates for such programs. However, even there this simple criterion does not consider the fact that exports of cash crops, such as tea and coffee from Kenya or meat from Argentina, are essential for the economy and care is necessary before considering these "agricultural surpluses".

iii) Other considerations

There are many developing countries where the land rents are high at specific places where the "alcohol crops" could be grown and there are others where the labour wages are high

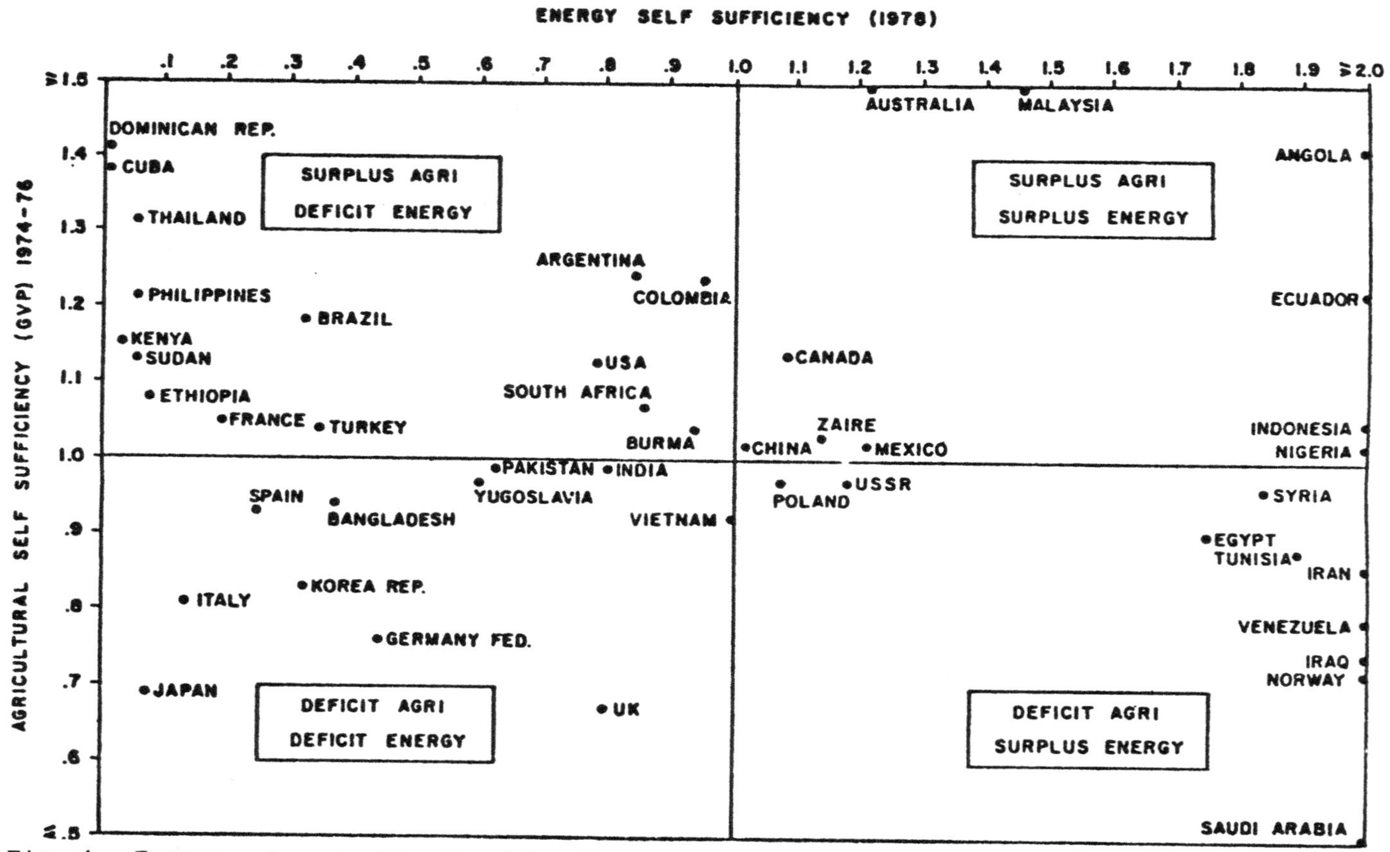

Fig. 4 - Energy and agricultural self sufficiency

Source: FAO (1980)

in these places. Moreover, there is a need for investment and technological infrastructure for these schemes to be successful.

4. THE ROLE OF ELECTRICAL ENERGY

Fuel has accounted for almost 98% of the total rural energy requirements so far and will continue to be important, but electricity is a superior form of energy in many respects. It is preferred for such applications as high-temperature furnaces, welding, lighting, television, deep pumping, grinding and crushing. Thus, the provision of electricity could give rise to new types of activities that otherwise might not be possible. On the other hand, electricity, which requires a great deal of investment, planning and effort to obtain, may not be required for all end-uses, and its indiscriminate use is not recommended. Its best role in rural development is the accomplishment of such goals as:

a) opening up new activities that may be economically productive;
b) improving education, health and welfare;
c) assisting family planning by creating conditions for engaging in a wider variety of activities;
d) helping rural skills to stay in the rural areas by providing better opportunities and thereby reducing rural to urban migration.

The main difficulties in achieving rapid rural electrification are that large capital requirements and the low load factor in rural areas make it financially uneconomical. In the typical developing country, the urban load factor is about 0.50; the rural load factor varies from 0.10 to 0.25. However, the load factor gradually increases with the intensification of agricultural and industrial activities. Thus, one would expect that rural electrification should, in the long run, give a spurt to rural development. This seems to be borne out by a number of recent studies of the agricultural sector in India.[+)] In spite of the problems, most developing countries consider rural electrification a desirable objective and a prerequisite for development. However, the need to cut down costs in providing electricity through appropriate strategy and techniques cannot be overstressed.

[+)] A report by the Government of India, National Commission on Agriculture, 1977.

4.1 Electricity Requirements and their Growth

Unfortunately, there is no magic formula for establishing norms for rural electricity requirements. Much depends on the potential for industrialization and agriculture, size of the population and its income distribution, availability of credit facilities for setting up enterprises, availability of natural resources and skills, and other similar factors. Typical rates of consumption of electricity in various industries and agro-industries are indicated in Annex I. In the following paragraphs is a summary of this aspect from various studies on rural electrification in India+) and Kenya (Parikh, 1980), and by the World Bank (1975).

Once the decision in favour of rural electrification has been made, the next step is to increase the load factor (which is usually 5 - 15% in the initial stages) so that the capital invested in rural electrification is recovered from the revenues obtained. However, the Indian studies indicate that the load factor does not improve much for the first five years and in subsequent years it grows by steps. In the initial stages, the demand grows in the agricultural and domestic sectors. The growth in the rural industrial sector, however, takes time and depends on the availability of credit facilities, extension workers (from whom ideas concerning the use of electricity and training could be expected) and other infrastructure. Setting up of agro-industries usually precedes that of other small-scale industries. Thus, only when one analyses villages that have been electrified for more than a decade, does one see any progress in industry. Concern for lack of industrialization in the electrified rural areas should be viewed in this light - that there is a gestation period.

Assessment of demand based on population and a per capita energy norm are simplistic and may lead to false estimations. The demand for service hook-ups must be estimated from the income distribution and the nature of activities that may come up in the cluster of villages. Having assessed the demand, the choice of generation method and the likely unit costs of electricity can be determined as discussed in the next section.

+) Various studies sponsored by the Government of India, 1971 - 1978.

4.2 Choice of Generation Method

4.2.1 Local generation through diesel stations or grid connection

The choice between local generation of electricity or connection to a grid supply depends upon the distance from the existing grid, the expected local and the costs of local generation with a diesel motor-generator set.
A World Bank report (1975) shows that, in rural areas, the capital costs of supply from a grid are much higher than those of local generation, but the fuel, operation and maintenance costs are much less. This is unlike large-scale power plants (100 - 3,000 kW), for which both the capital costs and the operational costs per kilowatt are less than for a small plant. A high utilization factor strongly favours the more capital-intensive and less fuel-intensive grid supply.
Table 3 shows the average costs of the two schemes. The grid is less expensive for load factors of 25% and above and for distances of 4 - 20 km from the main grid.

Table 3 - Comparison of the average cost of electricity supplied from a grid and locally (50 kW)

Load factor (%)	Cost per unit (cents/kWh)		
	Grid		Local
	4 km	29 km	
10	18	40	21
25	7	17	12
50	4	8	9

However, the critical distance varies with the local conditions. A comparison made for Kenyan conditions indicates that the break-even point is 45.1 km for having an annual consumption of 200 MWh, 63.5 km for 500 MWh, and 95.1 km for 1,000 MWh.

4.2.2 Other choices for local generation

The costs of local generation may be substantially reduced if local conditions are favourable for exploiting local resources. Of the technologies available or likely to be available in the near future, two seem to be of particular relevance: small hydroelectric plants and windmills.

4.2.3 Small hydroelectric plants

Rural areas can make use of small hydroelectric plants working on any nearby water resource with a fall of 10 - 20 m. If a natural fall is not available, it may be possible, in come cases, to create an artificial fall, providing employment for local labour. Small reservoirs or earthen dams could be used to generate electricity with small turbines.
Since 1966, for example, China has installed 60,000 small hydroelectric stations over 1,000 hilly areas. The total capacity is about 3,000 MW and is increasing annually by 400 - 500 MW. The stations are in the 40 - 250 KW range operating often on heads of less than 10 m.

4.2.4 Windmills

Windmills convert mechanical energy into electricity. The demand for mechanical energy should be satisfied first before any thought is given to using windmills to generate electricity. A detailed description of windmill designs appropriate for developing countries is given in the United Nations series on energy resources. It mentions that the village craft of making wooden mills with cotton or jute sails, which existed a few decades ago in many countries, ought to be revived to reduce costs and employ local labour.
Windmills are especially suited for coastal and mountainous areas. The wind speeds required are in the range of 8 - 25 km/h for mechanical energy for pumping and 15 - 40 km/h for electricity production. There are windmills in Crete that start pumping at a wind speed of 8 km/h and reach the optimum rotational speed of 25 rpm at 13 km/h. If left to themselves, they will increase to a maximum of about 40 rpm in higher winds. However, the sails are usually reefed at speeds above 25 rpm. This type of windmill pumps 3,000 litres/h, lifting water as much as 5 m, 10 - 12 hours per day, 4 - 5 months per year, and has an expected lifetime of about 20 years. Each pumping system costs about $ 480 (windmill $ 320, storage tank $ 120, pump $ 40). The cost of pumping is approximately 1 cent/m^3 (1974 prices).

REFERENCES

Earl, D.E. (1975). Forest Energy and Economic Development. Clarendon Press, Oxford.

Food and Agriculture Organizations's Production Yearbook (1975). Rome.

Food and Agriculture Organization (1980). Sixteenth FAO Regional Conf. for Latin America "Energy Cropping versus Agricultural Production", LARC/80/8. Rome.

Hall, D.O., Barnard, G.W., Moss, P.A. (1982). Biomass for Energy in Developing Countries, Pergamon Press, UK.

Khan, A.M. and Hölz, A. (1981). Evolution of Future Energy Demand till 2030 in Different World Regions - An Assessment Made for the Two IIASA Scenarios. Forthcoming Research Report. Int. Inst. for Appl. Systems Anal., Laxenburg, Austria.

Parikh, J.K. (1979). Renewable Options: What Could Developing Countries Expect from Them? Presented at the UN Univ. Conf. at the East-West Centre, Honolulu, Hawaii, 1979. Published in Energy, pp. 989-994.

Parikh, J.K. (1979). Planning of Rural Energy Systems: Issues and Perspectives. Published in Appropriate Ind. Techn. for Energy for Rural Requirements, UNIDO's Monographs on Appropriate Industrial Technology, pp. 119-140.

Parikh, J.K. (1980). Energy Systems and Development; Constraints, Demand and Supply of Energy for Developing Countries. Oxford Unv. Press., Delhi, India.

U.N. Energy Resources Development Series No. 16 (1976). Proc. Subtitle of the Meeting of the Expert Working Group on the Use of Solar and Wind Energy, ESCAP, Bangkok, Thailand.

United Nations (1979). World Energy Supplies 1973-78, Statistical Papers, Series J., No. 13, New York.

World Bank (1975). Rural Electrification. A World Bank paper. Washington D.C.

ANNEX I

INPUT NORMS FOR ENERGY USES IN RURAL INDUSTRIES

Industry	*Daily consumption of electricity (kWh)*	*Average daily consumption of raw materials and fuels*	
Sugar	81 000	Sugar cane	125 t
		Sulphur	10 t
		Limestone	20 t
		Firewood	25 t
		Coal	20 t
Coffee grinding	1	Coffee beans	33 kg
Rice milling	32	Paddy	2 t
Flour milling	36	Paddy	1.2 t
		Diesel oil	18 litres
Poultry	47	Water	1 000 litres
Dairy	50		
Sericulture	1.1		
Textiles	95	Cotton	1.4 t
		Coal	0.66 t
		Fuel oil	6.58 litres
Wool (carding)		Wool	1.64 kg
		Diesel oil	1 litre
Straw board		Waste paper	1.7 t
Press	3		
Tyres	3.3	Rubber	12.36 kg
		Firewood	0.46 Qtls
		Petrol	1.3 litres
		Water	3.3 t
Oil (ground-nut)	63.2	Ground-nuts	2.96 t
Painting	16	Paint	1.67 litres
Pottery	33.3	Water	·40 litres
		Kerosene	2 litres
Engineering	30	Welding rods	70
		Mild steel sheet	500 kg
Iron and steel	10 400	Furnace oil	8.22 litres
		Raw iron	0.4 t
Aluminium utensils	18.9	Aluminium	13 kg
Mechanical repair	3.3		

Source: Perspective plan for rural electrification in the Telegana region of Andhra Pradesh, India. National Council of Applied Economic Research, New Delhi, India.

ENERGY NEEDS AND URBANIZATION

B. Chateau
Institut Economique et Juridique de l'Energie,
Grenoble.

HABITAT AND SERVICES IN URBAN AREAS

As regards energy, urban economic life displays many specific features which radically distinguish it from rural life: different needs, uses, economic activities and economic and spatial content. In this respect, retrospective analysis of energy consumption patterns in industrialized countries is revealing since, it clearly shows that urbanization and the huge increase in energy consumption during the last 30 years were concurrent phenomena (see Table 1).
In this study we shall endeavour to identify and analyse the specific energy features of urbanized economic systems.

ENERGY AND THE URBAN ECONOMIC FUNCTION

Briefly, we can link the urbanization phenomenon with the process of population concentration caused by a certain type of industrial and economic development. In addition to the economic activities directly linked with this development, other activities appear which are derived from the necessity of organizing and controlling this human concentration. Thus the town becomes a privileged centre of multiple service and administration activities, the existence of which is itself linked to the phenomenon of urbanization.
As far as the geographical situation is concerned, the town is characterized by increasing spatial separation, as population grows, of various societal functions (such as residence, work, trades and leisure facilities) which create specific urban

H. Neu and D. Bain (eds.), National Energy Planning and Management in Developing Countries, 77–88.

TABLE 1 : URBANIZATION AND ENERGY GROWTH

	1950		1960			1980	
	hh. (9)	tce/ cap (7)	hh. (9)	pers. (8)	tce/ cap (8)	pers.	tce/ cap (8) (1978)
United States	0.63	3.17	0.72	0.67	8.17	0.73	11.37
France	0.55(5)	1.95	0.65(1)	0.62	2.47	0.78	4.37
United Kingdom		4.40	0.81(2)	0.86	4.86	0.91	5.21
West Germany	0.41(6)	2.57	0.45	0.77	3.70	0.85	6.02
Japan		0.553	0.69(3)	0.62	1.17	0.78	3.83
Brazil	0.32(4)	0.19	0.49	0.46	0.33	0.65	0.79
Korea		0.05	0.29	0.28	0.26	0.55	1.36
Others		0.12	0.33	0.29	0.15	0.41	0.29

(1) 1962
(2) 1961
(3) 1963
(4) 1940
(5) 1946
(6) 1948
(7) Source: J. Darmstadter et al., Energy in the World Economy, J. Hopkins Press, Baltimore, London, 1971.
(8) pers. = % of urbanized population
Source: World Bank, Report on the Development in the World, 1980.
(9) hh. = percentage of urbanized households
Source: Statistical Yearbooks 1951, 1966.

needs.
As far as energy is concerned, four main points should be made:

- certain activities, which exist only in embryo in rural areas, are developed and organized in urban areas, so generating a substantial increase in associated energy needs (for example all the town facilities);
- other activities, intrinsically linked to urbanization appear to generate new energy needs (e.g. various types of leisure activity, urban transport, etc.);
- the employment activities of city dwellers, radically different from those of rural people, lead to behaviour and life styles which are, in their turn, very different and much more energy intensive (e.g. office work);
- whereas renewable and non-commercial energy is the basis for satisfying energy requirements in rural areas, urban settlements mainly rely on commercial energy.

In studying the energy requirements of urban areas, these specific energy features lead one to propose a typology based on the following categories:

- households, sometimes classified according to incomes or social and professional categories, which coincide with "dwelling" units;
- the tertiary sector, distributed according to function (teaching, offices, trades, etc.);
- industries, specifying those which are intermediate, concentrated and energy intensive (e.g. iron and steel, non-ferrous metals, cement, etc.);
- urban transport;
- inter-urban transport, of persons and goods, on which the physical and economic continuity of the urban system depends;
- the rural sector, which looks increasingly like an economic and spatial extension of the urban area.

For obvious reasons of simplicity we choose to limit our analysis of "Energy and Urbanization" to the first two categories households and the tertiary sector.

THE ENERGY REQUIREMENTS OF HOUSEHOLDS

Typology of needs

The energy requirements of households are considered here as being those for useful energy which springs from

social needs arising from home life.
Although these social needs are independent from the urbanization phenomena, the ways in which they are perceived, their intensity and their mode of satisfaction are closely linked to it:

- The urban environment imposes both constraints on life style and opportunities to satisfy needs (e.g. open fire cooking may be impossible but electricity and running water are available). This is a first specific aspect;
- A second specific aspect of the urban environment is that it defines the style of social organization, including distribution of various domestic duties among family members. This is particularly important as regards the expression of needs and the conditions of their satisfaction (e.g. eating habits, use of homehold appliances);
- Finally, the routine and rhythm of work constitute a third important specific aspect of the urban environment, the effect of which is particularly noticeable at the level of intensity of needs (this is particularly true as regards such needs as thermal comfort levels or communication possibilities).

Thus we are led to define a typology of household energy needs which is specific to urban areas and reflects not only basic requirements but also their expression in an urban environment. Thus we distinguish between energy needs linked with

- thermal comfort (heating and/or air conditioning);
- availability of hot water for cleaning;
- cooking;
- lighting and the use of household electrical appliances (i.e. specific requirements from electricity).

Development of needs in industrialized countries

Energy consumption per dwelling in industrialized countries has developed as illustrated in Table 2.
Analysis of the development of these needs in industrialized countries - where the urbanization phenomenon is the oldest and most complete - brings to light the following facts:

1) During the last three decades, energy needs for heating have increased spectacularly almost everywhere, despite a slight improvement in thermal insulation levels. This is mainly due to widespread adoption of central heating, rather than to the general - but modest - increase in the unit volume of dwellings. In fact, by allowing comfortable temperatures to exist throughout the house, rather than just in the living room as previously, central heating techniques have led to a marked

TABLE 2 : FINAL ENERGY CONSUMPTION PER DWELLING (GJ/Dwelling)

	1955	1960	1970	1977
U.S.A.	153	166	192	–
U.K.	104	96	82	78
France	–	44*	–	84
F.R. Germany	–	57	87	–
Japan	14	14	30	–

* 1962

Source: B. Chateau, B. Lapillonne, Energy Demand, Facts and Trends, A comparative analysis in industrialized countries, Springer Verlag, Wien, New York, May 1982.

increase in energy requirements to achieve thermal comfort levels (thus changing the intensity and form of expression of this need). Undoubtedly it was the combined effect of rising income levels and low prices for hydrocarbon fuels which led to this phenomenon. We do not yet know to what extent the increase in the cost of commercial energy is capable of slowing down - or even totally avoiding - such a phenomenon in countries where it has either not commenced or has only just started.

2) A similar phenomenon occurred in the industrialized countries regarding hot water for cleaning purposes. This was, however, due to two different trends:

. one stems from the initial use, then the general need for hot water for cleaning and washing purposes;
. the other, quantitatively much more important, follows from the progressive installation of baths and shower rooms in new dwellings - a virtual necessity for new dwellings from the beginning of the Sixties onwards.

Whereas this phenomenon has followed the rise of living standards it seems to have little connection with the level and trend of energy prices, being more the expression of a new social need, probably linked to the urban life style.

3) Energy requirements for cooking have developed under contradictory influences which it is very difficult to separate:

. demographic influences such as "decohabitation" (decrease in the number of persons per household) and the corresponding increase in the number of households;
. changes in eating habits (meals eaten outside of the home, deep-freezing, etc.;
. changes in cooking habits, etc.

On account of the absence of statistical information and accurate surveys concerning these needs, we cannot know how and in what sense they actually were developed. This is of minor importance since the level of energy needs for cooking is low compared to other domestic energy needs.

4) Two main factors account for the development of household electrical energy requirements:

. the general adoption of electrical lighting and the substantial increase in illumination requirements that accompanied it;
. widespread adoption of major household appliances (refrigerators, washing machines, etc.).

Whereas the first factor is mainly of a technical nature (although not unconnected to the rise in living standards), the second is a direct consequence of increased incomes. In fact, we can discern a very high correlation between levels of household appliance ownership and income in all industrial countries. Nowadays the market for refrigerators, washing machines and television sets is close to saturation level in most industrialized countries, although dishwashers and deep-freezers are not yet widely owned by less affluent sections of the population. In addition to income there are other factors to be considered regarding widespread penetration of household electrical appliances: living or eating habits (deep-freezers), number of persons per household (dishwashers), existence of collective equipment (washing machines), easy access to food stores (deep-freezers), etc.

In conclusion, it would appear that the growth of demand for electrical energy has been little affected by the levels and trends of electricity prices. This is probably due to the large difference between the high cost of the actual household appliances and the relatively low costs of the energy required for their utilization. (Once you have bought a refrigerator you use it.)

<u>What are the lessons for the developing countries?</u>

Before attempting to transpose to developing countries the energy developments observed in industrialized countries during their urbanization phase, it is necessary to know to what extent there exist similarities in the growth of domestic energy requirements. Therefore the two following aspects of this question should be considered:

. What are the interactions between development of life styles and the developing urbanization pattern? (This question is particularly important when urbanization develops in a very concentrated manner around several very large cities - as in Mexico, Seoul or Sao Paulo);

- What are the interactions between urbanization style and the development of techniques designed to meet energy requirements?

On the basis of the answers to these questions an appraisal can be made regarding the connections between development of services and energy requirements, on the one hand, and between the development of energy prices and these requirements - together with equipment - on the other.
It is likely that, in those countries where development already is underway and resembles that in industrialized countries, developments similar to those described above will take place, with variations due to climate (for instance no heating requirements, potentially large requirements for air conditioning).

THE ENERGY REQUIREMENTS OF THE TERTIARY SECTOR

The tertiary sector, composed of service activities and administration, is for the most part intrinsically linked with the urban phenomenon. The energy requirements stemming from the activities of the tertiary sector are by their nature very similar to those for dwellings, at least in urban areas.
Generally speaking, the development of the energy requirements of the tertiary sector is primarily determined by the development of the population. It is mainly through this linkage that the concomitant phenomena of urbanization and economic growth, generators of tertiary activity and employment, bring about the growth of the energy needs of this sector. It is also through this linkage that the energy implications of greater or lesser tertiarization of the economy are assessed. According, however, to the nature of the tertiary activities, the weight of the various types of requirements and their development factors can differ greatly.
Energy requirements in teaching and training activities are influenced, first of all, by needs for thermal comfort and lighting and, secondly, by the need for hot water for sanitary purposes, by cooking and by the use of electronic and electrical equipment. Their intensity is derived from social standards and from the budget available for the teaching institutions. Globally, they also depend closely on the size of the school and university population, and space standards for teaching premises.
For offices and commercial activities thermal comfort needs and specific requirements for electricity (for lighting, lifts,

electronics, etc.) widely predominate. Intensities are generally determined by working conditions and productivity requirements.Moreover, they depend on the number of people employed in offices and trades, which depends in turn on economic development and the level of centralization, as well as on the levels of occupancy of offices and commercial premises.
Among other tertiary activities, we should mention the sectors of health (public and private hospitals), and tourism (hotels and restaurants), where energy requirements per person employed can be very high. This is true for every type of energy requirement, the total depending on accomodation and lodging capacities.
Finally, public lighting, typically an urban phenomenon, creates requirements for electricity at levels directly dependent on the size of the urban population and the illumination standards demanded.

Since information on the energy consumption of the tertiary sector is generally very fragmentary and imprecise, it is unfortunately not possible to draw reliable quantitative lessons from the past development of requirements in industrialized countries, nor, (even more so), is it possible to transpose their experience to situations in developing countries.

ENERGY NEEDS FOR PASSENGER TRANSPORT

Any journey made by an individual by means of a vehicle entails energy consumption - either directly, if the vehicle is used for the individual's sole use, or indirectly, if the vehicle is being used collectively in any case. The level of this consumption, and the type of energy required, depend directly on the length of the journey, the mode of transport used and the conditions in which it is used.
On a national (or regional) level energy demand for passenger transport is thus directly determined by the number of individuals travelling (the population of the country), by the frequency and length of their journeys, the distribution of these journeys among the various modes of transport and the technical characteristics and conditions of use of vehicles.

Needs for mobility

Within the range of mobility needs one can generally identify those corresponding to:

. home-to-work trips, the frequency of which is only determined nationally by the activity level of the population and the organization of work;

- private urban trips motivated by various social needs (shopping, pleasure, etc.) the frequency of which is determined, at the material level, by lifestyle and consumption modes, by urban organization (i.e. the level and structure of urbanization), and by the spread of car ownership;
- other private trips, for various purposes and various reasons (e.g. holidays, family visits, etc.) the frequency of which is also linked to lifestyle, to the spread of car ownership and to urban organization, but which also depends on household income;
- business trips of all kinds, made for purely business reasons, the frequency of which depends solely on the level of economic activity.

It should be emphasized that, in the case of most of these trips, distances are directly linked to dwelling zone conditions (urban or rural zone, size of towns, etc.) and depend moreover on the possibility and availability of motorized road transport.

With the exception of certain business trips by car and inter-urban private journeys by car, the distance of which does depend on the size of the country concerned, the length of trips is generally quite independent of country size. This can be confirmed statistically by comparisons of personal mobility within countries of varying sizes.

Car ownership and modal distribution in industrialized countries

Car ownership influences the modal distribution of trips in two ways:

- directly, since the availability of a private car offers an alternative which did not previously exist;
- indirectly, since car ownership influences both the frequency of trips (spread of specific car trips) and their length (complete freedom as to place of dwelling).

Historically, as we have seen, this phenomenon in industrialized countries resulted in a growth of personal mobility and an increased contribution of the private car to this mobility, without any corresponding decrease in the mobility ensured by public transport (see Table 3).

Comparison of car ownership in three very different industrialized countries (France, the U.K. and the U.S.) shows that:

- there is an extremely strong and general relationship between increase in household income and purchase of a first car;
- the relationship between increased household income and multiple car ownership is both much less apparent and also different in the three countries.

TABLE 3 : PER CAPITA PASSENGER MOVEMENT (km/capita/year)

	1955	1960	1970	1975	1977
U.S.A.		11050	14520	16320	
	(7300)	(10130)	(13160)	(15500)	(16600)
France		2970	6870	8750	
		(1670)	(5400)	(6900)	(7900)
F.R. Germany		4550	7500	8430	8850
		(2920)	(5780)	(6560)	(7000)
U.K.	4020	4870	7210	8000	8220
	(1705)	(2740)	(5510)	(6380)	(6640)
Italy	1880	2450	5750	6620	
	(820)	(1190)	(3950)	(5000)	(5100)
Japan		3840*	5560	6250	
		(400)*	(1730)	(2200)	(2330)
Netherlands		3050	7080	8970	
		(1380)	(5630)	(7530)	

* 1965

Figures in brackets: per capita movement with private cars.

Source: B. Chateau, B. Lapillonne, op. cit.

In addition, analysis of household income in the different countries leads to the following three important conclusions:

. the level of car ownership is almost entirely determined by the average price of cars, average income and income distribution: it appears to be quite independent of petrol prices;
. the price of petrol has an influence on car ownership and on the trend to multiple car ownership (once mobility permitted by the first vehicle has been saturated) without being a determinant factor;
. multiple car ownership is linked in part to income, but depends also on urbanization patterns and the development of public transport services.

Conditions of availability of public passenger transport, characterized by infrastructure, frequency of services, price and quality thus influence the modal distribution of journeys in two ways:

. by means of competition with private cars, whenever such

competition exists (for journeys between home and work, holiday trips, etc.);

. by limiting multiple car ownership to a greater or lesser extent.

Comparisons between countries and towns where public transport has developed at different rates indicate that similar levels of mobility may be ensured by quite different modal distributions.

From the point of view of energy the degree of freedom in modal distribution of journeys is most important since the difference per passenger-kilometre transported differs on average by a factor of three between public transport (except air transport) and private cars.

Despite minor technological differences cars differ considerably in their consumption (thus North American cars require, on the average, twice as much petrol per kilometre as European cars). Weight, power and conditions of use account for the greater part of these differences.

Two main factors can be used to explain consumption per passenger-kilometre transported for public types of transport: the level of consumption of vehicles and their occupancy rate. For all types of public transport the constant trend of the past few decades has been towards a regular decrease in consumption per passenger-kilometre available, due to increases in vehicle size, the use of increasingly light materials and technical improvement to engines. Rate of occupancy is directly linked to the use made of vehicles (urban, interurban) and to the particular services provided. Thus, in Europe, the rate of occupancy varies from 10% to 15% for urban public transport, from 35% to 45% for interurban rail transport and is around 60% for air transport.

CONCLUSION: WHAT LESSONS CAN BE DRAWN FOR DEVELOPING COUNTRIES?

In terms of the analysis developed above most developing countries have certain special circumstances which must be kept in mind before attempting any adaptation.

For a start, the level of car ownership is generally low and will certainly remain so for a long time to come, either for political reasons (for example heavy taxation of cars to protect the balance of payments), or for economic reasons (slow or uneven development of incomes) or even for cultural reasons (no particular social symbol attached to the car).

Secondly, the main determinant of the level of need for motorised trips - especially interurban trips - almost always

corresponds to the level of supply of public means of transport. These are generally used to the maximum of their capacity: the potential demand which could be met by public transport thus almost systematically exceeds the supply.
Finally, in a great number of developing countries, non-motorised trips (on foot, by bicycle or by animal transport) and trips on two-wheeled vehicles still account for a great majority of journeys. The two most significant aspects of the relationship between means of transport and energy forms thus become:

- the availability of a two-wheeled motor vehicle and its effect in terms of mobility;
- the consequences of an increase in the supply of public transport services for the shift from non-motorised to motorised trips and on the development of mobility.

Having made these remarks, one may assume that the analysis of the development of energy needs in the transport sector of industrial countries may generally constitute a sound basis for investigation of possible evolution of these needs in developing countries.

CONSIDERATIONS ON THE RURAL/URBAN DIMENSION IN PLANNING FOR ENERGY IN THE DEVELOPING COUNTRIES

Oliviero Bernardini

Centro Studi Energia (CESEN), Genova

This paper briefly explores the interrelations between agricultural and urban systems as reflected in the energy requirements of development. It underscores the need to seek alternative solutions to the energy problem of developing countries directed more at the deep-seated social and economic origins of development than to their outward energy-related manifestations. Energy, after all, is only the final linkage in the causal chain allowing increasing numbers of people to survive satisfactorily on diminishing land and water resources. For many developing countries there appears to be no solution to the energy problem that does not in some way attack the aggravating resource balance in rural areas.

THE ENERGY CRISIS AND THE COMING REVOLUTION IN ENERGY PLANNING

No earlier than a decade ago energy planning was largely conceived as synonimous with energy supply planning. Energy was generally cheap and abundant. Any increase in demand could be met with in real time, almost anywhere around the globe, by expanding oil supplies. Refinery capacity was in place or could be rapidly expanded to cope with any conceivable increase in liquid fuels requirements. The multinational oil companies were above all interested in accelerating oil production before expropriation from foreign concessions, an event which had been foreseen by their intelligence units at least since the '50s (1). Electric supply systems relying increasingly on oil and gas came in units still sufficiently small in size as to allow relatively short lead times for electricity supply expansion.

H. Neu and D. Bain (eds.), National Energy Planning and Management in Developing Countries, 89–117.

All this came to an abrupt end when Opec countries eventually layed claim to their oil resources in 1974, raising oil prices so as to reflect the long term scarcity value of the resource base. The rise in the cost of energy and limitations in supply which occurred during the last decade put energy on a much more equal footing with other factors of production, specifically capital and labour, sometimes to the extent of inverting previous trends in substitution. It was no matter of coincidence that the '70s also bore witness to the discovery of the existence of a double energy crisis in developing countries. Especially after the second shock in 1979 and the spontaneous return to woodfuels in many end use sectors, many oil-poor developing nations began to give serious consideration to their neglected traditional fuels sources, which at least did not imply deficits in foreign exchange. In many instances, however, the deterioration of forest resources and woodfuel supplies was too far advanced to allow any immediate hope of salvation coming from this direction (2).

The rediscovery of the traditional fuels gave rise to considerable activity in almost all developing countries to obtain the necessary cognitive base relating to woodfuels and other biomass patterns of supply and demand. In the early '70s information on this aspect of energy use patterns was almost non existent (3). In the early '80s almost all developing countries had either undertaken one or more major woodfuel and rural energy surveys, or were planning to undertake one (4).

The other major rediscovery spurred by the renewed interest in traditional fuels regards the non-monetary sector of the developing economies. Our conventional monetary measure of the wealth of developing countries refers almost exclusively to the urban modern sector. It is still common practice to adopt indicators relating the economic and energy using characteristics of the modern sector to the country's total population, as if the urban and rural sectors were part of the same tightly interknit world. According to World Bank estimates, per capita income in the low income countries averaged 230 US $ in 1979. If the rural non-monetary sector is excluded from these estimates (as it should be) this figure increases to the much more respectable 1353 US $. In the same way commercial energy consumption increases from 0.32 to 1.91 toe/capita (5). These effects are clearly much more evident for the African countries with largest rural population components.

In developing countries, no energy planning exercise can be considered serious and meaningful that does not address the traditional fuels and the non-monetary sector in their own right. Indeed for many countries in their very initial stages of development, the vast majority of economic and energy assessments based on conventional tools originally developed for industrial societies and transferred indifferently to the less developed countries (6), must be considered as largely useless to the extent that they focus only on the modern sector. Lumping together the two extremes of what is in reality a dual economy and treating them as if they were part of the same average system, implies missing some of the more important phenomena underlying social and economic change, and many of the most critical planning variables.

One of the more recurrent aspects of the new focus of energy planning in the developing countries relates to the technology, economics and social management of accessible and benign energy systems for the rural population either based on traditional resources or in substitution of them (7). The broader aspect which in the long run appears to be the most far reaching, however, is the integrated systems approach to national energy planning. In the developing countries, the major systems interaction is with agriculture and the rural non-monetary element of the population. It is to some of the major aspects of these interactions with the energy system and with particular reference to three countries, Kenya, Indonesia and Egypt, that this short essay is addressed.

Foremost it is necessary to get away from the conventional notion of economic growth as the ultimate objective. Most energy planning studies and the formal or conceptual models behind them, are built on the notion that the objective of human societies is the maximization of gdp. Brought to its extremes however, the negative environmental and social repercussions of uncontrolled industrialization and economic growth, which have become increasingly apparent in the developed countries during the '60s and '70s, have in due course led to a reconsideration of the value of the concept of economic growth as a measure of well-being (8).

Uncontrolled economic growth and accompanying urbanization have been charged with irreversible effects of urban and technological stress on the neuroendocrine system with eventual disruption of human biological rhythms, pathological breakdown of space and time consciousness (9), growing breakdown of the human communication process and the privatization syndrome (10), loss of spatial and social responsibilities followed by declining citizen participation in the local decision making process, rising crime rate and increased central intervention in social control, loss of local to central decision making power, vulnerability of complex centralized systems (11), vulnerability of specialisation, frustration and revolution of rising expectations, environmental breakdown through air, land and water pollution, hazard of mass destruction through nuclear and biological weapons, threats to privacy and freedom through surveillance technology and bioengineering (12), and many other.

In ultimate analysis, gross domestic product is only a measure of the monetisation of the urban system requirements, a reflection of the money flows that are necessary to lubricate the highly complex, myriad interactions necessary to support urban mechanisms of survival. Most of the multiplied urban earnings are required to satisfy wants that are eventually, only regrettable urban induced necessities related to (13):

- high freight transport demands due to increasing hinterland and area of sustenance,
- high distribution and storage demands, and increasing number of stages and storage levels from bulk delivery to point of sale,
- high personal transport demands to satisfy increasingly distant local life requirements and to recapture the living space lost in the rural/urban transition,
- the deployment of ever novel technologies to save premium time and space in normal household chores,
- increasing demands for products and services of urban societies living under increasing time and space constraints,
- rapidly increasing industrial and agricultural labour productivity related to worsening balance between active population and urban induced demand for products and services
- multiplied commercial energy requirements of urban societies.

On the other hand, the inevitability of urbanization in many population/resource situations around the world does require and is impossible without, rapid economic growth. Any growing urban system that does not have adequate economic growth is doomed to inanity and reversal in the long run.

The question is not whether to urbanize or not to urbanize but what type of urbanization to pursue. Recognising that the objective of societies is not economic growth but individual betterment and survival on an ever diminishing natural resource base, and that monetary flows are essential only to simplify and make possible transactions in increasingly time and space limited urban systems, then the role of economics in energy system planning in developing countries may well lose its time honoured position of centrality.

The major features of an energy planning model based on real phenomena rather than their derived monetary flows are:

- the causal determinism between population growth on limited natural resources and urbanization,

- the direct relation between economies of scale, regional markets, and the pattern of urbanization,

- the non-linear relation between urban concentration and commercial energy systems.

Population growth and urbanization

Though the micro-level decision of whether or not migrate, taken at the household or at the individual level, is a highly complex process, in developing countries, characterised by large rural populations, there can be no doubt that there exists an intimate macro-level correlation between population displacement and the decline in economic opportunities on the land (14). Thus while the ultimate destiny of migrants is determined by the differential pull effect exercised by the different attractive potentials of competing alternative settlements, which may be rural or urban, the underlying causality of migration flows in developing countries seems to be largely anchored in the push effect of the deteriorating balance between growing population and limited local resources.

Historically, population movements have been shaped by natural barriers such as mountain ranges, deserts, lakes and seas. Geographical and climatic, as well as ecological features delimit catchment areas within which the population is by and large contained. The appearence of urban catchment areas of national scope and the development of channels of physical communication, such as roads and railways connecting catchment areas, have naturally increased the leakage between them. But even in industrialized countries with well-developed transport networks, well-established cultural as well as economic ties tend to constrain population movements within boundaries which were historically of geographical and ecological nature, though interregional leakage has in time, increased considerably.

Historical accidents and geographical peculiarities are sometimes responsible for local situations but, on the whole, regions with the greatest population are those with the most fertile land, despite the fact that distance, poor communication, poverty and established cultural diversity have in the past provided almost insurmountable obstacles to migration. The continuing occurrence of these phenomena in Kenya are clearly revealed in a comparision of real and natural population growth in agro-ecological regions that have historically functioned as population catchment areas. Table 1 shows the historical changes in the magnitude and direction of migration flows that have accompanied the decline in land availability over the last two decades. From these results it would seem that in Kenya the turning point from net rural inflow to outflow sets in when the local per-capita equivalent high potential land availability declines below roughly 0.5 - 0.7 ha.

Under conditions of rapid population growth, however, migration to other agricultural regions provides only a short-term palliative. Rural/rural migration which had been increasing in Kenya prior to 1969 with the opening up of the lower Rift Valley areas, is even now quite clearly on the decline as indicated by an analysis of the most recent census returns (16). As the best lands are used up under current technology, migration patterns turn, in fact, from predominantly rural/rural to predominantly rural/urban.

Once the last agricultural frontiers are conquered the rural poor are forced to migrate increasingly to towns and cities. It is difficult to stop the mounting pressure of urbanization while inhabitants perceive a greater chance of a decent living standard in urban areas, even if this chance is minimal. There has been in the past a direct inverse relation between aggregate high potential agricultural land availability in Kenya and the rate of

Table 1 Agricultural land availability and interregional population movements in Kenya (15)

Region	equivalent high potential land fraction of total area	per capita (ha) 1962	1969	1979	Net regional immigration (% per year) 1962-69	1969-79
1	0.59	0.80	0.61	0.44	+0.5	-0.7
2	0.59	0.55	0.41	0.32	+1.1	-1.7
3	0.44	2.53	1.96	1.25	+0.6	+1.2
4	0.33	0.69	0.54	0.38	+0.4	+0.2
5	0.75	0.58	0.48	0.34	-0.7	-0.3
6	-	-	-	-	-	-
7	0.15	0.38	0.31	0.21	+0.1	+0.6
8	0.21	1.36	1.06	0.76	+1.3	-0.1
9	0.23	5.15	4.36	2.56	-0.6	+2.0
10	0.06	2.38	2.19	2.34	-1.4	-4.3
11	0.02	1.57	1.67	1.14	-3.1	+1.1
12	0.05	1.23	1.10	0.73	-0.8	+1.2
Kenya	0.15	0.99	0.78	0.56	0.0	0.0

urbanization (table 2). Regionalized migration models applied to Kenya (16) indicate that under current trends in agricultural technology, small-farm cultivation practices and patterns of land use, the urbanization trends in Kenya will continue to accelerate into the future.

ECONOMIES OF SCALE AND THE PATTERN OF URBANIZATION

The size of cities is related to the market regions inside which they are embedded and the population dynamics of the sub-regions which they represent. These in turn are determined by prevailing production, transport and communication technologies, and geographical and political barriers to population movements.

The natural process leading from population growth to the birth of new towns and later to the uncontrolled growth of single cities is well illustrated in the bimillennial history of Central Europe (13). Surpluses in agricultural production allowed by progress in farming techniques beginning with the millennium (improvements in the plough, the fallow and open-field system, new crops, the 4-crop rotation etc (17)) were accompanied by a considerable increase in population growth, to reach an unprecedented 0.5% per year in the early 14th century (18).

Whether increases in population pressure resulted eventually in aggravated economic opportunities on the land under the low-productivities of the times, or whether for reasons related to the prevailing feudal structure of society, the 300 year period stretching to 1300 was accompanied by the birth of hundreds of new towns, most of them not more than a thousand inhabitants strong, and each typically closed on itself by a city wall effectively limiting, or at least slowing down, the local point growth in population.

Table 2 Land availability and urbanization in Kenya 1948-94

year	equivalent high potential land availability (ha/inhabitant)	urbanization rate (% per year)
1948	1.58	-
1962	1.00	3.9
1969	0.78	4.5
1979	0.56	6.1
1994	0.31	9.4

The Black Death which abruptly ended this first urban revolution in the early 14th century has been frequently ascribed to the poor sanitary conditions prevailing in the new and burgeoning urban centres. More generally, its causes may be identified in the inability of urban-related technologies to be developed and deployed fast enough to cope with the requirements of the unprecedented growth in population in the preceding several centuries. Population remained stagnant, or increased only very slowly, for about 4 centuries, in a life struggle against war and pestilence, again reflecting a continuing general mismatch between resources, population growth and technology.

A town of 10 thousand souls could in principle be nourished, clothed and sheltered from the land contained within a radius of 5-10 km, a short enough distance that could be travelled by ox cart in less than a day's journey. But with the increase in number of towns and their growth accompanying the first urban revolution, the land required for each town's basic sustenance in food staples and material wants, began to overlap with that required for other neighbouring towns. The technologies were not immediately available to solve the resulting production and logistic problems which called for larger scale and speed over increasing transport distances (19).

Sanitary conditions in the urban settlements eventually improved and technologies in all fields became available to allow population to take off once again into sustained growth around the middle of the 18th century. Though this second urban revolution was initially accompanied, as was the first, by the birth of many new towns, this distributed local growth process appeared unable to cope with the unprecedented growth in population under existing land productivities. The new transport technologies of the time (canal, train and ship) allowed enormous increases in market regions to take full effect of the scale economies possible with the novel production techniques and technologies of the industrial revolution. As the growth of new towns became increasingly unable to cope with unsustainable population in rural areas a relatively new phenomenon appeared which was the multipied growth of existing settlements.

The natural sequence of events would not be very different in today's developing countries except for the displacement in time and the enormous changes brought to the world system over two centuries of European industrial development. Many of today's African peoples, for example, were not even remotely on the map a century ago. On the other hand, the existing infrastructure and technological boundary conditions (health care and sanitation, dimension and speed of transport and communication systems, scale and volume of production processes) are very substantially different from those that accompanied Europe during its urban and industrial revolutions. The pre-existence of this technological dowry is in fact having the effect of compressing into a short century one-thousand years of European development, with all the social and economic adjustment problems that this obviously implies.

A basic difference between the early urbanisation and industrialization of Western Europe and that presently occurring in the developing countries, is that for many products these countries are embedded in a world or region-wide market while their resource/population balance does not generally require advanced technology and broad market region solutions. For example, many countries are at a stage in which their resources/population problems could well be solved through the birth and growth of a large number of towns and cities with largely local markets, rather than the extended growth of single metropolises.

A number of perverse mechanisms related to balance of payments, fiscal and monetary policy come, however, to sustain one another in discouraging the multiplication of distributed growth poles and boost the growth of established urban settlements (20). More rapid population growth than experienced ever before, higher powered communication systems, greatly increased barriers to territorial migration have all aggravated the pressures on existing urban settlements. The consequent increase in the demand for modern urban-related products and services has inevitably reflected itself in balance-of-payments problems in many developing countries as yet unendowed with an adequate industrial base.

In the past, as frequently in the present, these problems have been attacked through policies of import substitution. However, the development of import substitution industries usually implies the same or similar capital/labour/technology mix as in the industrial countries. Since noone any longer wants "second best" even though it may be more "appropriate", developing countries frequently get among the more advanced technologies thus aggravating labour/capital ratios even further (21). For reason of size, infrastructural endowment, political relations, administrative requirements and qualified labour availability these industries are usually located in the proximity of major urban centres thus further reinforcing the trend to concentrated urbanization (22). Export-oriented industrial development, to the extent that its products have to compete in world markets, frequently leads to similar scale effects in production function and location, again privileging extreme levels of urban development.

While these policies have in the past apparently alleviated

some aspects of the balance-of-payments burden, the more lasting impression is that they have created enormous problems in labour relations and income distribution (23). Many developing country governments still discriminate against informal sector urban activities competing with modern sector counterparts. To the extent that they are consistent with a lower rural/urban wage differential, the promotion of small-scale traditional activities would reduce misdirected forms of aggressive urbanization and be more adapted to the locally prevailing population/resource equilibrium.

The need to allocate scarce resources to industrial development of frequently capital-intensive nature located in privileged urban centres, has inevitably resulted in neglect towards the rural segments of the economy, and particularly agriculture. This has ultimately resulted in increased differentials between rural-urban wage levels and, as a consequence, increased migration pressure on urban areas. In most instances it has also resulted in increased import dependence for food products, impacting negatively on the balance-of-payments and contributing yet another mechanism favouring concentrated urban growth.

In almost every case the solutions chosen to the developing country problem, fundamentally because they ignored the strong rural-urban coupling, have eventually given rise to a positive feedback response situation. During the '70s the situation has been immensely aggravated by the rise in the cost of oil imports, resulting ultimately in a balance-of-payments cul-de-sac which may now never be solved except on the basis of extreme hardships.

Largely as a consequence of such policies inspired to maximising monetary flows or gdp (24), rather than solving physical problems, most developing countries are now distinguished by a very few (sometimes only one) very large and rapidly growing cities, while other settlements grow more slowly and remain frequently small. This unbalanced urban growth is in dismal contrast to the more distributed growth that has occurred in most of today's industrial countries and particularly in Europe (table 3).

Even so, monoclear urban growth does not set in immediately and indiscriminately, but only once the economy has become more fully coupled to the world market system. The mechanism is, in fact, not clearly apparent in countries in the earliest stages of development such as, for example, Kenya, which are still only weakly linked to international markets. In Kenya the most dynamic growth poles have not been large cities but settlements in the order of 10-20 thousand inhabitants (16), constituting what are in effect still local market regions.

Of the 0.95 millions added to the urban sector in Kenya in the period 1969-79, 0.49 million went to cities with > 10^5 inhabitants, 0.32 million went to towns with 10^4-10^5 inhabitants and the remaining 0.13 million went to settlements with 10^3-10^4 inhabitants (25). If it should be possible to continue this distribution trend into the future, then under a 10% forecast annual urban growth, these settlements would see additions of respectively 3.7, 2.4 and 1.0 million inhabitants by 1995. Because of the increasing number of growth poles with decreasing settlement size the magnitude of the disproportion is in fact more pronounced than meets the eye.

In absolute terms, moreover, the absorptive capacity of one large city is greater than the combined capacity of several small centres. As Kenya becomes more heavily linked to the international system, as it will soon have to under present policy trends, then the increment in the size of the major growth poles will inevitably accelerate, and quite significantly. Much of the above forecast of 10% per year in 1979-94 is most likely to occur in the largest urban growth poles, Nairobi and Mombasa. Under these circustances Nairobi, for example, may well reach a population of 4-5 million by 1995 from a present level of less than 1 million. Such rates of growth are by no means unprecedented, as indicated by the city of Bogota whose population grew 5-fold in the space of a decade to reach 5 million in 1980.

URBAN CONCENTRATION AND COMMERCIAL ENERGY SYSTEMS

Economies of scale in supply of materials and energy inputs, in production, and distribution systems are all in favour of increasing urban dimension. External economies are also generated by clustering of suppliers and customers of intermediates within the same urban continuum leading to substantial savings in transport and communication costs. Economies are also apparent on

Table 3 Annual percentage growth rates of urban agglomerations by settlement density 1950-70 (26)

	settlement density		ratio
	10^2-10^6	> 10^6	
less developed countries	3.94	6.69	1.70
developed countries	2.54	2.25	0.88

account of the finer division of labour and a better match between supply of skills and job opportunities. Further advantages of large urban locations come from the indivisibility of supply of public and private services requiring minimum threshold city size (20).

Micro-economic and technical analyses, however, all too frequently ignore the macro-system energy implications of development policies, again a consequence of our inheritance from the easy oil and gas era, when practically any level of energy demand growth could be quickly and easily matched with supply. Urban development policies and related economic and agricultural policies clearly have significant energy counterparts. It is these interactions, as yet little perceived and ill-understood, that are slowly emerging into a significant systems effect likely to dominate the whole process in the future, even to the extent of inverting past trends in urban and economic development patterns (27).

Increased money flows and increased energy flows are associated, whether directly or indirectly, with all the time and space saving development induced by urbanization of societies. The most evident effect of the urban transition regards the direct and indirect demands that it has generated through transport and storage, housing and home activities on a great variety of different sectors of economic growth stretching from textiles, food processing and steel, to cement, chemicals, engineering, and electronics in manufacturing, and in countless areas of tertiary activities necessary to service the fundamental and derived demands of urban dwellers with ever decreasing time and space opportunities.

The increasing level of mobility of goods and persons induced by urbanisation, and the long distances involved, could be satisfied only at increasing levels of throughput of intermediate products and services each with its own characteristic lifetime. The increased energy requirements to procure food from farm to table includes not only the energy to power the ships, trains and trucks, and the energy to produce them, but also the energy to provide the steel and the steel plants and a complete cascade of manufacturing processes all along the line, each with its own value added and contribution to economic growth as well as energy consumption (13).

Urbanisation not only increased the number of steps between the field and the table, from production, to processing, transporting, packaging and selling the food, with a corresponding increase in the amount of direct and indirect energy use, but the extreme decrease of the agricultural population which accompanied it, also called for new methods of growing and collecting the food which were, quite naturally, energy intensive and large scale. Each of these developments gave rise to a branched chain of upstream activities which contributed to increasing,besides energy demand, also the national accounts measures of "well-being".

The explosion in quantity and variety of manufacturing and service sectors that developed as a consequence of the derived demands of the urban society, implied foremost, the division of human labour and secondly, but no less important, its tremendous increase in productivity, accompanied generally, by an increase in scale and energy consumption.

The number of products necessary to sustain the urban society in fact grew much more rapidly than the labour force. Specialisation through division proved only fleetingly useful in providing productivity increases. The process of induced multiplication of products accompanying urbanisation -and essential to it- and the high rates of throughput at which these were demanded, required much more than labour itself could supply. Mechanisation, in fact, appeared very early on in time and was indeed the principal mark of the industrial revolution. However, while in terms of intermediate demands, productivity gains may loom enormous and bear witness to the ingenuity of the human mind, on the other hand, in terms of the overall final demands which the various product chains satisfy, productivity gains appear pitifully small or even negative. As Leach points out (28), with reference to the food chain in the United Kingdom, it presently takes about 4 million workers -including support industries and services- to supply 6 million toe of food energy a year, giving an output of about 8000 kcal/man.hr "with a performance little better than that for pre-industrial subsistence farmers, albeit in the tropics", but with immensely increased energy requirements per food calorie delivered to the table!

The money flows induced by a given level of urbanisation, are, within limits, the same for any country around the world, as can be readily seen from a standard cross-section between the urban/rural population ratio and per capita gdp (13). Variations, apart from those due to inter-country differences in the definition of "urban" and "rural", must be ascribed to faster or slower rates of approach to the equilibrium condition. In early stages, urbanisation is an extremely dynamic process and it might be expected that some countries would follows the equilibrium growth path closer than others. Differences from the mean are particularly evident for countries with urban/rural ratios between 0.3 and 0.9 (respectively 23 and 47% urban population) which corresponds roughly to the take-off in gdp growth (29).

Resource constraints, misguided or ineffective economic

policies, political upheavals, all retard the approach to equilibrium, while particularly favourable conditions would accelerate it. For some countries, conditions are so unfavourable that urbanisation stagnates or proceeds only very slowly, limited by threshold rates of capital accumulation and build-up of monetary flows. In such countries, located today mostly in South-East Asia, rural-urban wage differentials are unattractive to the point that urban growth is frequently little more than natural population growth (30), with the prospect that reverse migration flows may one day well set-in as potential migrants prefer to survive on declining plots of land (with diminishing body weight) rather than face the dereliction of urban poverty.

The other side of the coin is energy. Greater money flows consequent to higher levels of urbanisation are reflected in increased levels of energy consumption. The overall energy content of each money unit flowing through the urban system, the <u>energy/gdp ratio</u>, is closely linked with the level of urbanisation and the degree of approach to equilibrium conditions as measured by achieved levels of per capita gdp.

The revelated historic experience suggests a marked similarity in the energy/gdp curves between all regions both with regard to their shape and to their positioning on the per capita gdp axis, though not their height. As countries urbanise, there is initially a slow, and later, a more rapid increase in energy/gdp ratio, accompanying the build-up of the urban-industrial system and the related changes in the structure of production and consumption. A maximum in the ratio, which may be as much as 3 or 4 times greater than the value reached in the earliest development, is achieved only in fairly advanced stage of growth (the United Kingdom in 1880, the Unites States in 1925, Italy in 1975). Thereafter, there is a slow but definite decline in the energy/gdp ratio which may be attributed partly to changes in the structure of economic activity but is mostly accounted for by structural turnover, improved logistics and organisation, and, most basically, new technology (31).

The growth process is intimately related to urban build-up, both of cities and communication networks. In the beginning, economic activity tends to have relatively low impacts on commercial energy demands. It is largely concentrated in agriculture and in services and the labour, as opposed to the energy intensive capital component, is high. In this period of growth the relation between income and energy consumption is highly variable between one year and the next. Single additions in capacity to the rapidly growing but still small urban system can make large differences to the overall picture. Averaged out over sufficently long periods of time, however, the relation is approximately one-to-one. That is income and energy consumption grow at about the same rates.

A large number of countries are today in this very incipient stage of development. Very rarely, however, is it recognised that these energy systems are on the verge of a much more rapid and accelerating period of growth than ever experienced in the past. There is a period in time, in fact, when this type of stability of growth is suddenly and irrevocably upset. Mechanisms previously diluted within the system, coagulate and reach a threshold level of influence. A type of chain reaction sets in, with one type of demand leading to another in a rapidly branching system. Considerable identity problems arise for many of these countries in intermediate stages of development, presently experiencing rates of energy growth greater than gdp growth.

By analogy with advanced industrial economies, energy system planning in many of these take-off, or pre-take off countries, is made on the assumption that energy savings induced by increases in the price of oil will quickly and substantially reduce the rate of growth of the energy/gdp ratio from current levels. In reality, however, there is little that can be done in this sense, if the system is to continue on its urban/industrial growth path.

Price is, in fact, only a minor element influencing energy demand growth in early stages of development, dominated as it is by urban and industrial build-up and the related income factor. It is only when the more dynamic phases of growth have subsided, as in industrial countries, that price changes become a significant factor in energy demand growth, even to the extent of dominating the income factor.

The case of Indonesia, illustrated in figure 1, is just one of a number of clamorous examples where official energy forecasts made largely by analogy with industrial country trends, can be dangerously misleading for energy policy planning. According to official forecasts made in 1976 (32), the income elasticity of demand was expected to have reached its maximum value in that year and to decline very substantially thereafter. The observed results up to 1980, were, however much closer in agreement with those obtained from a more refined model of growth taking into account the country's incipient stage of development and continuing rural/urban dynamics (33), which indicated that the maximum might not be attained until the mid or late '80s according to the conditions of growth. Recent forecasts made for the Indonesian Government are more realistic (34) but still lead to 50% lower levels of energy demand in the year 2000 than those obtained from the above model.

A factor aggravating the energy resource balance of developing countries into the future, is that the energy inputs required to satisfy increasing monetary flows will be increasingly commercial and modern. While cities remain small and still integrated with the rural hinterland, their energy requirements can be partly satisfied with traditional energy resources such as fuelwood and charcoal. As cities expand, however, their sustenance radius extends ever further away. Fuelwood is quickly cut out of the picture on account of its low energy density and consequently high transport costs. While fuelwood cannot stand distances greater than 50 km charcoal, with greater energy density, can accept overland hauls as great as 200 km. As cities extend in size, moreover, scale economies suggest the deployment of centralised modern energy resource distribution and production systems: from bottled gas, to electricity grids, to gaseous distribution networks. When such economies of scale are reached, charcoal is quickly displaced even from the poorest sectors of the economy.

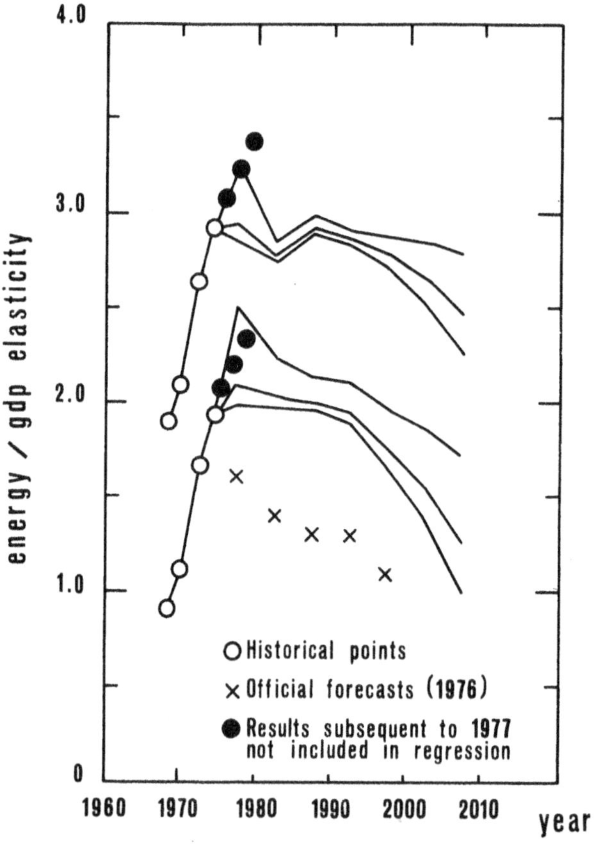

Figure 1 - Commercial energy/gdp elasticity projections for Indonesia: several scenarios (5 year average centred on mid point)

The historic data available for the period up to 1980, however, do not yet reflect the effects of the ravages brought to forest, land and water resources during the last two decades of unprecedented population pressure. The reduced availability of traditional renewable resources which, in many countries, will develop into an outright crisis already during the course of the '80s, will inevitably spur the growth of commercial and modern fuel demands relative to equivalent gdp conditions occurring in the past. This mechanism provides yet another component to increasing monetary flows with significant positive feedback implications for urban migration (through the balance -of-payments/import substitution/export development loop) and further growth of the urban-industrial energy system.

The future of the urban energy systems of developing countries remains uncertain, however, or at least difficult to forecast with any degree of accuracy, because of the delicate and unquantified balance between contributing and opposing factors. It seems certain, for example, that the unprecedented urban concentrations of developing countries will eventually give rise to macro-system chains of products and service demands that are greatly amplified relative to what has been normal for the smaller urban agglomerations of past industrialized country experience under equivalent conditions of urban population. This factor would tend to contribute yet another component of energy growth relative to the historic experience.

At the same time larger urban agglomerations allow greater economies of scale in energy production and distribution systems providing a negative component to energy growth. The most important factor, however, and the one that the expanding urban systems of developing countries should take full advantage of, comes from the improvements in urban and industrial technologies that will become available over the coming decades. This has been the one most important feature in reducing the maximum energy/gdp ratios of each new batch of countries reaching industrial maturity over the last century and doubtless will continue to be so in the future.

THE CRITICAL VARIABLES IN THE FUTURE ENERGY PLANNING OF THE DEVELOPING COUNTRIES

Model forecasts for the period up to 2000 made taking into account current trends in rural/urban dynamics, imply continuing high rates of growth of modern commercial fuels demand in the developing countries as a group, if these countries are to continue to experience satisfactory and consistent levels of economic growth (table 4). Under such assumptions, which imply average gdp growth rates of about 4.5% per year, energy demand in the developing countries in 2000 would be only slightly lower than all the energy consumed in the world today ($\sim$ 6000 Mtoe). Notwithstanding higher prices, oil will continue to be the preferred incremental fuel, assuming it is available in sufficient quantities and the developing economies can afford to pay for it.

There is today apparently no simple way, of going through the early and intermediate stages of development without increasing recourse to oil. Yet, unless drastic measures are taken to very substantially restructure patterns of development in the lesser developed countries, they will not generally be able to afford oil in sufficient quantities to satisfy their growing urban induced energy demands. Moreover, even if such restructuring were to take place (a very unlikely circumstance given the short time available and the extreme adjustment problems involved), incremental oil demands in the world as a whole, could not possibly be satisfied by foreseeable levels of supply. The situation is aggravated by the fact that before the end of the century, many of today's oil exporters will add to swell the ranks of the oil deficit countries contributing to multiply the demands on potential world suppliers, while, of course, there are technical and economic limits to expanding oil supplies into the future (35).

Solutions to the energy problem are generally sought in technical fixes such as energy saving, new fuels and new technologies. It is unlikely, however, that any of these solutions can be very effective for developing countries. Their capacity to save energy and to deploy new energy technologies is

Table 4 Demands for urban fuels in developing countries 1950-2000 (Mtoe)

	1950	1975	2000	growth rate (%)	
				1950-75	1975-2000
Latin America	45.7	237.7	1212.3	6.8	6.7
Africa, South and East Asia	41.4	238.2	1660.3	7.3	8.1
Middle East and North Africa	7.5	85.4	687.4	10.2	8.7
Centrally Planned Asia	31.9	423.7	1949.0	10.9	6.3
total	126.5	985.0	5509.0	8.6	7.1

largely limited in the short term by economic and balance-of-payments problems as also by the current continuing rapid rates of increase in energy demand. Some respite may be available from substitution with other conventional energy resources such as hydropower, but in quantities generally insufficient for the purpose.

Most of these solutions tackle the problem at the surface considering only its more outward manifestations: energy consumption, economic growth and its surrogates. The very complex nature of the developing rural-urban energy system, and the many critical variables involved suggest, however, a number of weak links having little or no direct bearing on energy itself. Energy, after all, is only the final link in the chain allowing increasing numbers of people to survive satisfactorily on diminishing land surfaces. As discussed in previous pages, the root of the problem lies fundamentally in the aggravating balance between population and agricultural land resources.

For many developing countries there appears to be no solution to the energy problem that does not in some way attack the problem of the aggravating resource balance in rural areas. There is widespread reluctance, however, to conceive the emotional issue of rural-urban migration and patterns of development as anything that is open to change or modification from the apparently inertial historic trends (36). In reality there is no experiential foundation in the common belief that rural out-migration has its origins in people's dislike for agricultural occupations and attraction for the bright lights of the city (37). It is widely recognized on the other hand, that people automatically flee the land when it is no longer able to provide adequate and acceptable forms of survival (38).

Economic opportunities on the land for the rural population can be improved in a number of ways. In many parts of the world land is kept idle through inequitable ownership patterns, absentee landlord ownership and outdated inheritance practices. Land productivity can be immensely improved through intensive solutions such as irrigation, use of fertilizers, both natural and artificial. In many developing countries, land productivity can gain enormously from fairly simple improvements in cultivation practices that have elsewhere been well-established for hundreds of years. There include deep tillage, lying fallow, four-crop rotation, animal husbandry. It is a commonly held belief that animal labour cannot be introduced in the East-African enviroment because there would not be enough land from which to feed the animals. Yet animal ploughing, because it reaches very much deeper into the soil than human labour, can improve land productivity by a factor of at least two.

Improved organization and agricultural technology, more

effective pest control, better crop treatment and cropping patterns, more efficient harvesting and storage, and many other areas of improvement, are available at relatively little cost to improve yields through reduction of wastage. Yet when there is any interest in the agricultural sector, most of the emphasis seems to be placed in mechanisation and other labour saving devices, that is labour rather that land productivity increases. This is not unnatural, given that most of the investment effort is left to private capital which clearly seeks the highest individual rewards. The result, however, is that the rural-population is expelled even more surely from agricultural occupations.

In those rare occasions when land productivity has been increased in the past, that is in favour of labour rather than capital, it has invariably resulted in a remarkable decline in rural out-migration lending credibility to the thesis that the push factor is more important than the pull (39). But perhaps the most remarkable evidence lies in the extremely short response times of such initiatives, implying that the rural population does not need to be coaxed into staying on the land once conditions are sufficiently favourable to allow this. The experience in Egypt following the great irrigation scheme of the mid - '60s, indicates a time lag of barely 1-2 years between increases in reclaimed land and reductions in the rate of urbanization (figure 2).

In the absence of such an initiative, urban population growth in Egypt would have continued to accelerate at least until the early '80s. Model estimates in this case indicate an urban population of around 23 million in 1980 as against 18.2 million, while commercial energy requirements might have had to be some 130% higher just to maintain the same or similar living conditions to those attained in that year (40). It would be difficult to imagine a more effective and relatively low cost, energy saving measure providing at one and the same time the added bonus of large increases in employment opportunities and reductions in urban slum development.

Now that the irrigation potential of surface water resources is practically exhaused in Egypt, increasing land availability for agriculture will need to be based increasingly on ground water potential. On this question, the Egyptian Government might consider the possiblity of devolving a substantial part of its ephemeral oil income to development and reclamation of new agricultural areas, essentially continuing the good record of past, rather than embarking on an increasingly energy and urban intensive import substitution policy. Solar energy, of which Egypt abounds, in photovoltaic applications to activate tubewells, is a major prospective option to this end. The point

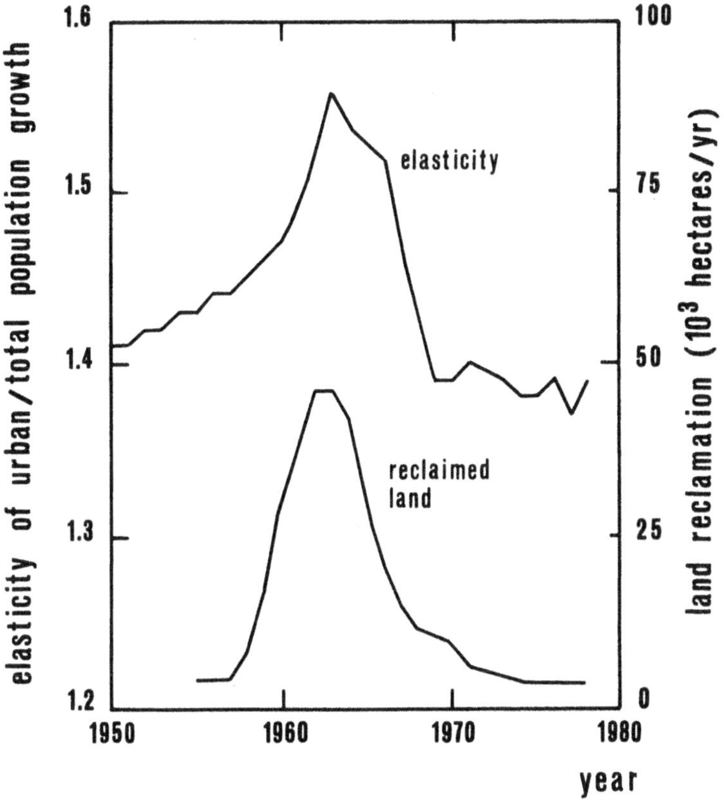

Figure 2 - Land reclamation and urbanization in Egypt (5-year moving averages)

to be made, hereagain, is that this apparently agriculture-oriented policy has significant and important positive implications also for energy policy and urban settlements development.

With the progressive exhaustion of agricultural frontiers, such extensive solutions to the population/land resource problem become increasingly costly, particularly as distances increase, or soil conditions require greater and greater capital outlays. Particularly high costs can be involved when the rural-rural migration requires crossing large stretches of water as in the case of the Indonesian transmigration programme. This programme, trasferring potential migrants from rural Java to the Outer

Islands, picked up momentum in the late '70s to reach an apparently very high level of 310 thousand persons/year in 1980-81, from barely 30 thousand a decade earlier (41). Despite the very large effort involved, this is,however, still about 5 times lower than will have to be reached if the country is not to run into unmanageable energy resource problems in the coming decades (33).

Considering that the costs of transmigration average around 5000 $/family, including the initial set-up investments, a programme of 1.5 million migrants per year (roughly the present population growth on Java) would cost around 30% of present yearly oil receipts, which the Indonesian Government is instead earmarking for urban infrastructure development and import-substitution oriented heavy industry.

Advising in favour of such a boosted transmigration policy is the large reduction in energy requirements (as much as a factor of 2 to 3 (42)) that would be achieved relative to a situation in which continuing population growth leads to an urban continuum on Java towards the end of the century (population density in 1980 already averaged 699 persons/km^2). Under current economic plans, Indonesian self-sufficiency in oil is almost certain to end abruptly in the late '80s, after which the country will begin to experience acute difficulties in fueling her increasingly energy intensive economy. Embarking immediately on a 20 year high-level transmigration programme of agriculture-based settlements on the Outer Islands, would result, instead, in extending Indonesian's self-sufficiency in oil out at least to the end of the century and beyond, while converting this resource into a self-perpetuating endowment for future generations.

While countries like Egypt and Indonesia presently have the economic potential, in the form of oil resources, to put to fruit their high cost land resources, there are countries, like Kenya, without any such endowment. With its exceptionally high population growth of 4% per year, its very scarce modern energy resouces, its relatively limited high-potential land, and its rapidly vanishing forests and deteriorating soils, Kenya is perhaps the extreme case in population/land resource management.

As already discussed, population pressure on the land in this country is so high, and population growth will continue inevitably to be so rapid, as to promise exceptionally high levels of urbanisation unless rapid and effective action is taken in the immediate future. There is, of course, need to seek a system response that is different from concentrated urbanisation, because intensive urban growth, if it is not to be poverty-ridden, requires rapidly increasing modern energy resources (figure 3), which Kenya apparently does not have and can ill afford to buy.

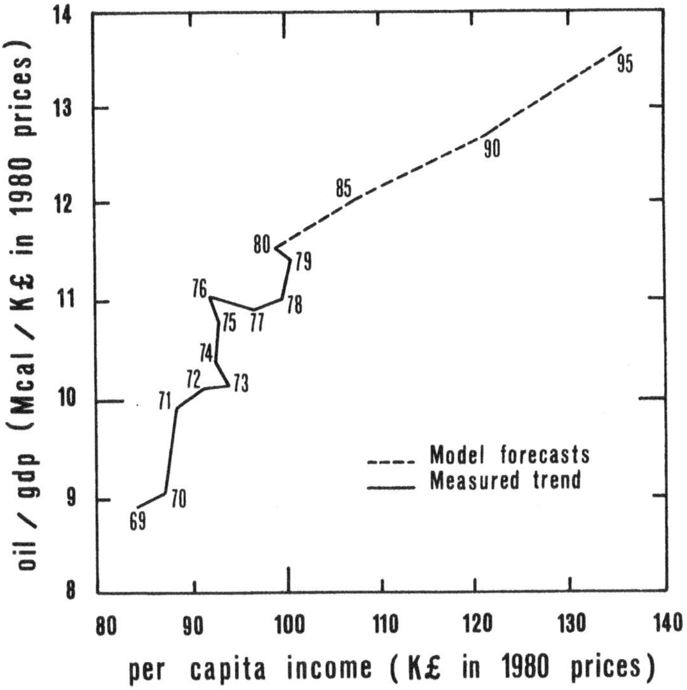

Figure 3 - The oil intensity of gdp in Kenya 1969-95

Eventually, the solution to Kenya's very difficult energy problem will have to be found in a very carefully balanced set of actions. Kenya will need to undertake extensive restructuring of her economy (more austere, more export oriented) in order to permit her to double and even triple oil imports by 1995. This is the minimum quantity of oil that is consistent with any satisfactory from of development in this country, and the need to safeguard forest resources for future generations. She will need to deploy far-reaching programmes aimed at improving her efficiency of woodfuel use and conversion. She will need to embark on a strong hydropower and geothermal expansion programme and later in the '80s, a coal programme to enable her to guarantee the increasing quantities of electricity which her system needs. At the same time she will need to undertake significant research, development and demonstration programmes in new and renewable energy technologies suited to Kenya and based as much as possible on Kenyan industry and man power in order that these will be increasingly available as the need requires beyond the '80s (43).

All this will be sufficient, however, to guarantee a bare doubling in overall primary energy supply (including woodfuel and non conventional resources) between now and 1995. Although this is significant because of the large savings obtained through woodfuel substitution by oil and electricity, and other major savings in woodfuel conversion and end uses, is does not leave very much space for the urban intensive type of growth that lies in store for Kenya if she does not make a special effort in favour of rural development and urban settlements policy.

In the former case, high potential land availability will need to increase at an annual rate of at least 0.6%, while in the latter case it will be necessary to devise policies favouring the growth of intermediate size towns at least 30% faster than large cities over the next 15 years. These are two major cornerstones of Kenyan development policy if energy supplies of the magnitude indicated, are to be sufficient to satisfy urban-induced incremental energy requirements into the future.

The Kenya Government seems to be open to the latter of the two options and there are clear indications of the willingness to create incentives to favour the growth of smaller settlements (44) though it remains to be seen what form this development will take. There is, however, a long way to go before well devised rural development policies of the magnitude and direction required are effectively and convincingly deployed (45).

CONCLUSIONS

The existence of a gap between supply and demand of energy, means in general, that many of the social and economic conditions of a scenario cannot be met. That is, the system has to fall back on some more modest line of growth and some less comfortable form of survival. Though lack of energy implies that social and economic targets connot be achieved, some of the physical conditions of a scenario will to some extent materialise. For example, it is difficult to stop the mounting pressure of urbanisation while rural inhabitants perceive a greater chance of a decent living standard in urban areas, even if this chance is minimal. The demand for products and services generated by rural migrants, insofar as they imply the use of energy, will not be fully satisfied however. Given the conditions of economic growth, it will not be possible to finance an adequate housing strategy, and rural migrants will inevitably flock into ever expanding slums bordering the cities. Given the insufficiency of modern sector activities, informal micro-economies will tend to emerge within these urban societies, based on services and small craft. These however, can only survive if it is possible to provide adequate sanitary conditions and above all, food.

There is no end to human ingenuity in devising ways of converting solar energy into human forms of survival even in circumstances of abject poverty. There exists no better example than the human-crab cycle of the favelas of the North East region of Brazil , in which the only external input is solar energy (46). Human systems left to themselves develop ad hoc organisations capable of locally optimising benefits under the given resource constraints. When these are too limiting, of course, the system collapses. One of the major roles of government is to remove or otherwise broaden, the resource constraints both by guaranteeing the conditions necessary to provide adequate or alternative resources, and by inducing or favouring structural modification within the system which are less avid of resources.

To this end the governments of the developing countries might begin to consider the possibility of focusing on the more deep seated social and economic origins of their energy problems, which lie not so much in the modern sector as in the rural sphere and, in particular, in the aggravating balance between the rural population and local resources.

REFERENCES

1. Mead W.G.,1979, **An Economic Analysis of Crude Oil Price Behaviour in the 1970s** , The Journal of Energy and Development, p. 212.

2. See for example, **Report of the Ad Hoc Expert Group on Rural Energy** , United Nations Conference on New and Renewable Sources of Energy, Nairobi, 1981.

3. Openshaw K., 1978, **Woodfuel: a Time for Re-Assessment** , Natural Resources Forum, vol.3, p. 35.

4. Among such countries it is possible to cite Bangladesh which undertook a comprehensive nationwide energy survey in 1975 (**Bangladesh Energy Study for the Government of the People's Republic of Bangladesh**), ADB and UNDP, project BGD173/038/B/01/4S; Kenya with two surveys focussing specifically on woodfuel and rural energy consumption (**Rural/Urban Huosehold Energy Consumption Survey : October - November 1978** , Central Bureau of Statistics, and Akinga W., 1980, **Woodfuel Survey** , Ministry of Environment and Natural Resources, Forestry Department); and Ethiopia which is presently undertaking a very broad spectrum energy survey of this type (**Cooperation Agreement between the**

Ministry of Mines and Energy and Cesen in the Energy Sector, february 1982). Interest in this field is of course indicated by the very large number of studies and publications emerging in the last 5 years.

5. **World Development Report 1981**, World Bank.

6. See as typical examples Hoffman L. and Mors M., 1979, **Energy Demand in the Developing World: Estimation and Projection up to 1990 by Region and Country**, Regensburg, and also Wolf C., Relles D.A. and J. Navarro, 1981, **Oil and Energy Demand in Developing Countries in 1990**, The Energy Journal, vol. 2, p. 1.

7. The vast majority of applications in this field refer to local and partial systems management as indicated, for example, in such studies as de Lucia R.J., 1980, **Energy from Organic Residues in Developing Countries: Present Use, Potentials and Observations on Feasibility**, Bio-Energy World Congress, Atlanta.

8. On this point see Nordhaus W.D.and Tobin J., 1973, **Is Growth Obsolete?** The Measurement of Economic and Social Performance, NBER Studies of Income and Wealth, vol. 38; Schumacher E.F., 1975, **Small is Beautiful: Economics as if People Mattered**, Harper and Row;and Bruckman G., 1977 **A Western Type Economic System of the Future, IIASA mimeograph, Laxenburg.**

9. **Houston J., 1976,** Promotheus Rebound: an Inquiry into Technological Growth and Psychological Change, **Technological Forecasting and Social Change, vol.9, p. 241.**

10. **Pawley M., 1974,** The private Future, **Random House, New York.**

11. **Hannon B., 1977,** Energy, Land and Equity. **Center for Advanced Computation, University of Illinois, Urbana.**

12. Changing images of Man, **a report of the Center for the Study of Social Policy, Stanford, Research Institute, 1973.**

13. **Bernardini O., 1981,** The Urban Determinants of Economic Growth and Energy demand, **Economia delle Fonti di Energia, vol. 13, p. 119.**

14. **The many-faceted nature of the problem is well brought out in De Jong G.F. and Gardner R.W., 1981,** Migration Decision Making, **Pergamon Policy Studies.**

15. **The composition by administrative districts of the 12 Kenya**

regions is defined as follows:

Region 1
Weston Pokot
Trans Nzoia
Elgeyo-Marakwet
Uasin-Gishu
Nandi
Bungoma
Busia
Kakamega
Siaya
Kisumu

Region 2
South Nyanza
Kisii
Kerico

Region 3
Barigo
Laikipia
Nakuru
Nyndarua

Region 4
Embu
Meru

Region 5
Nyeri
Kirinyaga
Murang'a
Kiambu

Region 6
Nairobi

Region 7
Machakos

Region 8
Taita
Kwale
Kilifi
Mombasa
Lamu

Region 9
Narok
Kajado

Region 10
Turkana
Samburu

Region 11
Wajir
Mandera
Isiolo
Marsabit

Region 12
Kitui
Garissa

For the justification behind the district clustering as also for the details relating to the discussion of interregional migration see reference 16.

16. Republic of Kenya, Ministry of Energy, Consultancy Contract Project no. 4505.016.32.29, Annex B.

17. See Slicher van Bath J.H., 1966, **The Agrarian Revolution of Western Europe, AD 500-1850** .

18. Mc Evedy C. and Jones R., 1978, **Atlas of World Population History** , Penguin.

19. See Bernardini O. op. cit. reference 13.

20. Beier G., Churchill A., Cohen M. and Renaud B., 1976, **The Task Ahead for the Cities of the Developing Countries** ,World Develompent, vol. 4, p. 363.

21. On appropriate and intermediate technologies see Gadgil D.R., 1974, **Appropriate Technologies for the Total Development Plan** , in Appropriate Technology for Indian Industry, Hyderbad; Schumacher E.F., 1974, **Social and Economic Problems Calling for the Development of Intermediate Technology: a Definitional Discussion** , in Choice and Adaptations of Technology in Developing Countries, OECD, Paris; and Jequier N., et al. 1976, **Low Cost Technology: an Inquiry into Outstanding Policy Issues** , OECD Development Centre, Paris.

22. See Bergsman, Greenston and Healy, 1972, **The Agglomerative Process in Urban Growth** , Urban Studies, vol. 9; Richardson

H.W., 1973, **The Economics of Urban Size** , Lexington Books; Alonso W., 1968, **Urban and Regional Imbalances in Economic Development** , Economic Development and Cultural Change.

23. On this point see the lengthy discussion under reference 20.

24. The conviction with which such views are propounded is well expressed in Balassa B., 1978, **Export and Economic Growth: Further Evidence** , Journal of Development Economics Vol. 5, p. 181; and Michaely M., 1977, **Exports and Growth: an Empirical Investigation** , Journal of Development Economics, 4, p. 49, who find, not surprisingly, in export growth the key to gdp growth.

25. Republic of Kenya, Ministry of Energy, Consultancy Contract Project no. 4505.016.32.29, Annex B, Appendix B6.

26. **City Projections 1950-2000** , United Nations, 1976.

27. Colombo U. and Bernardini O., 1979, A **Low Energy Growth 2030 Scenario and Perspectives for Western Europe** , Report prepared for the Commission of the European Communities and published as annex to **In Favour of an Energy Efficient Society** , EEC, Brussels.

28. Leach G., 1977, **Energy and Food Production** , IPC Science and Technology Press, Guildford.

29. Rostow W.W., 1978, **The World Economy** , Macmillan Press, London.

30. **Urban/Rural Projections from 1950-2000** , United Nations, 1974.

31. See Colombo U. and Bernardini O., op. cit. reference 27.

32. Arismunandar A., 1977, **Availability of Conventional Energy Resources in Indonesia to Meet Present and Future Domestic Demands** , 10th World Energy Conference, Instabul.

33. Bernardini O., Bodda L. and Ninni A., 1981, **Problemi di analisi e pianificazione energetica in un paese in via di sviluppo: il caso Indonesia** , Economia delle Fonti di Energia, vol. 15, p. 169.

34. **Indonesia National Report** , United Nations Conference on New and Renewable Sources of Energy, Nairobi, 1981.

35. The most lucid and outstanding declarations on the question

of the approaching energy crisis still remain, **Energy: Global Prospects 1985-2000** , Editor Wilson C.L., MacGraw Hill, 1977, and **Coal: Bridge to the Future** , Editor Wilson C.L., Ballinger, 1980.

36. See Doxiadis C.A. and Papaioannou J.G., 1974, **Ecumenopolis, the Inevitable City of the Future** , Athens Centre of Ekistiks.

37. May R.J. and Skeldon R., 1977, **Internal Migration in Papua, New Guinea** in Change and Movement, Readings in International Migration in Papua, New Guinea, Australian National University Press, Camberra.

38. Simmons A.B., Briquets S.D. and Laquian A., 1977, **Social Change and Internal Migration; Review of Findings from Africa, Asia and Latin America** , Ottawa International Development Centre.

39. Among numerous examples, see Renaud B., 1971, **Economic Fluctuations and Speed of Migration: a Case Study of Korea 1955-75** , World Bank, Special Working Paper; Riddel J., 1978, **The migration to the Cities of West Africa, Some Policy Conclusions** , The Journal of Modern African Studies, vol. 16, p. 241, and, also Arnold F. and Cochrane S.H., 1980, **Economic Motivation versus City lights: Testing Hypothesis about Inter-Changwat Migration in Thailand** , World Bank, Staff Working Paper no. 416.

40. **Prefeasibility Study for the Establishment of the Egyptian Renewable Energy Development Organization (EREDO)** , Cesen Consultancy to Qattara Hydro and Renewable Energy Production Authority, 1981.

41. **Indonesian Transmigration Programme Review** , vol. 1, World Bank Report no. 3170a - IND, 1981.

42. See scenario analysis in reference 33.

43. See extensive scenario analysis in Republic of Kenya, Ministry of Energy, Consultancy Contract Project no. 4505.016.32.29, Annex A.

44. **Republic of Kenya, Development Plan 1979-83** , Nairobi, 1979.

45. Mc Carthy F.D. and Mwangi W.M., 1982, **Kenyan Agriculture: Toward 2000** , IIASA Reports, vol. 5, no. 1, p. 131.

46. de Castro J., 1965, **Une zone explosive, le Nordeste du Bresil** , Editions du Seuil, Paris.

FOSSIL FUEL PROSPECTS IN DEVELOPING COUNTRIES

Paul S. Basile

Secretary of Management Committee
International Energy Development Corporation
and General Manager IEDConsultants, SA

The potential fossil fuel resources in developing countries have a potential which warrants a significantly greater exploration effect than has to date been the case. In 1979, some 540 exploratory wells were drilled in developing countries outside OPEC; in the same year roughly 15,000 exploratory wells were drilled in North America. Fossil fuels are essential for the development, modernization and economic growth in developing countries. In order to overcome some of the obstacles to accelerated exploration for and development of fossil fuel resources in developing countries, it will be necessary to mobilize new sources of capital and possibly new mechanisms for financing. International development assistance agencies, consortia of private firms/investors, and public institutions in the host country itself all will play a part.

1. BACKGROUND

It has become respectable and even fashionable to urge on developing countries the development of new and renewable sources of energy. While it is certainly true that there is in some developing countries significant potential for the expansion of hydropower and other renewable energy resources, it is by no means clear that all renewable resources taken together could ever amount to a substantial portion of developing countries' legitimate energy needs. More important, perhaps, than the ultimate long-term potential of renewable resources is attention to the immediate opportunity to exploit fossil fuel resources, which almost certainly exist in many developing countries but which, to date, are inadequately developed.

H. Neu and D. Bain (eds.), National Energy Planning and Management in Developing Countries, 119–136.

For industrialization, urbanization, and modernization most developing countries need increased supplies of fossil fuels, and at affordable prices. In those countries where there are potential fossil fuel resources, the income generation from developing those resources can be quite substantial. It is, therefore, worth examining the potential of fossil fuel resources in developing countries, the reasons why activities to exploit these resources have not been as great as they might, and the possibilities which exist for accelerating their development.

Oil importing developing countries face a dual energy crisis. Many of them are already experiencing the first crisis, vanishing supplies of fuelwood. Nearly all of them are dependent on oil for continued economic growth. Only a handful of oil importing developing countries have the capacity to generate foreign exchange in amounts anywhere near adequate to pay for their growing oil imports, service their often large foreign debt, buy essential food from abroad, and have anything left over for the import of capital goods.

Of the 74 developing countries which now import oil, 48 of these are dependent on oil for more than 90% of their commercial energy needs. In fact only 4 oil importing developing countries are dependent on oil imports for less than 50% of their commercial energy needs. These 4 -- India, Korea, Pakistan and Zambia -- have significant indigenous coal, gas or hydro resources at their disposal. In spite of recent complacency in industrialized countries, where conservation and new supplies are reducing the threat of energy shortages or economic pressures from oil import prices, the situation for oil importing developing countries is getting worse, not better. In developing countries oil imports continue to grow (even if at reduced prices) and balance of payments deficits widen.

The treatment of petroleum in this paper is considerably greater than for other fossil fuels, in part because of oil's dominance and importance in the global energy mix, in part because some oil examples are generalizable to other fuels, and in part because of space limitations here.

2. OIL

2.1 Resources

The consensus range for ultimately recoverable oil resources in the world is between 1,700 and 2,300 billion barrels. Of course a few estimates are much higher and one or two are somewhat lower.

Of the roughly 2,000 billion barrels ultimately recoverable, about 430 billion barrels have been produced and an additional 620 billion barrels or so are known, identified reserves not yet produced. This leaves then some 650 to 1,250 billion barrels of producible oil remaining to be discovered. It is expected that up to two thirds of that amount will come from extensions to existing fields and enhanced recovery techniques, while the remainder will come from newly identified fields.

One of the factors tending to bias towards more modest estimates of oil remaining to be found is the dominance of giant fields in current reserves and production. At the end of 1975 there were 272 known giant fields in the world (fields with greater than 500 million barrels of recoverable reserves), 36 probable giants and 117 possible giants and 33 super giants (fields with recoverable reserves in excess of 5 billion barrels). Some 80% of current recoverable reserves are in giant and super giant fields; 50% of all reserves are in super giant fields. At an even higher level of aggregation, there are some 22 oil provinces in the world (with greater than 10 billion barrels of recoverable reserves each) that contain 86% of the known recoverable oil resources. All the super giants and 95% of the recoverable oil of the giant fields are contained in these provinces.

Richard Nehring in "Giant Oil Fields and World Oil Resources", F-2284-CIA, June 1978, Rand Corporation, estimated that at the end of 1975 there were, remaining to be discovered:

4- 10 super giants	:	30-100 billion barrels
125-175 giants	:	100-175 billion barrels
400-800 large non-giants	:	60-160 billion barrels
Small non-giant fields	:	73-120 billion barrels
TOTAL	:	263-555 billion barrels

Of course, no one knows where these fields will be found. It is fair to assume, probably, that all of the super giants and most of the giant fields will be found in presently producing regions. If so, there still may be on the order of 100 to 300 billion barrels of oil to be found in currently non-producing regions. The table below from Nehring's report is not inconsistent with this generalization, although neither does it substantiate it.

TABLE 1

ESTIMATED ULTIMATE CONVENTIONAL WORLD CRUDE OIL RESOURCES BY REGION (at the end of 1975)

(In billions of barrels)

Region	Known	Potential	Total
North America	179.8	100-200	280-380
South America	68.4	52-92	120-160
Western Europe	24.6	25-45	50-70
Eastern Europe/ Soviet Union	102.4	63-123	165-225
Africa	75.6	45-94	120-170
Middle East	509.9	350-630	860-1160
Asia/Oceania	50.8	54-104	105-155
TOTAL	1011.5	688-1288	1700-2300

Based on this table one could, very heroically, speculate that some 10-15% of ultimately recoverable resources remaining to be found in the world will be found in currently non-exporting developing countries.

In many ways, of course, this highly aggregate observation is not very helpful. It is nice to know that there is some basis for believing that there are substantial quantities of oil remaining to be discovered in oil importing regions, but it is not terribly comforting to recognize that the finds will be predominately in small fields (probably very costly) and of course there is no information on precisely where the oil is to be found.

And, it must be noted that the resource figures cited above are by no means free from controversy. Some resource specialists believe the figures to be substantially higher; the discovered reserves to date being low due to the much lower drilling intensity in LDCs than in, for example, the United States. Others believe the figures above are too high; U.S. geology is unique, they claim, and commensurate drilling intensity should not be expected to reveal large quantities of new reserves in other places.

Still, for our purposes it is probably safe to accept that there is some (probably significant) potential for further development of oil resources in developing countries which is not being adequately realized.

2.2 Exploration Trends[(1)]

There are three major indicators of petroleum exploration activity:

1. The amount of acreage under license.

2. The number of seismic party months.

3. The number of exploratory wells drilled.

For all three indicators, readily available data are poor and must be interpreted carefully.

In developing countries as a whole, the amount of acreage under license (concessions, joint ventures, production-sharing contracts, risk contracts, etc.) appears not to have risen significantly since 1970, although there may have been some increase very recently (in 1981). While many new concessions have been awarded in the past few years, there have also been numerous relinquishments of acreage holdings. In 1976, there were about 21.4 mn. square kilometers in the world outside centrally planned economies (CPEs) under license, compared with 24.7 mn. square kilometers in 1970, a drop of 13%.

Exploration activity as indicated by number of seismic party-months has not shown much better performance. In 29 non-OPEC developing countries (encompassing all of the countries with significant activity), the number of seismic party-months rose between 1970 and 1974, only to fall back again in 1979. Table 2 below summarizes the data.

---

(1) The analysis summarized here and in the next two sections is reported more fully in "Exploration Trends in the Developing Countries: Trends in the Seventies, Outlook for the Eighties", a paper presented by Francisco R. PARRA, Executive Director of IEDC, at the International Petroleum Seminar sponsored by the Institut Français du Pétrole, Nice, March 1981.

TABLE 2

SEISMIC PARTY-MONTHS

	1970	1974	1979
10 OPEC Countries	755	921	877
29 Non-OPEC Developing Countries	1,402	1,666	1,561
TOTAL	2,157	2,587	2,438

The data, which show only a modest increase in activity over the period, are dominated by a small number of countries (of the 39 countries examined, five of them -- Indonesia, Algeria, India, Argentina and Mexico -- accounted for 58% of the total in 1979). However, the pattern of increase in the mid-1970s followed by a falloff in the late 1970s is generally representative of most of the countries. This may not be as discouraging as it seems, because more marine seismic was probably shot in the later period. (One marine seismic party-month is equivalent to 10 or 20 party-months on land in terms of kilometers of seismic line shot).

Exploratory drilling has increased in the Third World, but only modestly, rising from a total of 871 wells in 1970, to 906 in 1974 and 914 in 1979. This is a mere 0.5% per year over the whole period (0.2% per year from 1974 to 1979). In two areas - Africa and Latin America excluding OPEC countries - exploratory drilling has actually dropped since 1970, and only in Asia has there been a significant increase. Table 3 below sets out the data in summary form:

TABLE 3

EXPLORATORY WELL COMPLETION IN DEVELOPING COUNTRIES

	Number of Wells		
	1970	1974	1979
OPEC	350	398	375
Other LDCs in:			
Africa	117	81	112
Latin America*	334	327	298
Asia	70	100	129
Total Other LDCs	521	508	539
Total All LDCs	871	906	914

* Excluding Mexico

Note: Data is partly estimated

There is some logic in presuming that exploration activity in developing countries has increased since 1979, in response to higher prices and in conjunction with the dramatic expansion of drilling activity in North America. While adequate data are not yet available, there are some indications that this is true -- although the increase, if any, appears to be quite modest.

Exploration licenses were granted at an increasing rate in 1979-80; of the 428 license groups or consortia holding exploration permits in developing countries at the end of 1980, 145 (34%) were awarded during the two years 1979-80. Yet of the 71 countries where all of these permits are held, in only 40 were there new field wildcat wells drilled during 1980; 70% of these wells were drilled in just 9 countries. While there are certainly a good many reasons for this, it may be one indication that, although interest is up, the level of exploration work is not (yet).

2.3 Constraints on Exploration in LDCs

While the above data on acreage, geological and geophysical exploration and exploratory drilling is far from comprehensive in measuring the exploration effort in LDCs, it is nevertheless clear that no great upsurge in exploration occurred as a result of the crude oil price increases of 1973/1974. This is indeed peculiar, especially since that oddly popular form of exploration and production agreement -- the production-sharing contract -- usually allows the foreign contractor to keep the whole of the benefit of any price increase on his share of the oil. The simplest, most general and plausible explanation for this phenomenon is that the normal time-lag for exploration in LDCs to respond to price increases in crude oil is in the order of several years, and that, with the second shock-wave of price increases in 1979/1980 adding to the lagged reaction from 1973/1974, a rapid stepping-up in the pace of exploration could not have been expected before 1980. And, if any upsurge did get underway in 1980 and 81, it likely has been short-circuited by the soft oil market conditions of 1982.

Governments of several developing countries which have maintained a policy of keeping foreign companies out of petroleum exploration have had to alter their positions under the pressure of increasing oil import bills. This has been a painful and slow process, first in reaching the decision at all, and second in implementing it in such a way that exploration is stimulated without a complete sell-out of the country's resources.

In Third World countries, non-price constraints include violent political upheaval, border disputes, unacceptibility of foreign companies, inadequate legal framework, and inappropriate model exploration and production agreements. In addition, in some countries, exploration may be inhibited because a single company or consortium of companies has under old agreements tied up all of the available prospective acreage without commensurate work obligations or adequate relinquishment provisions.

In addition, there are also constraints on the availability of suitably financed technical capacity to undertake exploration in Third World countries. For example, the very large oil companies, including the majors and large independents, desire to maintain their main line business in the OECD area; the state oil companies in OECD areas have only recently shown interest in exploring in the Third World. For the larger companies security of supply for their refineries and distribution systems and closeness to the home market are important. For the smaller companies activities closer to the home market are important. For the smaller companies activities closer to home are much easier to deal with than the serious infrastructure problems caused by expanding activities into remote developing countries. In the U.S. and Canada the infrastructure for oil exploration is in place. Supplies are easier to obtain, leases on private lands are more easily negotiated, political risk is perceived to be lower, prices are higher and taxes sometimes fewer. The net result is that the available resources are overwhelmingly attracted to the U.S. For example, in 1979 10,500 exploratory wells were drilled in the U.S. alone, compared with 1,200 in the whole of the Third World. In 1980 the number of exploratory wells drilled in the U.S. rose by 24% to 13,000.

2.4 Outlook for the Eighties

The constraints noted above imply that exploration in the Third World will likely be limited to areas with prospects for export oil and that the pace of exploration in these areas will be relatively slow. Naturally, greatly increased prices for oil in 1979 and 1980 will have an effect on activity, but the fundamental constraints have not been removed. Policy-type constraints in developing countries reflected in petroleum legislation, model exploration and production agreements, general political risk, and an aversion to foreign capital appear to be diminishing. The outlook for further relaxation of these constraints is good. Yet it is not, at this point, clear whether the net effect of all the factors will point to a significant increase in exploration activities in the Third World or not. What does seem clear is that there is an urgent need to accelerate hydrocarbons exploration in developing countries, including exporting ones, and that because of the non-price constraints examined above, this need will not be adequately met through the simple working of the market.

Despite incentives which may be high, exploration in the 1980s in areas yielding or likely to yield export crude may not accelerate significantly above its present modest pace, while exploration in the medium-size and smaller countries where prospects are only fair may continue to be woefully neglected.

The only visible alternatives seem to be increased exploration by the national oil companies of the developing countries themselves, supplemented by financial aid from the outside. A few national oil companies already undertake and carry out vigorous exploration programs, but in most countries, resources available for exploration by the national companies are several limited, and although they will undoubtedly be increased as priorities shift under the pressure of oil import prices, these companies will for the most part lack both funds and technical capability to increase exploration sharply.

The two major outside sources that could help are the World Bank and OPEC countries. The World Bank has proposed the creation of an energy affiliate which would permit it to double its lending for the development of energy in the Third World. At present, this proposal appears to be lifeless. OPEC countries, jointly and individually, are another possible source of funds for exploration in other LDCs. The now-stymied long-term strategy formulated by a Ministerial committee last year recommended a sharp increase in aid specifically oriented towards petroleum exploration. Progress has so far been held up on this by the severe internal political difficulties through which the Organization is passing. But these two sources are at present the only ones likely to be able to make a major contribution to accelerated exploration in the Third World.

3. NATURAL GAS

3.1 Resources

During the rest of this century the role of natural gas in meeting the world's energy requirements will be greatly expanded in both size and importance. Many new opportunities will arise for countries to utilize their natural gas resources to meet their own energy and raw materials needs. Many countries will also be able to earn significant foreign earnings income through natural gas exports and others will expand and diversify their energy sources by importing natural gas.

The world's natural gas resources are estimated at approximately 10,000 trillion cubic feet (275 billion m^3), of which present proven reserves are about 2,600 Tcf (75 billion m^3). While the largest reserves are in the USSR, the Middle East and North Africa, many smaller but commercially viable reserves of natural gas are located in numerous countries throughout the world. Over 60 countries have some level of proven reserves, but only about half of these countries presently utilize their natural gas resources, and in some of these utilization is quite small or well below the ultimate potential.

Table 4 shows growth in natural gas production and use in developing countries and Table 5 lists the various LNG projects now on the drawing board.

In many places gas resources associated with oil production are being wasted through flaring. Approximately 20 billion cubic fee of gas are presently lost each day (200 billion m^3 per year). About one third of all gas reserves are associated with oil.

TABLE 4. Growth of Natural Gas and Consumption in Developing Countries. (Quantities in thousand teracalories).

COUNTRY	Production 1972	Production 1976	Production Inc/Dec	Consumption 1972	Consumption 1976	Consumption Inc/Dec
Algeria	47.5	93.9	98%	30.4	45.9	51%
Egypt	0.6	11.0	1733	0.6	11.0	1733
Nigeria	2.5	5.9	136	2.5	5.9	136
Argentina	52.2	64.0	23	60.5	79.9	32
Chile	12.4	11.1	(11)	12.4	11.1	(11)
Colombia	17.2	16.6	(4)	17.2	16.6	(4)
Mexico	108.2	120.3	11	109.2	122.2	12
Trinidad	17.5	15.8	(10)	17.5	15.8	(10)
Venezuela	96.0	118.3	23	96.0	118.3	23
Bahrain	10.5	20.3	93	10.5	20.3	93
Iran	166.9	211.3	27	89.8	124.1	38
Iraq	8.7	18.5	113	8.7	18.5	113
Kuwait	47.8	52.0	9	47.8	52.0	9
Qatar	10.3	13.7	33	10.3	13.7	33
Saudi Arabia	26.1	55.0	111	26.1	55.0	111
United Arab Emirates	6.3	8.2	30	6.3	8.2	30
Afghanistan	27.0	23.7	(12)	0.4	0.4	-
Bangladesh	5.4	7.5	39	5.4	7.5	39
Brunei	4.5	77.1	1613	2.8	16.8	600
India	5.9	10.6	80	5.9	10.6	80
Indonesia	11.4	30.5	167	11.4	30.5	167
China	23.8	36.5	53	23.8	36.5	53
Others*	59.6	89.3	50	31.4	42.2	34
TOTAL	768.3	1111.1	45%	626.9	863.0	38%

* Libya, Morocco, Tunisia, Congo, Gabon, Angola, Rwanda, Bolivia, Brazil, Ecuador, Peru, Barbados, Cuba, Syrian Arab, Burma, Sarawak and Pakistan.

Source: World Energy Supplies 1972-1976, Series J, No. 21; Department of International Economic and Social Affairs, United Nations; New York, 1978.

TABLE 5.

LIQUEFIED NATURAL GAS EXPORT/IMPORT PROJECTS

Exporter	Importer	Initial Deliveries	Daily Volumes Million C.F.	Contract Term Years
Operating:				
Algeria	France	1964	50	25
Algeria	U.K.	1964	100	15
Alaska	Japan	1969	135	15
Libya	Italy	1969	235	20
Libya	Spain	1969	110	15
Algeria	U.S.A.	1971-1978*	43.6-115	20
Algeria	France	1972	350	25
Brunei	Japan	1972	737	20
Algeria	Spain	1974*	50	15
Abu Dhabi	Japan	1977	450	20
Indonesia	Japan	1977	1050	20
Algeria	U.S.A.	1978	1000	20
Under Construction:				
Algeria	Spain	1979	436	20
Indonesia	U.S.A.	1979	550	20
Algeria	France	1980	515	20
Algeria	U.S.A.	1980	420	20
Algeria	Belgium	1982	340	20
Proposed:				
Alaska	U.S.A.	1979-80	200-400	20
Chile	Chile & U.S.	1980	400	NA
Iran	Japan	1980	400	--
Iran	Japan or W. Europe	1982-83	400	--
Nigeria	U.S.A.	1980-85	1600	20
Algeria	Canada & US.	1982	960	20
Algeria	U.S.A.	1983	1000	25
Iran	U.S.A.	1983	300	--
Sarawak	Japan	1983	1000	20
Algeria	Germany	1984	400	20
Algeria	Germany and Netherlands	1984	816	20

* In 1978 the daily volume was increased to 115 million c.f.
** Full delivery will begin in 1979.
Source: Energy Topics, Institute of Gas Technology, July 31, 1978.

TABLE 6.

ASSOCIATED GAS FLARED BY REGION

Africa	80%
Asia/Middle East	59%
Latin America	31%
Eastern Europe	21%
Western Europe	75%
North America	5%

Source : World Bank

Over the past decade the new additions to gas reserves have been much larger than those for crude oil and future gas prospects are even brighter. Further, economic developments have greatly increased the value attached to natural gas and dramatically increased the incentives for its utilization.

3.2 Elements of Typical Gas Development Programs

A major natural gas development program normally includes three principal phases: definition, development and evaluation and implementation. These phases are required in virtually all programs, whether in a developing country with substantial gas reserves looking to a major LNG export project, or in countries with very modest reserves needing to analyse the best use, pricing, and development of the reserves.

Definition Phase

1. Assess gas reserves including quantity, deliverability of reservoirs, location and composition;

2. survey country energy needs and energy programs;

3. identify potential gas utilization projects;

4. define general project development program and plan.

Development and Evaluation Phase

1. Evaluate and select site or route for gas facilities;

2. define a comprehensive and totally integrated facilities concept through feasibility and design studies (wellhead to consumer);

3. prepare accurate cost estimates covering all components of system (wellhead to consumer);

4. prepare construction and implementation schedules;

5. perform economic analyses to assess viability and to define incentives and risk;

6. develop an all-encompassing financing plan;

7. define project structure and organization, including all necessary contractual arrangements;

8. define project participants and ownership.

Implementation Phase

1. Consumate financing agreements;

2. evaluate and select contractors for engineering and construction of all facilities;

3. obtain all requisite governmental approvals and authorizations;

4. evaluate and select process technology;

5. select and procure materials and equipment;

6. arrange for management and technical support for engineering, construction and start-up phases;

7. organize training of start-up and operating personnel;

8. develop marketing plans and organization.

Not all of these detailed steps are necessary on all projects or evaluation programs.

4. COAL.

4.1 The Need for Coal[2]

It is becoming increasingly apparent that the global energy transition away from cheap and easily accessible sources of oil and gas to more expensive forms of energy and ultimately towards more renewable sources, will necessitate a return to large-scale and widespread use of coal. Few analysts see coal as a long-term option, yet the vast resources of coal in the world in its relatively (compared to oil) even distribution around the world argue for its major role as a bridge during the long transition in energy supplies.

As the World Coal Study notes:

> "A tripling of coal use and a ten-to-fifteen fold increase in world steam coal trade could allow the energy problems of the next two decades to be faced with confidence. With such increases in coal use, coupled with a mobilization of other energy sources and vigorously promoted conservation, it becomes possible to see how to meet the energy requirements of moderate economic growth through the world. But without such a coal expansion the outlook is bleak.

In the industrialized countries the potential for coal use is real and large, both in major industries such as cement making and steel for heat raising, but also as a source of electric power generation, and, of course, the potential for turning coal into useable liquid fuels is being evaluated. In developing countries the same points hold, but perhaps on a lesser scale. The modern industrial sector in these countries and electric power generation could indeed be fueled by coal. Yet the fundamental need in a great many developing countries is for expanded transportation systems which, baring some unexpected major technological breakthrough, will require liquid fuels.

Still, the potential for expanded coal use in developing countries is real. The major impediments to expanded coal use may be the need for infrastructure, railroads, handling facilities, storage sites and coal using machinery. Given the uncertain long-term future of oil supplies, and the rapid using up of fuel-wood and the increasingly high cost and difficult accessibility of charcoal, it may be wise for developing countries to seriously consider importing even modest amounts of coal for near-port powerplants or cement industries. This also begins the development

(2) Coal: Bridge to the Future, Carroll L. Wilson, Project Director, Ballinger Publishing Co., Cambridge, Mass. 1980, gives more details.

of the necessary infrastructure for coal development and use, should large-scale coal use be warranted sometime in the future.

4.2 Resources

The world possesses vast resources of coal as indicated in Table 7 below, far more than any other fossil fuel and sufficient to support greatly expanded world-wide use well into the future.

TABLE 7.

WORLD COAL RESOURCES, RESERVES AND PRODUCTION

	Resources (billion tce)	Reserves (billion tce)	% of world reserves	1977 Production (million tce)
USSR	4,860	110	16.6	510
USA	2,570	167	25.2	560
China	1,438	99	14.9	373
Australia	600	33	5.0	76
Canada	323	4	0.6	23
West Germany	247	34	5.1	120
United Kingdom	190	45	6.8	108
Poland	140	60	9.0	167
India	81	12	1.8	72
South Africa	72	43	6.5	73
Others	229	56	8.4	368
TOTAL	10,750	663	100.0	2,450

Although coal deposits occur in more than 80 countries, almost 90% of the world's known geological resources are concentrated in four countries: the USSR, the United States, China and Australia.

Coal deposits in developing countries are poorly known due to lack of intensive exploration. Increased exploration in the southern hemisphere could well result in significant new discoveries. Increasingly the world's coal resources and reserves are found to be larger and more widely distributed geographically than had been previously thought.

Exploration for coal in developing countries has been much less widespread and less intense than that for oil and natural gas, and the exploration intensity for the latter fuels has not been as great (as has been noted) as the resources might indicate. Some 50 developing countries have known coal resources, and 30 of these are currently producing coal, but their combined output accounted for only 5% of total 1977 coal production. Of these India is by far the largest coal producer and consumer. Columbia, and also Indonesia and Botswana, may have significant potential for exporting coal in the future.

4.3 A Role for Coal in Developing Countries

To be sure, the factors to be weighed by LDCs in substituting coal for oil in modern sectors such as electricity, cement, pulp and paper are very complex and action to use coal would require capital investment, a scarce commodity in most developing countries. Cement seems to be the clearest possibility; conversion of a cement factory from oil to coal can cost in the vicinity of $15-20 million. But, again, significant coal use could be achieved only through major investment, which may be viable since coal fuel costs are often half the costs of fuel oil.

Many developing countries can and should consider the possibility of importing coal to replace oil. Coal, even at $95 per ton of coal equivalent is equivalent only to $20 per barrel of oil. $30 per barrel of oil is equal to coal at $142 per ton. Of course, coal handling and using equipment is more expensive than equipment for oil, so the capital cost of a coal-fired powerplant at the receiving point is higher than an oil-fired plant or port. Coal conversion in various industries is well established, the technology is well developed, steadily improving and equipment is available from firms in many countries. Substitution of coal for oil is accelerated in many developed and industrialized countries.

The potential savings of using coal instead of oil can be illustrated. For example, a new 250 mega Watt powerplant would use about 500,000 tce of coal per year. At current CIF prices of $60/ton (12.50$ per barrel), the annual fuel cost would be $30 million. Assuming the same thermal efficiency in burning coal and residual oil at $30 a barrel, the annual fuel cost, using coal, would be $71 million, or a difference in favour of coal would be $14 million per year in annual fuel cost. This differential would be reduced as a result of various environmental control measures, which would add to the cost of burning coal. Even with very strict standards, there would be an annual saving of in excess of

$20 million for the 500,000 tce of coal used in the powerplant.

The critical need may be for structuring such coal use in place of oil projects in such a way that investors will be willing to finance the project and see a reasonable return. Such projects could easily benefit both the investors and the developing countries concerned.

PROBLEMS OF IMPLEMENTING ALTERNATIVE ENERGIES IN THE DEVELOPING WORLD

Markus Fritz
Visiting Scientist at the Commission of the European Communities, Joint Research Centre, Ispra Establishment*)

*) New address: Institute for Energy Analysis, Oklahoma State University, USA

ABSTRACT

Despite the fact that the Third World regions are naturally predestined to use "alternative energies" (i.e. basically solar energy in direct and indirect forms) - and further that they have an urgent need to do because of the oil price burden - there are four basic obstacles: in particular:

1) lack of appropriate and marketable technology;
2) lack of economic affordability on two levels: governmental level (R and D and demonstration) and consumer level (to acquire alternative technologies which are, apart from renewable energy itself, by no means "free");
3) social (as well as cultural) hurdles, combined with reluctance to accept changes in lifestyle;
4) organizational (i.e. political) deficiencies on an international as well as a national basis.

1. INTRODUCTION

There is a worldwide consensus on the urgent need to harness new and especially renewable sources of energy not only for environmental reasons, but above all to fill the gap of depleting and expensive oil reserves. This is especially true for the Third World, where two thirds of the sovereign and non-sovereign LDCs are covering 90 to 100 percent of their

H. Neu and D. Bain (eds.), National Energy Planning and Management in Developing Countries, 137–150.

commercial primary energy consumption by oil imports. However, if we look at the energy policies which have been elaborated by only a few LDCs so far - the majority, for different reasons, remained in a kind of "apathy" since 1973/74 - we are faced with a picture that demonstrates more or less the contrary. Not only this, despite the fact that these few national energy policies had been worked out individually, there are striking parallels. To reach the main aim, reducing the dependence upon oil import, those countries are striving for:

1) an increased use of domestic conventional energy sources (primarily coal and hydropower);
2) an only marginally increased use of alternative (renewable) energies.

To be sure, there are certainly many verbal expressions about intentions of harnessing alternative energies on a large scale, but what actually counts are the official plans and forecasts on future energy demand/consumption. The gulf between theory and practice is evident, e. g. in Thailand; although the country has drafted an ambitious programme for the use of non-conventional energy, this is not visible in demand planning before the year 2000. To give another example, although India is engaged in both theoretical as well as practical activity in the spheres of solar, wind and biogas energy, and although these activities have been designated as being "encouraging", it is nevertheless "unlikely that by 2000 AD these non-conventional sources of energy will contribute a significant proportion towards meeting the national energy needs" /1/.

As the author's comprehensive study on the future energy consumption of the Third World, published in summer 1981 /2/ reveals, there is not much chance remaining for alternative, non-conventional energy sources. This point is made even more vivid by the fact that most of the energy plans and forecasts have been made specifically for electricity demand, to be taken care of exclusively by commercial energy.

The five reasons for such a conventional policy are evident:

1) Conventional energy sources can be used more effectively than non-conventional energy (the basic reason why they are used).
2) The technology for using conventional energy efficiently is already available, and this on a commercial basis.
3) Conventional energy forms have been the basis for industrialization in the modern industrialized countries. It is out of the question for most of the developing countries not to take the same road to success and thus not to use

these same sources of energy.

4) Despite the fact that the Third World is naturally predestined to use renewable energy forms (most of the region is located in the so-called "sun belt" north and sourth of the equator), there are not only climatological implications that often hinder successful implementation, but also two basic implications in particular:
   . lack of appropriate and marketable technology;
   . social (as well as traditional) hurdles, especially in the highly populated rural areas, where the renewable alternatives are thought to be of special advantage.
5) Often overlooked in connection with plans for an increased use of non-conventional energies is the fact that the latter "do not displace imported fuels to a significant degree... The problem is that it is very difficult for a non-conventional energy program focused solely on village-level applications to make sizeable reduction on oil import figures."/3/.

In the following, special attention is given to 4) above, which is the most effective barrier against accelerated harnessing of new and renewable sources of energy in the developing world. Before that, however, we should clarify a semantic question. We are talking here about "new" or "non-conventional" energies. Most of them do not deserve this status at all. Wind, being man's earliest energy source, drove the Egyptian ships over the sea millennia ago. Sea water desalination, considered today as a modern technology, had been employed already some 100 years ago on a solar basis in 1872 at Antofagosto, in the northeast of Chile. Even OTEC (Ocean Thermal Energy Conversion) - a process utilizing the difference tween temperatures of water at the surface and the lower depths -, being considered a "sophisticated technology" today (after the failure of a pilot plant off Hawaii no OTEC plant is working worldwide now!), was invented half a century ago. Not only this, the first pilot plant was working then off Cuba with a capacity of 22 kW. In summer 1930, the American Society of Mechanical Engineers conferred on Georges Claude, one of the two inventors, its Fifth Anniversary Medal for "having only a few weeks since succeeded in solving a problem which has held and continues to hold the interest of the engineering world... " /4/. However, when Claude described his experiments, conducted with his friend Boucherot, in "Mechanical Engineering" in December 1930, he was actually (too) optimistic, when finishing his lecture with the words: "and there is already much, I think, to warrant us in stating, as I do state today, that humanity has from now on the certainty that its industries will never lack the precious energy that actuates them" /5/. He also stated a very important criterion being valid for all activities aiming of harnessing new (non-conventional) energies, which is unfortunately often overlooked: "in projects of this kind, however, only the hundreds of thousands of kilowatts are worth considering..." /6/.

2. IMPLICATIONS

As already mentioned above, the developing world is well equipped with a natural basis for harnessing renewable energy sources, i.e. solar energy directly or indirectly, but the natural activity of sun in itself is not actually the decisive criterion; rather it is climatical, technological, economic and social availability, as well as the culture of people.

2.1 Climatical Availability

Although the Third World is favoured by the sun in general, the insolation is by no means equally - and adequately - distributed. There are, for example, African regions along the Tropic of Cancer in the North with by far the highest average insolation. On the other hand, African countries on or near the equator, where actually the highest insolation activity is expected, like Gabon and the south-east of Nigeria, have approximately the same amount of sunshine hours during the whole year as New York /7/.
The situation is similar with Kinshasa, the capital of Zaire, equally being located near the equator and receiving 1.670 sunshine hours per year - average of insolation for the Federal Republic of Germany, a country not especially favoured in solar energy terms.
The climatic differences and constraints of solar insolation are not different in other regions of the developing world, as Asia reveals. The southern Chinese Sichuan Basin, with a warm climate ideal for biogas production but being located behind high mountains "and influenced by monsoons,... has 250 cloudy days a year. Dense fogs come, on the average, every third day and it rains nearly 150 days a year. Total duration of sunshine is merely 1.100 to 1.200 hours per year, with winters averaging 1.5 hours of sunshine a day (less than in Germany). LANDSAT satellites have been monitoring the earth at 9.30 a.m. local time for over seven years and have recorded only one cloud-free, unfogged morning over the basin" /8/.

The author is in the position to stress this problem by his own experience made in Kathmandu, Nepal, when attending the

"Seminar Workshop on Utilization of Renewable Energy Sources in Nepal" in 1981. The first three days there was no hot water available in the hotel because of dependence on solar collectors in that cloudy region.
Climatic obstacles also apply to the utilization of other renewable energy forms such as wind energy, as the example of India may illustrate, where this kind of indirect solar energy is available only seasonally; by and large, the winds occur mainly in the pre-monsoon period for a couple of months.
Also the average Indian wind speeds(10 to 20 km per hour) are comparatively much lower than those in the Netherlands, Australia or Canada which are about 25 km per hour and above.
Several consequences result from these characteristics of Indian winds. "Firstly, the seasonal character of the wind implies that the utilization factor can be as low as 15 - 20%" /9/.
Secondly, since "the power produced by a windmill is proportional to the cube of the wind speed and the square of the rotor diameter (see Table 1) in order to derive the same power from the (Indian) wind... the rotor diameters have to be 11.5 times larger" /10/.

The economic consequence of this problem is that power produced in India will be at least several times more expensive than electricity derived from an existing grid at the windmill site/11/.
This problem is certainly not limited to the Indian subcontinent.
Botswana - to give an example from another Third World region, is in principle interested in using wind energy to power water pumps in rural areas. At present this operation is "highly dependent on diesel... However, parts of Botswana turn out to have insufficient wind..." /12/.

To be sure, there is also the diametrically opposite problem - winds which are too strong. In the Philippines, for example, "most of the places with high mean wind speeds are also those with tropical cyclones. On the average, about 19 tropical cyclones pass through the country each year, some of which possess centre winds exceeding 200 km/h. Ingeneous 'typhoon resistant' designs are clearly needed, including perhaps systems which can be dismantled conveniently before the typhoon strikes and reinstalled easily after it has passed" /13/.

Looking at sugar-cane as feedstock of alcohol production, to give another example, favoured not only by Brazil, this plant requires not only good soil (not needed for food production), but also abundant rain fall, which is scarce in the vast arid zones of the Third World.

TABLE 1 - Windmill power output in Watts (assuming 70% of theoretical efficiency)

Propeller diameter in feet	Wind velocity kmph 8 / mph 5	16 / 10	24 / 15	32 / 20	40 / 25	48 / 30
2	0.6	5	16	38	73	130
4	2	19	64	150	300	520
6	5	42	140	340	660	1150
8	10	75	260	610	1180	2020
10	15	120	400	950	1840	3180
12	21	170	540	1360	2660	4600
14	29	230	735	1850	3620	6250
16	40	300	1040	2440	4740	8150
18	51	375	1320	3060	6000	10350
20	60	475	1600	3600	7360	12760
22	73	580	1940	4350	8900	15420
24	86	685	2300	5180	10650	18380

Source: H. Clews: Electric Power from the Wind. Solar Wind Co., Maine, 1973,p.4.

Severe climatic implications (constraints) can also be seen in biogas production. This is considered as an almost ideal form of energy, especially for the rural areas of the developing countries, because it supplies energy (methane for heating and lighting) and fertilizer, and destroys bacteria, thus having a very important hygienic effect as well. However, anaerobic fermentation needs high mean temperatures above 20°C. This is the reason why some 5 million biogas plants in the People's Republic of China of the country's total of 7.2 million are located in the southern province of Sichuan.

To come back finally to OTEC once again: besides the fact that "engineering problems have hampered the introduction of this technique" /14/, it is only practical in tropic regions (as already mentioned by the inventor Georges Claude /15/), as has been stressed by a study, conducted by order of the U.S. Department of Energy. Ten sites have been identified worldwide - all of them in the equator region, where surface water temperatures are the highest (see Fig. 1).

To be sure, despite these climatic limitations, a number of countries use already renewable energy, basically solar energy in direct form and this primarily for water heating, crop-drying, sea-water desalination or for pumping water. Israel has probably had the most success in the practical application of solar energy; by the mid-1970s there were "150.000 appartments using solar collectors; this probably represents the largest collector area per head of population (c. 0.15 m^2) in the world" /16/. Nevertheless, although the number of such water heaters doubled in the late 1970s, only around 1% of the total energy consumption of Israel is currently met from solar energy /17/.

2.2 Technological Availability

Solar energy application in Israel does not only stress reason 5) being responsible for less hope in achieving high impact by alternative energies, but also reason 1) - the lack of efficient technology (cf. 1). In other words, the exploitation of renewable energy sources either takes place in simple devices (primarily for heating water) with less impact on the whole energy balance or in advanced demonstration projects with no real impact either.

One typical example of the latter is Mali's multipurpose solar device at Diré, commissioned in November 1979; equipped with a collector capacity of 80 kW, the plant is used for pumping water (for drinking and irrigation), and in addition supplies electricity to a local agricultural co-operative. The facility merely represents a drop in the ocean. However, its capacity is equal to 0.2% of the total installed generating capacity of the country. The costs of the project, around US$ 1 million, are being borne by the French government as part of their programme "New Energy for the Sahel Countries" /18/. With this relatively small project, Mali nonetheless numbers among those developing countries with a relatively large amount of experience in the solar field.

Similar to this case is the installation of a solar power plant in Saudi Arabia with a capacity of 240 kW - approximately one two-thousandths of the total installed generating capacity of the country. (In other words, if a solar power plant with this capacity would be set up annually - which is not the case actually - it would take 500 years to replace the fossil today.)

The most spectacular project is probably SONNTLAN, intended to supply the Baja California (Mexico) village of Las

Barrancas, with a population of 250, exclusively with solar energy /19/. Set against the millions of villages in the Third World (India alone has 567.000), such demonstration and prestige projects figure as mere drops in the ocean.

The problems of lack of appropriate technology are exacerbated in those fields which require complex technologies, such as OTEC, which has already been mentioned above. The situation is similar in the case of tidal power; this source of energy (already used in France and the USSR) is only under active consideration by one Third World country, the Republic of Korea, where a 400 MW station may be built either in the Bay of Sosan, or Chonsuman. No power station yet exists to tap wave energy, and there are no plans for such a station, despite the suitability of this form of energy for islands lacking a continental shelf. Oil (tar) sands, the largest deposits of which are located in Venezuela, are still regarded as a reserve in view of the economic and technical problems involved in working them. The same applies to Madagascar, where the absence of a "suitable technology prevents the industrial exploitation of crude oil from these deposits"/20/. (Only Brazil, which has the second largest deposits in the world after the USA, has set a date for the commencement of recovery; using the nationally developed PETROSIX process, the first processing plants are expected to be in operation by 1983.)

2.3 Economic availability

Inappropriateness of technologies certainly hinders their market penetration. However, there are still more reasons, out of which the economic criterion is crucial. In Ghana, for example, the prices for "industrial power types of solar furnaces suitable for say large-scale irrigation in the North are still very high as imported, sophisticated tracking mirror systems and associated heat engines/dynamos are needed" /21/. The same factor makes the future of non-conventional energy uncertain in Malawi; only with "international assistance to enable implementation of the proposed plans will solar energy, e.g. have "a significant contribution to make to the economic development of the country" /22/. The same applies to the use of solar energy in Kenya, which is located directly in the equator region; "high capital costs of solar heaters, which at the moment vary from the equivalent of US$ 275 to 600 for a unit having 0.85 m^2 of collection surface depending on quality, prevents their widespread utilization" /23/. The same is true in the sphere of solar sea water desalination; "though the technology is rated simple, the cost is still prohibitive for widespread use" /24/. The implications mentioned are also evident in Kenya - which is taken here as an example for most developing countries - in the use of solar energy application for driving water pumps, a favoured area of solar energy application in the Third World generally. These are "very expensive as their technology is in its developmental stage. In Kenya there is only one solar pump and this was donated by the Republiv of France to the Republic of Kenya. The pump ... costs about US$ 66.000 ..." /25/. In other words, "due to the high costs of the pumps and the fact that the technology of solar water pumping is in its nascent stages, widespread use of solar water pumps is not anticipated before the turn of this century" /26/.

The widespread use of wind energy is also constrained by the same factors: "with the exception of water pumping and milling of grain, other applications of windmill technology, such as electricity generation, are not likely to arouse any interest in the near future because of high production costs ..." /27/.

The fact that countries such as Kenya possess local firms which produce solar water-heating equipment, does not overcome the cost problem. "The high capital costs of solar heaters ... inhibit their widespread utilization" /28/.

2.4 Social Availability

A breakthrough in market penetration can be only achieved by changes in the social pattern, as the example of biogas use in the People's Republic of China (the only country with extensive experience in this field) shows. There, biogas has been the subject of experiment and trial since the 1950s. The production and use of "marsh gas" begin in 1958, the year of the Great Leap Forward, when Chairman Mao gave his public support for the popularisation of this energy source /29/. By the late 1970s, 7 million units were in operation (mostly family units). The spread of this technology in China is assisted by three factors which are not operative in "western" oriented developing countries: the basic principle of social equality, organisation at village level and "the use of local materials and methods for local conditions " /30/. The latter brought down the price of the brick built Chinese digester to a quarter of the Indian industry built steel tank.
There is another important factor in comparing these two countries as far as biogas production is concerned - the above mentioned principle of social equality. There is a nimimum number of 4 - 5 domestic animals needed to supply a digester on family level, a number which is available only for some 15% of the

rural families in India, the country with the largest cattle population in the world. However, despite a rigorous policy of redistribution many peasant families in China also still lack the minimum stock of animals. However, there are approximately 10 times as many units operating in China than in India, the latter having 40% less population.

The introduction of digesters on village level is, on the other hand, also severely constrained for social reasons. "There is in India a time-honoured socio-economic rural structure assuring that dried dung is available to all villagers of all castes and means, regardless of cattle ownership ... once dung becomes the indispensable supply for a system whose final product is to be distributed to all, cattle owners will assert their latent property rights and, having power and status on their side, will get them; poorer families, whose women and children are the main collectors of dung, will lose" /31/. "Without equitable and enforceable distribution, that is, without a very major socio-economic change in the Indian villages, community-size biogas plants would become mostly another convenience for the better off segment of the population, hardly a way to "contribute promptly and dramatically to world equity" /32/.

It has certainly to be stressed that this example is not an exception, but is rather representative for the situation in many developing countries, where alternative societal systems are even more important than alternative energy sources.

2.5 Cultural acceptability

The power, or barrier of tradition, especially in the rural areas of the Third World is not only often overlooked, but also rather unexplored. This is even true where simple techniques designed to tap renewable sources of energy are concerned. Take, for example, the solar cooker; although "technically practical, the use of solar cookers has encountered psychological problems. Their use upsets the habits of the population and therefore careful preparation and the education of the people is necessary before the device is introduced. This introduction must not be hurried to avoid the possibility of the people rejecting the principle of the solar cooker" /33/.

3. CONCLUSION

The outline above shows that, although solar energy has - by direct and indirect application - the potential for liberating especially the tropical countries of the Third World from many of their energy problems, these countries "do not need solar energy, at the stage it is been developed so far", as the Mexican Energy Minister, Juan Eibenschutz, stated at the 7th Energy Technology Conference and Exposition in Washington, D.C. "Although there is theoretical potential for the application of solar technology, the technology is not yet here - the costs and amounts of materials required are just too high, even with the scenario of international prices of other energy resources going higher", Eibenschutz said."The World economic system" in his view, "has raised the prices of solar components faster than the price of the oil they are supposed to replace" /34/. Not In addition, social and cultural hurdles, along with climatic constraints, are hindering effective market penetration of alternative, i.e. new and renewable sources of energy.

This discouraging statement, however, does not mean that alternative energy sources do not have any chance in the developing world. Even electricity production - a special aim of the developing countries, as mentioned above - might be considered on a non-conventional basis, as, e.g., by wind energy. "First, (conventional) electric power is not and will not be available in many areas due to high costs of generation and distribution to small dispersed users. Secondly, there is a possibility of reducing the cost of the windmills by suitable designs" /35/. This means, however, that windmills (certainly, the same applies in principle to all other technologies for arnessing new and renewable energy sources) "have to be tailor-made to suite local wind conditions, and that wind-studies should not include only macrostudies of different regions, but also microstudies within a particular region" /36/.

So far, however, there is a tremendous lack of such studies. In April 1982, the author completed a comprehensive outline for a programme on solar data collection in the Arab States region in his capacity as energy consultant to UNESCO. The consequences of these deficiencies can be learnt, e.g. from a self-critical leaflet, published by the Danish Ministry of Energy in December 1981, containing an article entitled "Why the Solar Water Pump Did Not Work in Ppera" (a small village near Dar es Salaam, the capital of Tanzania). The following explanation was given: "The solar collector was not good enough, and the insulation very bad. Another problem was the cloudy weather from 10 a.m. to 3 p.m. most of the day. After the test period the well ran dry and further experiments were cancelled" /37/.

This case is by no means an exception, neither for development aid projects governed by experts from the developed world (often lacking intimate knowledge of the conditions in situ), nor for those conducted by the Third World on its own.
Therefore - and this is the final message of this paper - a comprehensive systems analysis, considering all relevant criteria as outlined above, has to be conducted for each project individually, before the first step of harnessing an alternative energy source can be taken. Only then will at least moderate success be achieved in the developing world.

REFERENCES

1. S.K. Kukherjee, R. Jagannathan: Energy Consumption in India; Recent Trends and the Problem of Demand Forecasting. Indian Institute of Management; Working Paper No. 198; February 1978.
2. Markus Fritz: Future Energy Consumption of the Third World; With Special Reference to Nuclear Power; An Individual and Comprehensive Evaluation of 156 Countries. Pergamon Press, Oxford, New York, Toronto, Sydney, Paris, Frankfurt, 1981.
3. Ernesto N. Terrado: A national programme for solar energy development in the Philippines. In: SUN, Mankind's Future Source of Energy. Proc. of the Int. Solar Energy Soc. Congress, New Delhi, India, Jan. 1978, Vol. 1, edited by Francis de Winter and Michael Cox, Atlas Corp., Santa Cruz, California, p. 101.
4. Georges Claude: Power from the Tropical Seas; Process Utilizing Difference Between Temperatures of Water at the Surface and at the Lower Depths - Expedients Employed in Locating and Submerging a Line of 6-Ft Pipe a Mile and a Quarter in Length - Results with Exeprimental Plant - The Inventor's Predictions for the Future. Mechanical Engineering, Vol. 52, No. 12, Dec. 1930, p. 1039.
5. Ibid., p. 1044.
6. Ibid.
7. Vaclav Smil: Energy Flows in the Developing World, American Scientist; Vol. 67, Sept.-Oct. 1979, p. 524.
8. Vaclav Smil: Renewable energies: How much and how renewable? The Bulletin of the Atomic Scientists, Dec. 1979, p. 16.
9. Amulya Kumar N. Reddy and K. Krishna Prasad (Indian Institute of Science, Bangalore): Technological Alternative and the Indian Energy Crisis. In: Nat. Sem. on

Energy, March 5-7, 1976, Hyderabad, Vol. II, Papers for presentation at seminar, Adm. Staff College of India/Inst. of Asian Studies, Hyderabad, p. 81-82.
10. Ibid., pp. 82-83.
11. S. P. Govinda Raju and R. Narasimha (Dept. of Aeronautics, Indian Inst. of Science, Bangalore, India): Windmills for Rural Use. In: SUN, Mankind's Future Source of Energy..., p. 1796.
12. Draft of the Energy Chapter for the 5th Nat. Development Plan; was sent to the author from the Ministry of Finance and Development Planning, 2 Aug. 1979.
13. Ernesto N. Terrado ... p. 102.
14. The non-conventional energies. The Courier, No. 51, Sept.-Oct. 1978, p. 73.
15. See ref. /4/.
16. Harry Tabor: Solar Heating and Cooling in Buildings (SHCB) in Israel. In: Description of the Solar R and D Programs in Many Nations; Based in Part on Presentations and on a Panel Discussion held on July 28, 1975 during the Conf. of the Int. Solar Energy Soc. at UCLA; Final Report; Edited by F. de Winter and J. W. de Winter, Atlas Corp., Santa Clara, California, Date published Feb. 1976, prepared for the United States Energy Research and Development Administration, Div. of Solar Energy; under contract No. E(04-3)-1122, p. 98.
17. Government of Israel, Ministry of Energy and Infrastructure, letter to the author, 10 July 1979.
18. P. M. Nyoike: Kenya's Energy Situation at Present and in the Future. Paper presented at the Forum "Third World Energy Strategies and the Role of Industrialized Countries", London, 20 - 22 June 1979.
19. Kurzdarstellung des Projekts SONNTLAN. Dornier System GmbH (DS-SO/NTE-53); 13 Dec. 1978.
20. Extract of the National Plan 1978 - 80; sent to the author from the Federal German Embassy, Tananarive, 27 Feb. 1979.
21. F. W. Lukey: Energy in Ghana (unpublished material).
22. W. K. Kennedy: Plans and Prospects for Solar Energy Utilization in Malawi. In: SUN, Mankind's Future Source of Energs ..., p. 127.
23. P. M. Nyoike ...
24. Ibid.
25. Ibid.
26. Ibid.
27. Ibid.
28. Ibid.

29. Popularisierung der Anwendung von Sumpfgas in den ländlichen Gebieten. In: Peking Review, 15 August 1975, p. 21.
30. Tai Mei-Then: Home-made gas for China's countryside. China reconstructs, May 1977, p. 27.
31. Vaclav Smil: Renewable energies ..., p. 17.
32. Ibid.
33. Cheichné Traore: Development and Use of Solar Energy in Mali. In: Solar Energy, Proc. of the Unesco/WMO Symp. (Geneva, 30 Aug. - 3 Sept. 1976); WMO No. 477, Secretariat of the World Meteorological Organization, Geneva, Switzerland, p. 256.
34. Latin American Energy Report, March 27, 1980, p. 47.
35. S. P. Govinda Raju and R. Narasimha ..., p. 1796.
36. Amulya Kumar N. Reddy and Krishna Prasad ..., p. 83.
37. New and Renewable Energy Sources; An Outline on Danish Research, Development and Industrial Application Activities. The Ministry of Energy, Denmark, 1981.

ENERGY FLOW IN NATIONAL ECONOMIES

Helmut Neu

Commission of the European Communities, Joint Research Centre - Ispra Establishment, Systems Analysis Division, 21020 Ispra (Va) - Italy

ABSTRACT

The paper gives an introduction to the basic concepts of energy chains, national energy balances and national flow diagrams. It outlines in particular the method elaborated at the Statistical Office of the European Communities on how to classify statistical energy data and bring them into a consistent scheme. This scheme is also applicable to developing countries with the introduction of traditional and non-commercial energy. Using flow diagrams derived from the energy data of four Third World countries it is demonstrated how energy flow diagrams can contribute to a comprehensive and global view of the principal features in national energy systems of developing countries and serve as a basic energy planning tool.

1. INTRODUCTION

There is a need for an integrated concept of national energy planning. What does this mean? And why?
Most energy planning in the industrialised and developing countries has been done until about a decade ago by each of the various energy branches such as electricity, hydrocarbons, coal and gas more or less independently of one another. Each branch has tried to gain more customers and maximize its output and influence and, if in commercial competition, its profits.

H. Neu and D. Bain (eds.), National Energy Planning and Management in Developing Countries, 151–174.

Following the first oil price explosion in 1973/74 and the related awareness of possible oil shortages, national governments have started to develop national energy policies which are based on an integrated view of the whole energy sector. Most specialists today agree that only by using all available energy sources in an optimum way can we avoid a serious and imminent energy shortage affecting a large part of the world. Moreover, this optimum way can only be found by an integrated approach.

Concepts for describing national energy flow systems (or energy balances) have been developed and are widely used now in the industrialised countries but not everywhere in developing countries. It is the purpose of this lecture to report on what national energy flow concepts are and what they are good for, with emphasis on the methodology.developed at the Statistical Office of the European Communities.

2. ENERGY IN THE ECONOMY

In national economies, in addition to such sectors as industry, transport, agriculture, households, there is usually also defined an energy sector which provides energy to the other economic sectors and includes activities such as mining of energy sources as well as transformation, transport, distribution, import and export of energy.

Generally, economists regard energy as a commodity which is subject to demand and supply. One may, however, consider energy also as a mean to provide a service. In fact, those who buy energy want to get a service: to cook a meal, drive a motor car or light a room.
Whether commodity or service, in principle one could measure energy in monetary units. But it is much more useful to measure energy in energy units.

When establishing an integrated concept of the energy flow in the economy it is indispensable to find a method of denominating the various forms of energy in common energy units. Usually this method is the calculation of the caloric equivalent content (e.g. of a kWh electricity, a barrel of oil, 1 kg of charcoal). No general agreement has emerged up to now on which unit to use. Preferably it should be an energy unit which is not too small, but results in figures which are memorizable.

The Statistical Office of the European Communities,

EUROSTAT, which has to harmonize the statistical data of the EC Member Countries, selected in the beginning the million of ton of coal equivalent (Mtce) (1 kg coal equivalent equals 7000 kcal) and later the million ton of oil equivalent (Mtoe) (1 kg oil equivalent equals 10000 kcal). Other statistical institutions prefer to use other units such as the physical unit 1 Joule (W sec), which, however, is so small that a high multiplicate, e.g. 10^{15} or 10^{12} Joules is necessary in order to get reasonable figures. We use in this lecture the unit 10^3 toe, which is about $4.2 \cdot 10^{13}$ Joules or 1.2 millions kWh*), as being more appropriate for energy flows in developing countries, particularly where they are small in size.

3. ENERGY CHAINS

To establish a national energy flow diagram or an energy balance it is necessary to distinguish between three different categories of energy:

i) Primary Energy refers to energy in the form of natural resources which have normally to be transformed into other forms of energy in order to be usable (an exception is e.g. natural gas). Table 1 gives the most important primary energy forms for DCs.

ii) Secondary Energy (sometimes called "Intermediate Energy") is a category corresponding to a commodity which is purchased by consumers. To be precise, a distinction has to be made between "secondary" energy as the end product of performed conversion processes (e.g. in a power plant or in a refinery) and "final energy" which is the energy delivered to consumers, not comprising losses in transportation and distribution, and the consumption of the Energy Sector itself. Although the difference is mostly small in quantity it has to be considered when using statistical data (see chapter 5).

iii) The End-use Energy (sometimes called "useful energy") corresponds to the energy needed for the service which the consumers wish to have. As Table 1 indicates there are three forms of end-use energy of importance, the mechanical energy (sometimes called "motive energy"),

---

*) See also the energy conversion table in the Appendix.

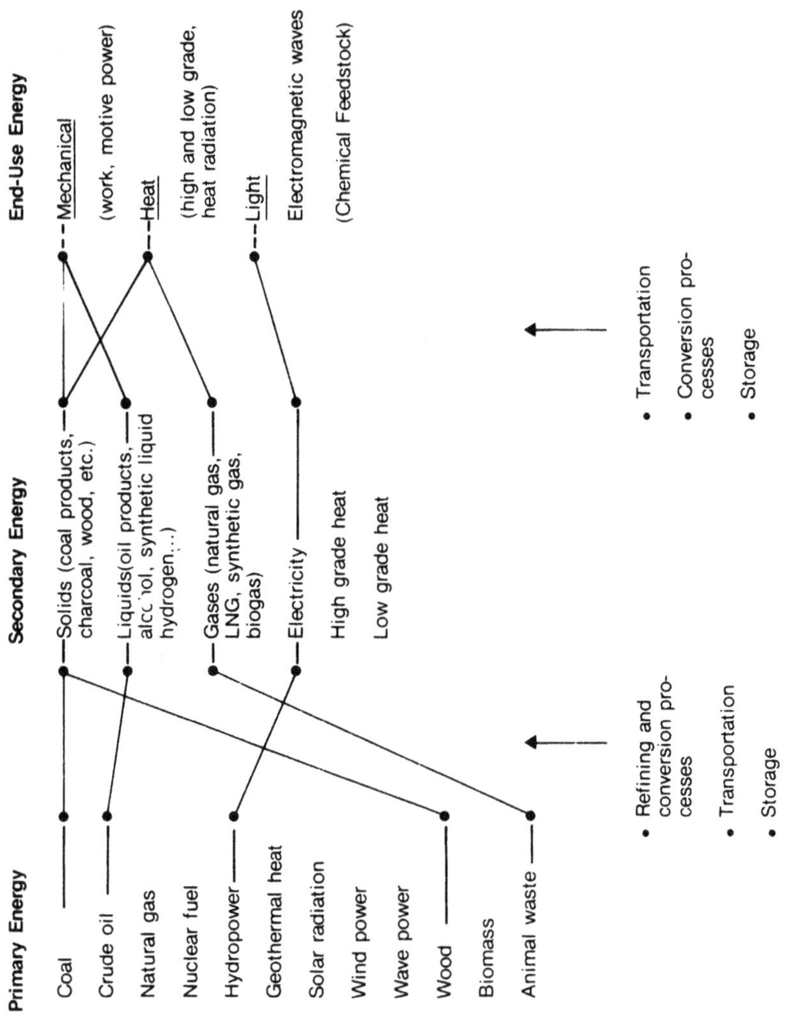

Table 1 Energy Categories and Examples of Energy Chains

light and heat energy. Examples of end-use energy are the mechanical energy driving a water pump for irrigation or the wheels of a motor car. End-use heat is e.g. the heat necessary for melting ores or heating water for cooking (without the losses dissipated into the environment).

In energy statistics chemicals used for non-energetic purposes (e.g. in the chemical industry) are usually included as well, whenever they are produced from solid, liquid and gaseous fuels.

Tracing the path from primary to secondary energy and to end-use energy, a number of processes can be identified as belonging to an energy chain (see Table 1). There are refining processes to bring primary energy into a more valuable form (e.g. crude oil into gasoline or wood into charcoal) and conversion processes changing completely the nature of energy (e.g. heat into electricity, solar radiation into heat). Furthermore transportation (e.g. by trucks, trains, ships, pipelines) belong to an energy chain.

In an energy chain some of the energy content is lost. There are conversion, transport, distribution and storage losses. This means that in an energy chain the end-use energy can be substantially smaller than the content of the primary energy. However, according to the conservative law of energy, the total energy passing into the environment must be exactly equal to the primary energy input. This is demonstrated in Fig. 1. What happens in the consumer device (motor car, stove, bulb) is that the end-use energy (mechanical, heat, light) is converted into "waste" energy (mostly low grade heat).

4. THE QUALITY OF ENERGY

Some of the conversion processes in the energy chain are linked with high losses. It is well known that because of the second law of thermodynamics, conversion of heat into mechanical energy is a rather inefficient process (vapour turbines, internal combustion engine). But the burning of fuels for cooking can also be very wasteful. On the other hand, there are very efficient conversion processes, such as transforming electricity into mechanical energy or heat.

This brings us to another characterization of energy which unfortunately cannot be done scientifically in a satisfactory way: one can attribute to energy a certain quality, which

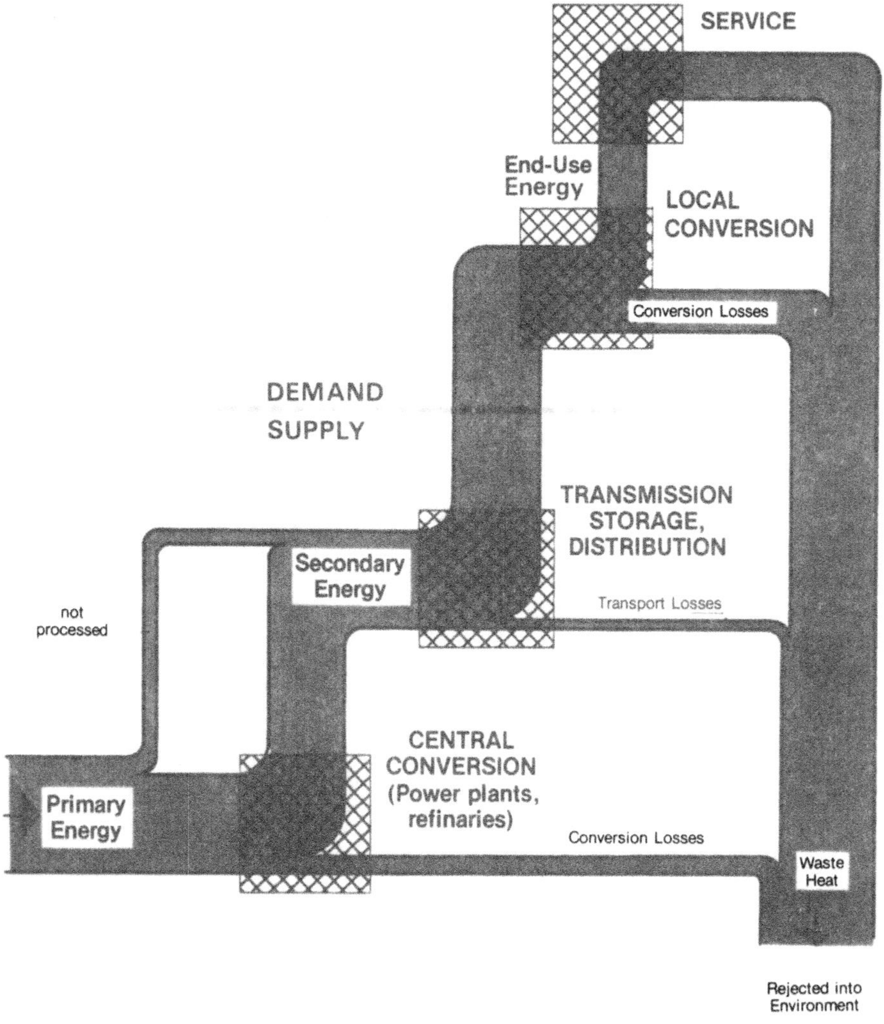

Fig. 1 Energy Conversion Scheme for National Energy Flow

plays an important role in energy management and planning.

First, one has to define in what respect quality has to be seen. One may consider criteria linked to the potential for convertibility into other energy forms, the transport properties and the potential for storage. Table 2 gives some indications on how to rank secondary energy forms according to these criteria. From this one may draw the conclusion that electricity can be regarded as by far the most valuable form of energy so long as transport over long distances and storage potential are disregarded. Low temperature heat is the least valuable energy form. As was possible in the last three decades in the industrialized part of the world, electricity and cheap oil products were an ideal combination and to a great extent responsible for the prosperity developed in these regions after the Second World War.

To attribute a quality to primary energy forms is still more difficult. One of the possible criteria which has been proven very important in the past is the energy density. From this point of view nuclear energy (energy density in the range of 10^4 kW/m^3) is a leading candidate and this was responsible for the high expectations which have been associated with nuclear energy (which up to now could not be fulfilled for other reasons). It gives, however, also an indication of the prospects for some new alternative energy forms such as solar power, wind power and wave power.

It is well known that the power density of solar radiation without concentration is very small about 1 kW per m^2 under clear sky conditions. The thermodynamic potential is sufficiently high (about 6000°C), but it needs concentration devices in order to get an energy form of a high quality. Concentration devices are highly material and capital intensive. This is the main reason why solar energy was successful up to now in the industrialized part of the world only for secondary energy of low quality (e.g. low temperature water supply for households and industry, crop and food drying, water desalination). It will also be important for its prospects in the developing countries.

This problem of quality has to be regarded carefully when designing an efficient energy system. E.g. the use of electricity for producing low temperature heat (e.g. for space heating) is a highly inefficient process; producing it with unconcentrated solar radiation is an efficient process from the standpoint of energy quality (there may be special conditions where other criteria are more important). Combined processes use this quality concept, e.g. in the co-generation of electricity and heat.

TABLE 2 - Ranking of secondary energy forms according to their potential for convertibility into end-use energy, transport and storage

	Convertibility	Transport	Storage
Electricity	very high	medium (low for long distances)	low
Solid fuels	medium	medium	medium
Oil products	medium-to-high	high	high
Synthetic liquid fuels	medium	medium	high
Gaseous fuel	medium	high (medium for sea transport)	medium
High temp. heat	medium	medium	low
Low temp. heat	low	low	low

5. THE NATIONAL ENERGY BALANCE

The energy balance-sheets of the member countries of the European Community are published regularly by the Statistical Office of the EC. In these balance-sheets all the operations in an energy chain (see Table 1) are recorded on the basis of the calorific energy content of each energy form. They make it possible to record the losses occurring in the course of processing operations and give the quantities of energy available to the final consumers for the period of one year. This type of balance-sheets ensures harmonisation between the Community's methodology and that of other international organisations, e.g. the OECD.

The columns of the sheets refer to the energy forms, the lines indicate the aggregates related to the operations which the different energy forms have undergone. At present, the columns of the sheets contain the following primary and secondary energy forms (see /1/):

- hard coal and derived solid fuels
- lignite and derived fuels
- crude oil
- petroleum products
- natural gas
- other gases (e.g. coke-oven gas, blast furnace gas, works gas)
- other fuels (e.g. wood waste, gas and heat recovered by industry)

. heat (including heat from nuclear energy)
. electrical energy

Not included in the balance-sheets are: solar heat, wind energy and energy from biomass.

The structure of the sheet lines is given in Table 3. It shows three different sections:

i) The first section (lines 1 to 7) indicates the actual supply and the overall consumption. Primary products relate to the indigenous extraction of energy from a natural source. Recovered products can be e.g. slurries and recycled energy products recovered in industry. The key aggregate of this part is the energy available for gross inland consumption.

ii) The transformation section (lines 8 to 13) show transformation inputs and outputs whenever the different energy forms are subjected to physical or chemical modifications. The difference between inputs and outputs constitutes transformation losses. The consumption of the energy sector covers the consumption of own-produced energy and of energy purchased by energy producers and transformers in operating their installations. Distribution losses are losses due to transport or distribution of electrical energy, natural gas and derived gases.

iii) The uses section (lines 14 to 16) shows final non-energy and (secondary) energy consumption of the different economic sectors and subsectors. Due to inaccuracy in the statistical data there is always a statistical difference.

The EC balance-sheets have been developed primarily for the industrialised nations. There is, however, no reason to change the items of columns and lines in order to adapt them to the special conditions in developing countries.

A possible structure of the columns for the use in developing countries may be:
. coal
. commercial firewood and charcoal
. crude oil
. oil products (this column could be split up into gasoline, naphta, kerosine and diesel oil, if suitable)
. natural gas
. other gases (e.g. from biomass)
. other fuels
. hydropower and geothermal heat

Table 3 - Line structure of the EC energy balance sheets

1.	Production of primary sources
2.	Recovered products
3.	Total imports (primary sources, derived products)
4.	Variation of stocks
5.	Total exports (primary sources, derived products)
6.	Bunkers
7.	Gross inland consumption (1 + 2 + 3 + 4 – 5 – 6)
8.	Transformation input
8.1	Conventional thermal power stations
8.2	Nuclear power stations
8.3	Patent fuel and briquetting plants
8.4	Coke-oven plants
8.5	Blast-furnace plants
8.6	Gas works
8.7	Refineries
9.	Transformation output (same substructure as 8)
10.	Exchanges and transfers
11.	Consumption of the energy sector
12.	Distribution losses
13.	Final energy available for consumption (7 + 9 + 10 – 8 – 11 – 12)
14.	Final non-energy consumption (chemical industry, other sectors)
15.	Final energy consumption
15.1	Industry (iron and steel industry, chemical industry, other industry)
15.2	Transportation (rail, road and air transportation, inland navigation)
15.3	Other sectors (households, commerce, public services, agriculture, fisheries a.s.o.)
16.	Statistical difference (13 – 14 – 15)

. electricity
. non-commercial energy.

Changes in the line structure (see Table 3) may be:
8.1 and 9.1: Public service thermal power stations
8.1 and 8.2: Hydropower stations
8.3 and 9.3: Refineries
8.4 and 9.4: Gas plants (e.g. biomass)
8.5 and 9.5: Self-use power plants (e.g. diesel and mini hydro plants)
16: Not identified, and statistical differences.

Table 4 shows, as an example, the energy balance of Kenya for the year 1977. The efficiency of hydro power stations is assumed to be 28%, representing the average thermodynamic efficiency of thermal power stations. Other concepts use the efficiency to transform hydro energy into electrical energy. +)

It should be noted that there are other concepts of energy balances, e.g. the concept developed by OLADE (organisation Latino-Américaine de l'Energie) /2/, which has been developed particularly for developing countries with two separate balances for primary and secondary energy.

6. ENERGY FLOW DIAGRAMS

Energy balances of the EC or OLADE type are, from the standpoint of statistics, fully satisfactory to describe the complex structure of energy flows. For the analysis of national energy flow systems, however, graphical flow diagrams, which have been developed about ten years ago simultaneously in the USA by the Brookhaven National Laboratory (BNL) /3/ and in Europe by the "Institut Economique et Juridique de l'Energie" (IEJE) in Grenoble, France /4/, are considered to be more suitable. They allow a quick overview of the general situation of a country and contain the same data as energy balances.

Figures 2 to 5 illustrate the BNL Reference Flow Diagram adapted to the situation in developing countries. An energy flow diagram can be seen as a network of energy chains, but usually

---

+) Our concept allows a better comparison of primary energy supply data between countries with dominant hydro and those with dominant thermoelectric electricity production.

Table 4 - Energy balance of Kenya, 1977 (10^3 toe/y)

	Coal	Crude oil	Petroleum products	Hydro power	Electricity	Total commercial	Non-commercial
1. Indigenous production				224		224	2810
3. Imports	38	2603	189		23	2854	
4. Variation of stocks		18	-63			-45	
5. Exports		-1439				-1439	
7. Gross inland consumption	38	2621	-1313	224	23	1593	2810
8. Transformation input		-2621	-124	-224		-2969	
8.1 Thermal power stations		-2621	-124			-2745	
8.2 Hydro power stations				-224		-224	
8.3 Refineries							
9. Transformation output		2501		90		2591	
9.1 Thermal power stations				27		27	
9.2 Hydro power stations				63		63	
9.3 Refineries		2501				2501	
11. Consumption energy sector							
12. Distribution losses							
13. Final consumption	38	1064	1064	113	113	1215	2810
14. Non-energetic consumption			31			31	
15. Final energy consumption	38	1033	113		113	1184	2810
15.1 Industry	1		263		51	315	60
15.2 Commercial sector			169		27	196	}
15.3 Residential sector			6		26	32	}
15.4 Public service			25			25	} 2750
15.5 Agriculture			61			61	}
15.6 Rail transport	37		174			211	}
15.7 Road transport			300			300	}
15.8 Air transport			24			24	
15.9 Other transport			11			11	
16. Not identified					9	9	

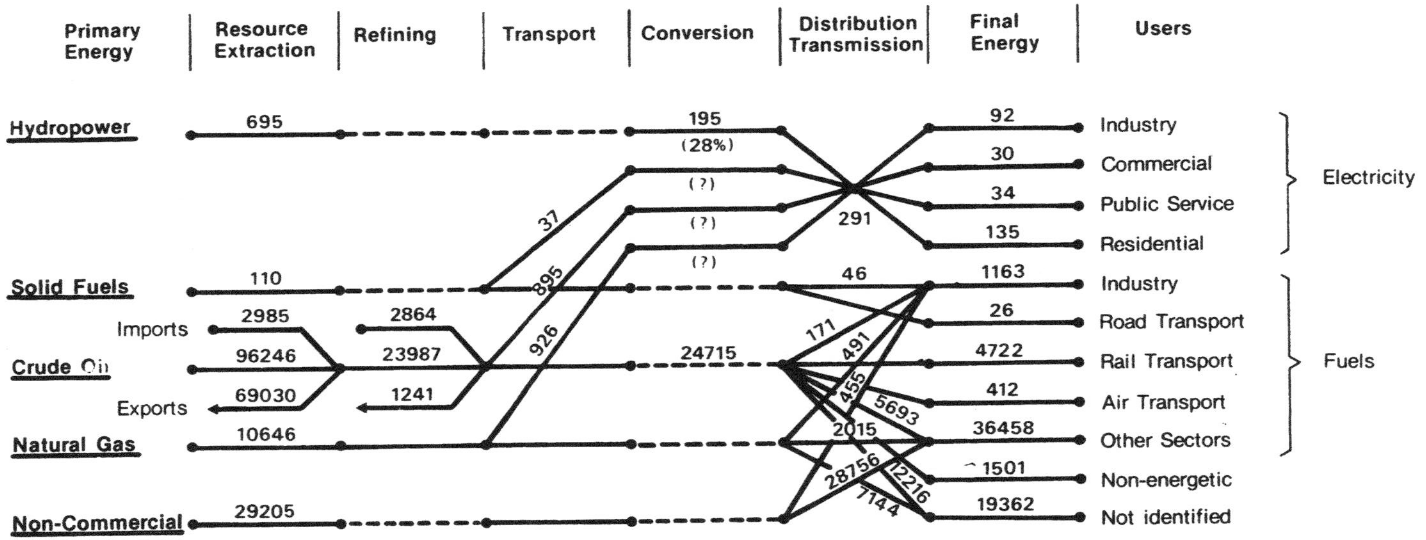

Fig. 2 Energy Flow Diagram for Indonesia, 1977 (Flows in 10^3 toe/y)

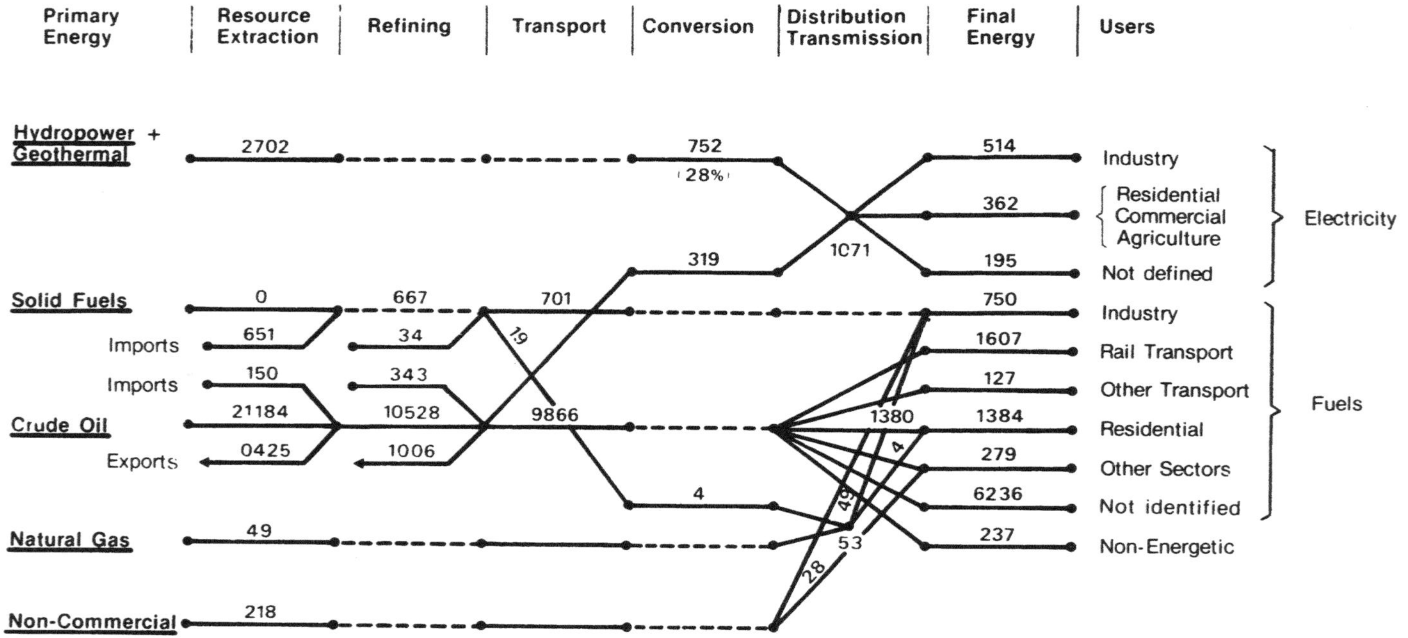

Fig. 3 Energy Flow Diagram for Egypt, 1977 (Flows in 10^3 toe/y)

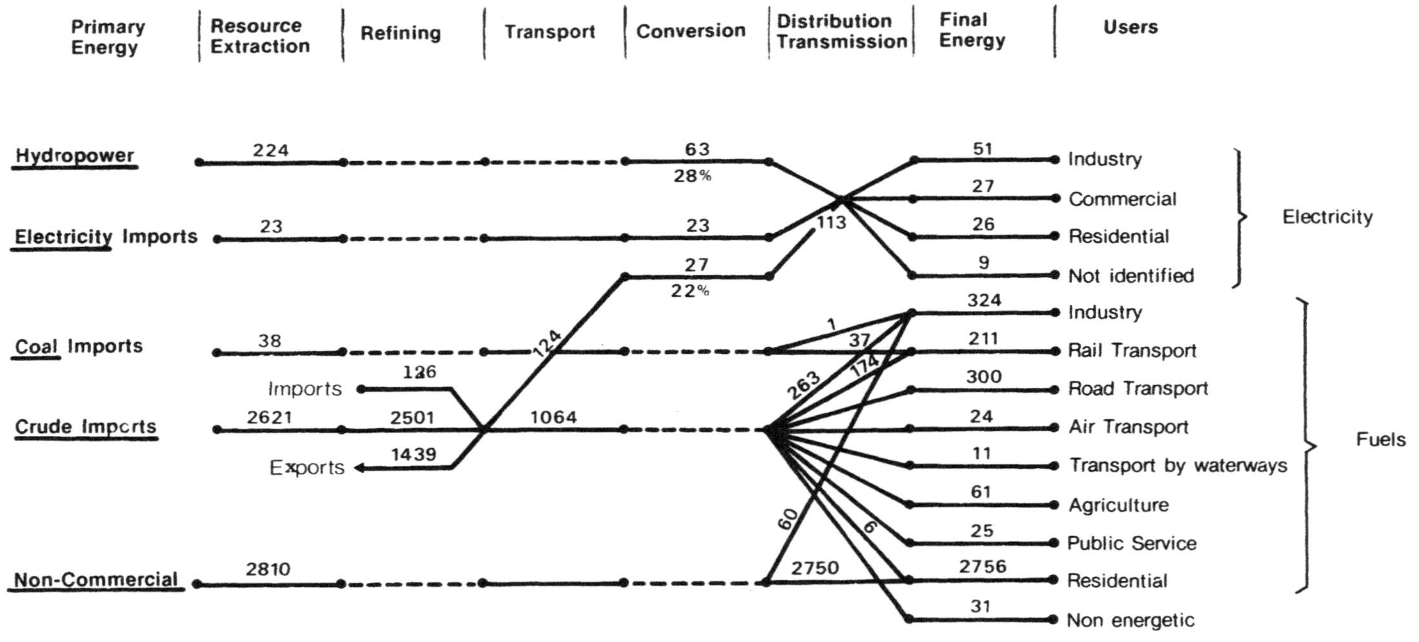

Fig. 4. Energy Flow Diagram of Kenya, 1977 (Flows in 10^3 toe/y)

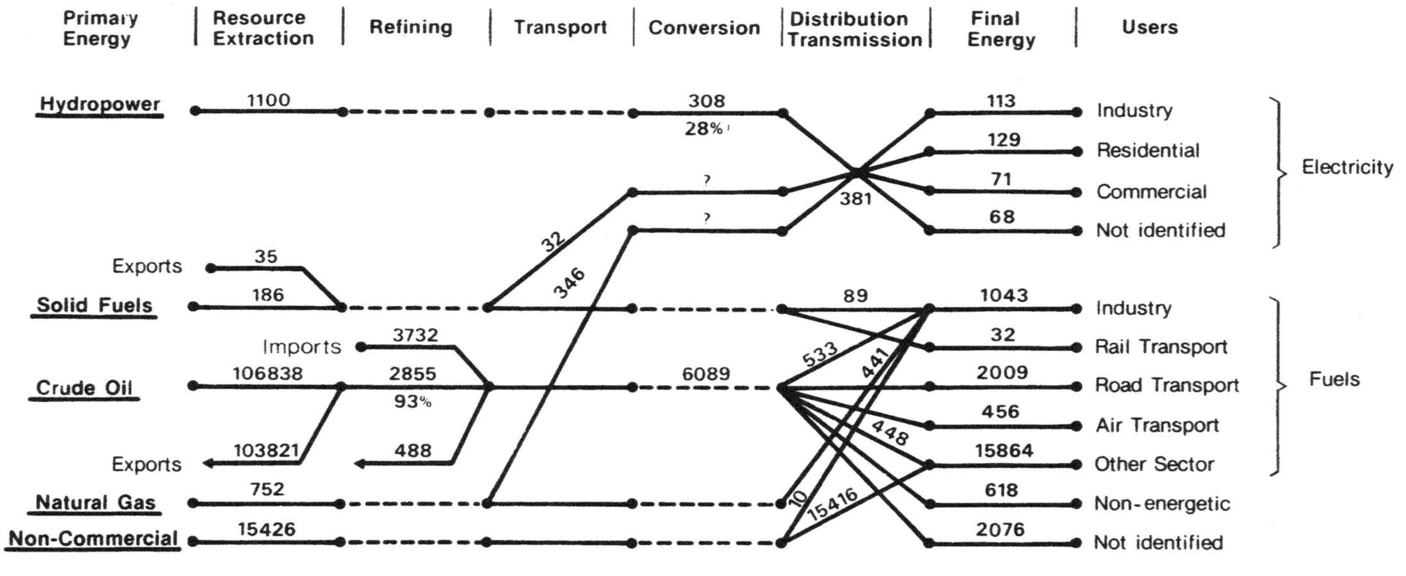

Fig. 5. Energy Flow Diagram of Nigeria, 1977 (Flows in 10^3 toe/y)

excluding end-use energy. In these diagrams the solid lines correspond to real processes and the nodes to distribution in the energy flow. (In the IEJE approach it is just the opposite.) The actual energy contents are given at the end of each process, taking conversion and transport losses into account.

In practice, the losses for refining, transport and distribution including storage are fairly small in comparison to the calorific content and are therefore mostly negligible in terms of the accuracy of statistical data. Only the thermodynamic conversion processes are associated with high losses, as mentioned earlier.

The level of aggregation on the demand side can be adapted to the availability of statistical data. It determines also the degree of information about the final energy mix on the users' side. It would be useful to distinguish between demand in rural and in urban regions.

The statistical data in Figs. 2 to 5 were made available to me by courtesy of the International Atomic Energy Agency (IAEA), Vienna. Note that the data in Fig. 4 are identical with the data in Table 4.

7. SPECIFIC ENERGY AND COMPARISON STUDIES

It is useful for the analysis of national energy systems to relate the energy figures to other parameters such as population, total output in the economy (GNP), or in part of it (sectoral output) to get a better insight. This also allows comparison of data for different countries.[+)]
Table 5 gives some specific energy data for the four countries and the EC related to population and GNP.

I would like to emphasize that the following conclusions drawn from these figures (assuming that the data are reliable) are only illustrative.

[+)] Because most of the data consist of aggregates of different energy forms which have a different "quality" (see chapter 4), comparisons can be only approximate.

Table 5 - Comparison of energy flow data related to population and GNP, for 1977

	Egypt	Indonesia	Kenya	Nigeria	EC	
Total primary commercial energy	13.7	43.3	1.6	8.3	915.8	Mtoe
Non-commercial energy	0.2	29.2	2.8	15.4		Mtoe
Percentage non-commercial	1.6	40.2	63.8	64.5		%
Population	38.9	143.2	14.4	66.4	259	millions
Pirmary energy /cap	0.357	0.505	0.306	0.358	3.586	toe/cap
Electricity consumption/cap	30.8	2.7	10.5	7.7	504	Watts/cap
Electricity as percentage of total secondary energy	9.2	0.8	9.3	5.4	10.5	%
Energy consumption of industry/cap	37.4	8.8	26.0	17.4	1130	kgoe/cap*
Energy consumption of transportation/cap	44.6	36.0	37.8	37.5	523	kgoe/cap*
Energy consumption of residential, Agr./cap	52.8	256	213	242	1258	kgoe/cap*
Not identified/cap	165.3	135.3	0.7	31.5		kgoe/cap*
GNP/cap	470	311	312	619	5345	US$/cap
Primary energy/ GNP	760	162	981	578	497	toe/$

* 1 kgoe = 10^{-3} toe

i) The primary energy consumption per capita is not so very different between the four countries and is roughly one tenth of that of the European Communities, when non-commercial energy, which is a small portion only in Egypt (or is this a statistical error?), is included.

ii) Electricity consumption per capita is, however, very small against that of the EC with the exception of Egypt (Asswan). It should be noted again that a distinction between rural and urban population would allow a better analysis.

iii) The picture changes when relating electricity consumption to total secondary energy consumption. This percentage for three of the four countries is not so very different from the EC percentage.

iv) The figures for industry reflect the low level of industrialisation in developing countries, particularly in Indonesia.

v) It is interesting to note the almost equal consumption per capita for transport of the four countries. Again a rural/urban distinction would give further information

vi) For the residential, agriculture and commercial sector the figure for Egypt is astonishingly low (possibly an error in non-commercial energy?).

vii) The analysis for Egypt and Indonesia is hampered by the high portion of energy consumption which could not be associated with a consumer sector.

viii) The primary energy consumption per GNP (including non commercial energy) is only for Indonesia much higher (about three times) than that for the EC (no explanation of the reasons is given). The correlation between energy consumption and economic output is very well known and one of the large concerns for development.

8. PROJECTIONS INTO THE FUTURE

Up to now we have looked on the energy flow in a national economy at a certain time as a "snapshot". Energy Planning means looking into and designing of possible "futures". Energy flow diagrams can be a helpful instrument.

I want to emphasize that I am by no means claiming that this rather simple instrument is adequate for understanding fully the complexity of national energy systems or adequate for national planning in general. It is, if you want, a "top down" approach to energy planning which has to be complemented by a very detailed analysis "to the bottom".

Furthermore the energy flow system is strongly influenced by the availability of energy and financial resources and by the economical, social and political situation of a country. A scheme of the most influential parameters is given in Fig. 6 . Planning needs to consider all these aspects.

I would like to emphasise one point in order to lead into the planned debate on "Is energy demand the correct starting point for energy planning in developing countries?".
The usual approach to projecting energy flows and energy balances into the future is to link energy demand, e.g. sectoral demand, to the growth of the sectoral output of the economy. Such linkage can be done using mathematical modelling (see e.g. /5/).

Figure 7 gives an illustration of how, in the past, energy demand projections have been made (on a more global way and without mathematical modelling) /6/. It was made in the framework of a Joint Egypt/USA study /7/. The shown scenario assumes a certain annual growth rate of secondary energy consumption and the related fuel mix. From this the growth rates of the different fuel consumptions can be calculated. The question arises if such a scenario is realistic.

It is not my intention to go more in the details of the problem of energy demand, but I want to rely on the principle. Projecting energy demand from economic activities is called a normative approach - normative in the sense of goal setting. Before the 1973/74 oil price shock most planning in industrialized countries consisted simply of extrapolating trends of economic activities and related energy demand. However this method of forecasting has failed completely in most cases since that time: planners speak now more consciously about scenarios (making different assumptions of economic growth) and strategies (which can be seen as a qualitative goal seeking process, see e.g. /8/).

An interesting question is, if for developing countries demand scenarios and strategies are useful and realistic. Because the demand concept implies that demand can be satisfied by supply, taking into account the economic interplay of supply and demand, and its regulation by prices and price elasticities. In fact, this still holds for most of the industrialized countries

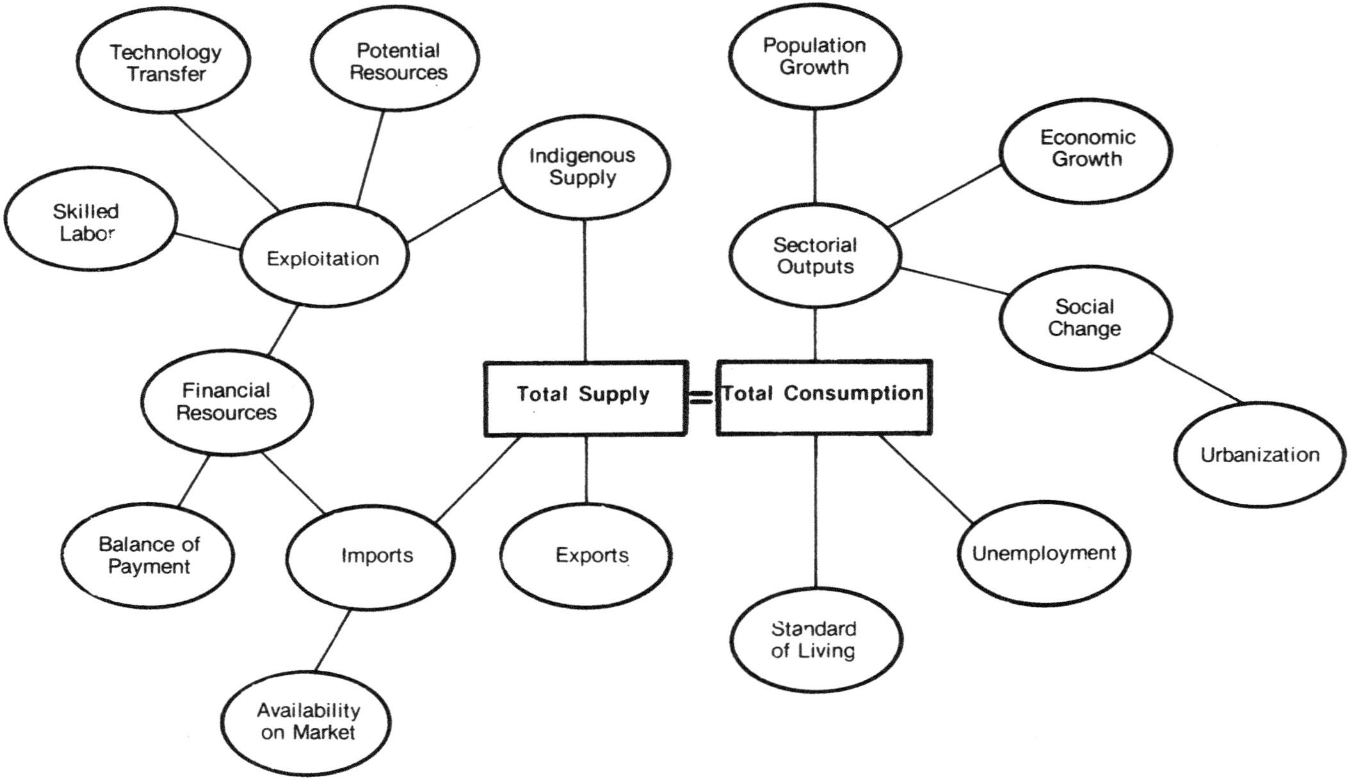

Fig. 6. Components and Parameters of the Energy System in DCs

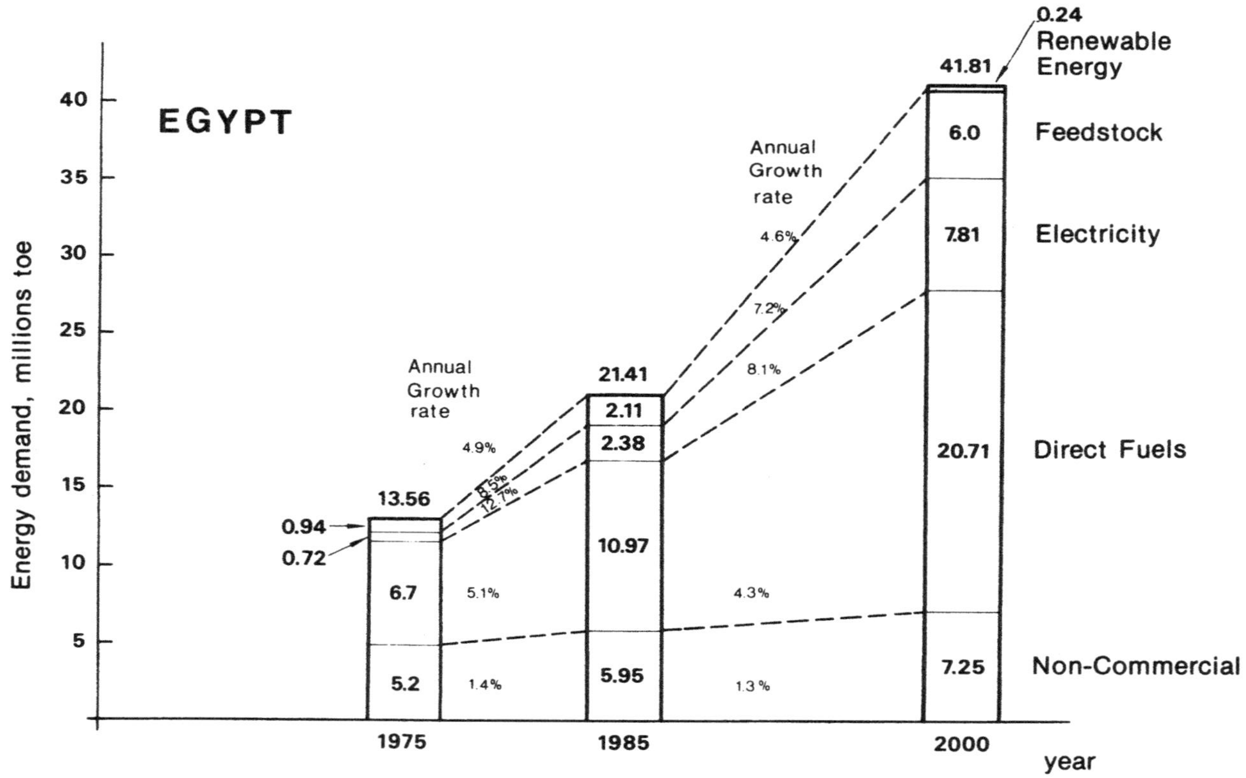

Fig. 7 Secondary Energy Demand Scenario (6)

(particularly in the free market economies) and some DCs, even if nearly all suffer more or less from balance of payment problems.

In DCs however, even if consumers can afford to buy energy service, the state can pay only for limited imports or lacks of funds for exploiting domestic resources. In these cases we have a supply limited situation. The problem of planners is now very different from that of demand driven planning: how best to use available supply for the good of the country and its people may become the real problem.

REFERENCES

1. Energy Statistical Year Books, Statistical Office of the European Communities, Luxemburg-Kirchberg, BP 1907.
2. OLADE Methodology for the Elaboration of Energy Balances, Latin American Energy Organization, Quito, Ecuador, 1980.
3. Hofmann, K.C.: The United States Energy System - Unified Planning Framework, Doctoral Dissertation, Polytechnic Institute of Brooklyn, 1972.
4. Finon, D.: Optimization Model for the French Energy Sector Energy Policy, 1974.
5. Energy Models for the European Community, Energy Policy, special issue, IPC Science and Technology Press Ltd., Surrey, UK, 1979.
6. Cirrillo, R.R. et al.: The Assessment of a Country's Energy Posture, Argonne National Laboratory, 1979.
7. Joint Egypt/United States Report on Egypt/United States Cooperative Energy Assessment (5 volumes), US Department of Energy, Report No. DOE/IA-0002/01, 1979.
8. Energy: Global Prospects 1985 - 2000, Report on the Workshop on Alternative Energy Strategies, Mc Graw-Hill, 1977.

APPENDIX

Energy Units

For national energy balances the following units are used:

tce (tons of coal equivalent)
toe (tons of oil equivalent)
GJ (Giga Joules, 1 Joule = 1 Wsec)
BOE (barrel of oil equivalent)

M = Mega = 10^6
G = Giga = 10^9
T = Terra = 10^{12}

1 BOE/day = 50 toe/year

Conversion Table

	kcal	GJ	tce	toe	BOE	10^3kWh	10^6 BTU
1 tce	7 x 10^6	29.3	1	0.7	4.8	8.14	28
1 toe	10 x 10^6	41.8	1.47	1	6.8	11.6	40
1 ton firewood	3.2 x 10^6	13.4	0.46	0.32	2.2	3.7	12.8
1 ton charcoal	7.5 x 10^6	31.4	1.1	0.75	5.1	8.7	30
1 ton dung	3.4 x 10^6	14.1	0.48	0.33	2.2	3.8	13.2
1 GJ	0.24 x 10^6	1	0.034	0.024	0.16	0.28	0.95
1 BOE*	1.46 x 10^6	6.1	0.21	0.146	1	1.7	5.8
10^6 kcal	1 x 10^6	4.18	0.147	0.1	0.68	1.16	4
10^3 kWh	0.86 x 10^6	3.6	0.124	0.086	0.58	1	3.4
10^6 BTU	0.25 x 10^6	1.055	0.036	0.025	0.17	0.29	1

* assumptions, not defined.

ENERGY FLOW OPTIONS

Gerald Leach

International Institute for Environment and Development

Reasonably detailed information about the structure of energy demand is necessary for planners to appreciate and act upon the substantial opportunities for energy conservation and fuel substitution in developing countries. The paper outlines the advantages and problems of developing such an information base and provides an overview of opportunities and issues in the area of energy conservation and fuel substitution.

The previous lecture discussed national energy systems in terms of fuels flowing from a few sources to a variety of final consumers and end-uses. This lecture goes much further down-river to examine those consumers and end-uses at the distant reaches of the energy flows.

The reason for doing so is that the 'top down' perspective of national energy systems has proved inadequate for today's planning needs. Symbolised by Energy Ministries with their fuel divisions, it belongs to a time when the main objective of energy industries and of planning was to meet demands for different fuels as they arose through economic growth and new patterns of production and consumption. In most countries, competition among fuel industries added to this emphasis on increasing and balancing supplies of fuel. There was little interest in how or why these demands for energy originated.

Today priorities are different, especially in

H. Neu and D. Bain (eds.), National Energy Planning and Management in Developing Countries, 175–193.

developing countries. For many reasons, including high oil prices and worsening terms of trade, national economic objectives are threatened. Attention has therefore turned, as in industrial nations, not only to developing domestic energy supplies, including new and renewable sources, but also to the better management of energy demand. The more effective use of energy ('conservation') and switching to alternative sources from scarce or high price fuels such as wood and oil ('substitution') are now high on the energy planning agenda.

Nearly all these objectives rely essentially on changing the structure of final energy demand. Although in principle there are thousands of methods of doing this, in reality the suitability of each one depends on the specific circumstances of each of millions of consumers: on his or her attitudes and practices, the forms of energy required, and the nest of social and economic conditions surrounding every use of energy and every energy-using technology.

None of these 'ingredients for change' appear anywhere on the standard national energy flow maps. They can only be seen, and their importance to energy planning appreciated, if one looks further and deeper than we have done so far into the fine structure of national energy systems.

Obviously, this mapping of the almost infinite variety of energy demands can never be completed. Move from sectors to sub-sectors to final consumers and the number of actors in the game proliferates alarmingly. Move from fuels to end-uses of energy and one has left the territory of sales statistics, fuel bills and energy meters. There are no longer any records of the energy flows. Even for a single class of consumer such as 'schools', a range of fuels may be employed to meet a large variety of end-uses of energy - for space and water heating, lighting, cooking, machines and so forth - that can be established only by intensive site visits or questionnaires. Clearly, compromises must be made and cut-off points established for the analysis of end-use demand, depending on manpower resources, the availability of data and difficulty of assembling it, and the utility of the information for planning purposes.

The major aim of this lecture is therefore limited to presenting some mapping guidelines, using case studies wherever possible. These examples will be taken mostly from the urban-industrial sector to keep the arguments focussed and because rural energy problems are the subject of special sessions during the seminar. What kinds of information does one get at each level of analysis of end-use demand? How difficult is it to get there? How useful is the information for planning, especially in selecting and implementing alternative energy options? Hopefully, these guidelines can be explored further by the seminar working groups using their own case studies as starting points.

MEASURING CHANGE

No store keeper would flourish if he knew only his weekly income and outgoings. He needs to know his stock and which lines are selling well or poorly. Similarly with energy. The aggregate figure of primary fuel consumption per unit of GDP - so often used to compare nations, or national performance from year to year - misses almost everything of importance, including an answer to the question 'why?'.

The first step towards mapping energy demand is to break down this aggregate into components by relating fuel consumption to other factors of importance. Fuels are not used without purpose but to provide for other ends, whether to make a ton of steel, create a dollar of economic output, provide for so many ton-kilometres of freight transport, or keep the schools and hospitals going. Similarly, conservation does not mean reducing fuel consumption as such, but reducing it per unit of output or service.

By measuring fuel consumption per unit of output at a reasonably fine-grained sectoral level, one can learn much about the tuning of the various parts of the energy economy: for example, how performance in each sector has been changing (are policies working?), how performance compares to that of other similar countries (are we doing as well as we might?), or where the largest potentials for the growth of consumption or for efficiency improvements lie (where should policy aim next?).

In developed countries hundreds of studies at this first level of sectoral fuel mapping have been carried out. For example, the European Commission is now assembling sectoral fuel performance indicators for each member country to compare national performances and the extent to which they have changed since the first oil price shock of 1973/74. A special session at the International Conference on Energy Use Management, in October 1981, compared such indicators for the first time for all OECD countries. Industrial associations and independent bodies such as the NATO Committee on the Challenges of Modern Society programme and the UN Economic Commission for Europe have also done much work analysing and comparing fuel consumption per unit of output for industries and industrial processes.

The following three examples must suffice to underline the benefits of these first level studies. They involve different levels of sectoral mapping and give a remarkable range of perspectives on change within the energy economy.

The first example involves a coarse sectoral breakdown. While researching IIED's Low Energy Strategy for the United Kingdom, we discovered by using only the aggregate UK fuel statistics, that from the mid-1950s to mid-1970s road transport accounted for 70 per cent of the total increase in UK secondary fuel consumption. This was largely because of a huge and once-only rise in the number of cars from just under 3 million to over 14 million in two decades. The main implication for the UK is that as this growth slows and levels off due to saturation in car ownership, the growth of total oil and energy consumption will also slow appreciably - even if nothing is done to make cars more fuel efficient or switch people from cars to public transport.

Saturation effects of this kind are obviously missed by any forecasting method based on aggregate energy consumption and economic data, which can therefore easily overestimate future demands. Such methods are still widely used in developing countries, at least to provide baseline forecases of future demands - often for as long as 20 years ahead. Although saturation effects in these countries may be some way off, it is worth noting the general rule that has been found to apply to almost every set of forecasts for developed countries: when you disaggregate by sectors you get a lower (or much lower) total demand than when you don't.

The second example involves a finer level of analysis. Since World War II, fuel consumption by agriculture in industrialised countries has risen rapidly owing to increased mechanisation and fertilisers. These changes have been roundly attacked for their 'energy profligacy' and held up as warnings against the modernisation of agriculture in developing countries.

However, detailed analyses have shown that the arguments are not so simple. For example, while annual fuels going into US corn production rose between 1945 and 1975 from 19 to 30 GJ per hectare, yields rose from 2 to 5 tonnes. Thus to get 5 tonnes of corn, energy inputs dropped by 17.5 GJ and land requirements fell by 60 per cent. Manpower decreased five-fold. Over the country, 20 million hectares of corn land were released, mostly to grow higher value, lower energy input soya beans. What at first sight seemed like energy waste, on closer scrutiny is revealed as an astonishing increase in agricultural productivity and energy productivity. Sound planning might protect such achievements by allocating sufficient oil to agriculture and attempting to save it elsewhere (although, of course, there are many other considerations in such an agricultural policy).

The third example involves a still finer level of analysis and reveals the potential for substituting oil by coal in industry. The UK National Coal Board recently completed a study for the IEA on the stock of industrial boilers in OECD countries. At present these consume annually some 530 million tons of coal equivalent (equal to 7 million barrels of oil per day). Only 15 per cent of this fuel is coal; and consumption is expected to increase to 800 mtce by the year 2000. The study found that with expected future oil and coal prices, in the year 2000 the use of coal might rise to 40 per cent through vigorous marketing and to 80 per cent with the addition of government subsidies for coal equipment.

This projection of a major switch from oil to coal is itself interesting. But it is also important to note what kinds of information about the boiler stock were needed to arrive, by careful calculation, at this answer. Assessments had to be made of the numbers of small firetube or shell boilers and large watertube boilers; the load factors of each type; the fuels used in each type; the lifetimes of each; and the costs of replacing different sizes of each type of boiler by new oil and new coal burning equipment. Only with reasonable estimates of all these items of information could the final assessment be built up.

INTO END-USE DEMAND - A CASE STUDY

These examples have hinted at some of the benefits of sectoral energy analysis but have said little about the difficulties. To examine the benefits and problems more fully, the most useful approach is to track one energy flow from the coarse national fuel statistics to its conclusion in a specific product, with various excursions on the way.

The example I have chosen is the factory-drying of vegetables. The starting point is the eight-fold breakdown of fuel consumption

Industry sector	Percent of industrial energy	Percent of industrial output	Energy intensity (MJ/£)
Iron and steel	27.3	4.9	340
Building materials	9.2	3.3	166
Chemicals	14.9	6.5	138
Textiles etc	5.5	4.3	78
Paper, printing etc	5.5	6.8	49
Food, drink etc	8.5	14.3	36
Engineering	17.6	35.8	30
Other trades	11.5	22.7	29
All industry	100	100	60
Services	(27)	(83)	20

Table 1. Secondary fuels and output in UK industry (1976)

within industry given in the UK energy statistics, as shown in Table 1. The service sector is also included for comparison.

Even at this first level of disaggregation (Level 1) there is much information of importance to energy and macro-economic planning, notably in the final column. Energy intensities within the industrial sector vary by nearly 12:1. The intensity of the services sector is only a third that of industry as a whole. While the economic output of services is over 80 per cent that of manufacturing, its fuel consumption is little over one quarter as much.

Clearly, data of this kind are valuable for translating macro-economic targets into fuel forecasts and in guiding macro-economic policy itself. Annual data can show the changing energy intensity of each sector. Such changes could be due to any combination of an altered mix of products in the sector, changing prices for products, changes in the mix of fuels, and more efficient use of fuels. To tease out the effects of these various factors demands much more detailed sectoral analyses.

Yet already, at this first level, most countries have run into considerable data problems. In many countries a breakdown of the kind shown in Table 1 may not be easy to construct because fuels such as petroleum products are supplied by different organisations, using different sectoral breakdowns for their sales records, with no government requirement to co-ordinate and submit their sales statistics. In some European countries the national fuel statistics do not even separate the residential and commercial sectors, although combined they account for 30 to 50 per cent of total energy use and are obviously quite different in kind so far as energy policies are concerned.

In developing countries the data base may be even more sketchy. Lee Schipper, who has recently completed a three-year end-use analysis of Kenya, found that he had to supplement government industrial fuel statistics by nearly 50 visits to individual firms and factories, but even so was unable to assemble fuel consumption data for all the major industrial sub-sectors.

Continuing our journey, we come to a finer breakdown of energy use (Level 2) which shows how fuel consumption in a sector is divided up by type of fuel and main end-uses. This is done in Table 2 for the UK food and drink industry and provides much valuable information for assessing the scope for conservation and fuel substitution for the whole industrial sector. Perhaps the most important examples are:

How much co-generation? This is the production of electricity at the factory level so that waste heat can be utilised for processes and space and water heating. Total energy efficiencies are usually in the 70 to 80 per cent range compared to 30 per cent or less for electricity generation and transmission from a power station. In principle, relatively cheap coal can be used instead of oil (or natural gas). The Table shows that some 25 units of fuel were used to produce 1.7 units of electricity and 16.5 units of heat by co-generation - an average efficiency of 75 per cent.

How much direct-fired and process heat? Processes using direct-firing - an industrial-scale pottery kiln is a good example - normally require 'clean' heat as supplied by electricity, gas and highly-refined oils. Usually the temperatures are high so that large energy savings are in principle possible while the waste heat can often be used in other processes ('heat cascading'). However, the fuels are not easily substitutable. Process heat refers mostly to steam-raising in boilers. Again, energy can be saved by cascading, by returning heat to pre-warm the boiler water feed ('recycling') and simply by plugging steam leaks, etc. As shown by the IEA boiler study mentioned above, there may be large potentials for using coal instead of other fuels. In the UK food industry only 16 per cent of process heat is for direct-firing, which suggests good opportunities for fuel substitution. By contrast, in the UK building materials sector, 83 per cent of process heat is for direct-firing.

What are the temperature ranges? This gives a different view on the potentialities for energy savings, fuel substitution and 'new' fuel-saving technologies which work only within certain temperature bands. Examples of the latter are heat pumps and solar water heating, both of which are cheapest (although not cheap) at

Petajoules

	Solids Coal	Other	Oils	Gas	Elec	Heat	Total
Delivered energy	25.6	1.1	108.2	45.3	24.5		204.7
Private elec.	-5.6		-13.2	-6.6	1.7	16.5	-7.2
Net energy	20.0	1.1	95.0	38.7	26.2	16.5	197.5
Process energy via							
steam - 80C	9.1	*	28.2	8.1	*	6.5	52.6
80-120C	5.3	*	32.6	9.9	*	6.4	54.9
120C	*	*	6.5	1.7	*	1.1	9.5
	14.6	1.1	67.3	19.7	*	14.0	117.0
Process energy direct-							
fired - 200C	*		3.1	3.7	2.2		9.6
200-400C	*		2.4	5.7	*		8.3
400C	1.3		*	2.2			3.8
	2.1		5.8	11.6	2.2		21.7
Total process	16.7	1.1	73.1	31.3	2.2	14.0	138.7
Space & water heating							
Central steam	2.8		15.8	5.2			23.8
Local boiler	*	*	1.4	*			2.2
Direct-acting	*		*	*	1.1	2.4	5.5
	3.2	*	18.1	6.5	1.1	2.4	31.5
Motive power							
Stationery			1.5	*	20.3		21.7
Vehicles			1.9		*		2.2
Other uses	*		*	*	2.3		3.2
Total net energy	20.00	1.1	95.0	38.7	26.2	16.5	197.5

* small quantities

Table 2. Energy use in the UK food and drink industry (1976)

lower ranges below 100-120°C. Space heating and water heating for washing and cooking are special cases of the lowest grade of heat which with good system design can often be supplied almost entirely from other factory heat processes.

How much motive power? Electricity for lathes, drills, cranes, etc., and petroleum for factory vehicles such as fork lift trucks, are not easily substitutable. But studies in developed countries have shown that, without affecting performance, energy savings of 20-30 per cent or more can be made in equipment of this kind.

Data on energy end-uses can be collected only by surveys of individual and representative units, whether these are factories, offices, schools or households, etc. These may be made by government-sponsored experts or, as is often the case in developed countries, by 'energy managers' employed by medium to large size firms. In either case the requirements for skilled manpower are substantial. The data shown in Table 2 were assembled for all eight UK industrial sectors from the records of site visits over many years by a large industrial energy management organisation and also by two government schemes which for the past five years have been systematically auditing energy use in industrial sub-sectors. All this information was put together by a branch of the UK Department of Energy using a good deal of experienced guesswork. Some of the figures probably have an error of 20 per cent either way.

Nevertheless, the benefits of these end-use surveys are considerable. Keeping to the sectoral level (Level 2), they allow broad assessments to be made of the scope for energy saving, fuel substitution, and the introduction of new technologies. Priority areas for research and development, potential markets for existing technologies, and leverage points for government policy can thus be identified. As one illustration, the surveys just described have led in the UK to government estimates that some 25-30 per cent of energy can be saved per unit of output in different industrial sectors by the year 2000. With some processes and technologies the potential savings are consideably greater. These are now the subject of further assessments of economic potentials for energy saving.

A second illustration of the value of this type of analysis is the energy source/energy use matrix in which the possible suitability of energy supply and conversion technologies are matched against different types of end-use demand. An example, for the application of renewable energy sources in isolated developing country villages, is given in Table 3. The 'merit stars' are the merest indications of technical suitablity and say nothing about socio-economic or cultural constraints and opportunities. Bearing the latter in mind, seminar working groups might like to take this table and re-write it for their own countries.

ENERGY USE

SOLAR TECHNOLOGY	Water pumping	Water desalting	Transport, tractor, etc.	Fertilizer	Food grinding	Food drying	Food cooking	Food cooling	Lighting	Communications	Heating space	Heating water	Stationary (small machines) power
Solar cells	X					X		X	X	XX			X
Flat-plate collectors	X	XX				XX		X			XX	XX	
Concentrating collectors	X	X					X	X					
Solar engines	X				X			X	X	X			X
Wind (mechanical)	XX				XX								X
Wind (electric)	XX				XX			XX	XX	XX			XX
Water (mechanical)	XX				XX								XX
Micro hydro	XX				XX			XX	XX	XX			XX
Bioconversion: pyrolysis			X	X		XX	XX				XX	XX	
Biogas	X		X	XX		X	XX	X	XX				

Table 3. A village energy source/use matrix

To appreciate the further benefits of end-use surveys we have to reach down to the level of the individual firm or factory, village or household and its variety of energy end-uses. This is done in Figure 1 for one energy stream in the industrial food sector.

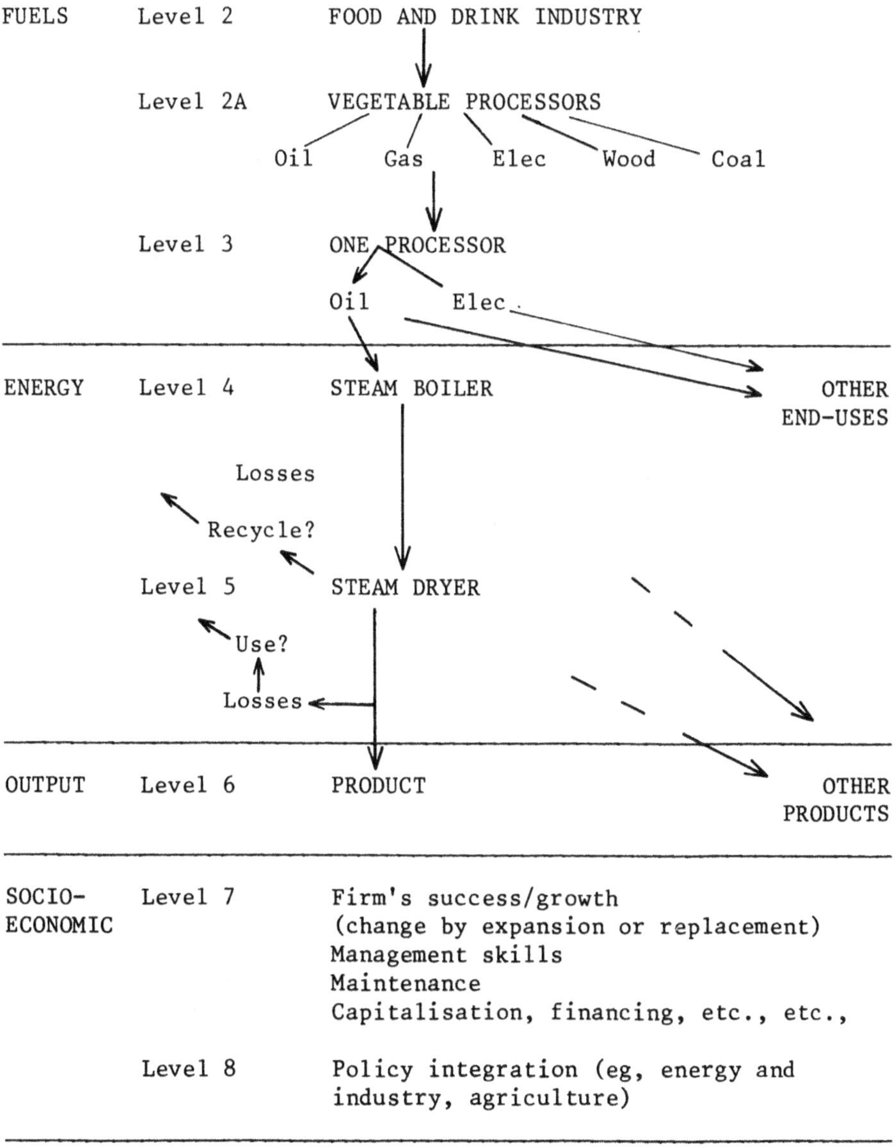

Figure 1. An end-use energy flow in industry

Clearly we have reached a region of very considerable complexity. In this case, of several possible fuels, oil is used to raise steam by a particular boiler plant to dry a particular product. There are two stages, each with its own losses, efficiencies and opportunities for improvement. Other products are processed in the factory, which has other end-uses of energy. Surrounding all these energy activities is a host of potentiating or constraining factors which are merely indicated by the items marked 'socio-economic' environment at Level 7. Last but not least, we have gone so far down the energy flow that we have left energy policy and entered the territory of other ministries. These may or may not welcome interference and almost certainly have other priorities than 'energy'.

Indeed, we have gone so far that the only useful perspective is not to look down the energy flows but to look up them. The world of each individual energy user starts at - indeed is - Level 7. This is where he or she stands when considering what might be done to improve energy use. From this standpoint - the view from the bottom - energy choices also form a branching network of possibilities and impossibilities, options and cul-de-sacs, that are often quite as complex as the view of energy flows 'from the top'. Furthermore, each perspective may reveal choices that are not apparent to the other.

The true benefit of the most detailed site-by-site energy surveys is now apparent. They provide specific information on possible energy choices that may be open - or not - to people within their real environment of constraints and opportunities. There are countless examples from developed and developing countries where effective change in energy systems has begun only when this specific information has been provided. Most people are too preoccupied with more immediate concerns to work out their best energy strategies and lack the expertise to know what can possibly be done in their particular circumstances.

This leads us to the final level of end-use analysis, that of precise specification of the task to be performed. The world is littered with examples of the failure of energy technologies to perform as required or as circumstances permit. The replacement of the 'inefficient' three-stone cooking fire, which can burn almost any available fuel, by stoves with a greater technical efficiency which can only burn sticks of a prescribed size, is one well-known example of such failures. The replacement stoves are often quickly abandoned.

Another example, which illustrates the point clearly at a purely technical level, is water pumping with solar cells (photovoltaics). A recent World Bank/UNDP survey of the performance and economics of such pumps found that success or failure was acutely sensitive to the combined efficiency of the motors and pumps, which varied from 11 to over 50 per cent. These efficiencies were in turn very sensitive to the correct matching of the pump/motor design to the head of water to be pumped. The key problems facing the introduction of solar water pumps in developing countries were identified as the accurate specification of the pumping task to be performed and the ability of manufacturers to meet these specifications from their range of products.

With this example we have reached the farthest ends of the energy flows - and boxed ourselves into a very tight corner indeed. What has matching a pump to the precise depth of water to do with the planners in the Ministry of Energy? Of course, very little. Planning at government level is about steering broad courses through a future of great uncertainty, not about minutae such as this. So let us draw back a little and look from a different perspective at the energy flows and options for changing them.

ENERGY CONSERVATION: OPPORTUNITIES AND PROBLEMS

The technical opportunities for energy conservation are so many, various and specific to each set of conditions that no single lecture can begin to cover them adequately. The same is true of the political and institutional questions of how conservation (and fuel substitution) can best be implemented.

What I propose for the remainder of the lecture, largely as a basis for working group discussions, is to provide a brief check list of the main types of option for altering energy flows and their attendant problems that need to be considered in energy planning. For the sake of brevity, I shall not deal with familiar issues such as investment costs, fuel prices, economic appraisal methods, and the like.

Starting at the bottom of the energy flows, nine broad topics are perhaps worth considering.

1. Altering the service or product

The use to which energy is put may not be necessary or may be acquired by other means. Mobility is a good example: the need
The use to which energy is put may not be nece
for it can be reduced by land-use planning and it can be provided by bicycles and buses instead of cars, although perhaps not with the same convenience or 'status'. Similarly, it may be better for foreign exchange balances and capital expenditure constraints to import goods such as nitrogen fertilisers from a low cost source than to import (or develop indigenous) fuels to make them domestically. These issues clearly go far beyond energy policy into the area of integrated macro-economic planning and the allocation

of foreign exchange to import priorities, etc. These decisions are interdisciplinary and inter-ministerial.

2. Reducing obvious energy losses

Many firms in developed countries have reduced their fuel bills by 10 to 15 per cent almost overnight, at little to no cost, by attention to details such as plugging steam leaks, mending holes in furnace doors, or insulating pipes and furnaces. Turning off unwanted lighting or idling machinery also comes under this category. The main barriers to such simple and effective measures are lack of awareness, and concern with more pressing problems of management. There is little that government energy planning can do in these respects except create awareness through the mass media, help firms by training energy management personnel - and increase energy prices. This is the domain of individual firms.

3. Redesign of the end-use device

Generally speaking, this is where the largest opportunities for energy saving lie. Furthermore, most of the really big opportunities are in just those sectors which in developing countries are either large energy users now or have high actual or potential growth rates, including transport vehicles, lighting, cooling and other electrical appliances. Let us examine some of these very briefly:

Cars. Many technical studies have shown that the specific fuel consumption of the European-type car normally used in developing countries could be reduced by a factor of roughly two to two and a half by the mid-1990s. For example, in France the average consumption of the 1975 passenger car fleet was 9 litres/100 km in 1975, but the targets for new models agreed to by government and the motor industry using known technologies are 7.3 l/100 km in 1985 and 6 l/100 km in 1990. Prototype cars such as the Renault VESTA and Peugeot ECO 2000 are expected to be mass-produced in the 1990s with specific consumptions of 3-4 l/100 km, largely due to weight reduction, reduced air drag, new engines and transmission systems. A similar performance of 3-4 l/100 km is expected from the small (31 kw) Volkswagen Rabbit prototype with a supercharged diesel engine; while larger VW prototypes, similar in size to present family saloon cars, are designed to give 3.4-4.5 l/100 km. In the USA, the informal motor industry target for 1995 model cars is 3-6 l/100 km, a reduction of up to five times on the 1977 new car average of 15 l/100 km.

Trucks. Similar studies have suggested that the fuel consumption of road freight vehicles could be halved by 1990-2000 by a number of separate measures. Using an American example, the following individual savings are thought to be technically possible and economic: engine improvements (bottom cycling, elimination of water cooling) 30 per cent; aerodynamic drag reductions 20-25 per cent; full shift to radial tyres 5-10 per cent; exhaust driven turbine for supercharging 5 per cent; lower loss rubber compounds in tyres 4-5 per cent.

Lighting. Lighting equipment now in use around the world spans an astonishing 600:1 range of efficiencies. Using as a unit the lumens of light given out per watt of energy consumed, moving from a kerosene wick to a kerosene mantle lamp gives a 10-fold improvement from 0.25 to 2.5. A standard electric incandescent bulb scores about 15. The recently-produced range of 'mini fluorescent' bulbs that fit standard light sockets and have a lighting quality close to that of daylight score around 50. Standard fluorescent tubes have a range of 85-110, while low pressure sodium tubes have a performance of about 155. The theoretical limit is 680 lumens/watt. Even within the range of standard incandescent/fluorescent equipment, the range is about six to one.

Cooling. In the USA the annual electricity consumption of the average refrigerator rose from 600 to 1600 kWh between 1945 and 1975, but recent designs could bring consumption down to as little as two-thirds of the 1945 figures. Thus one University of California prototype with the same storage volume as the current US average should have a performance of 400 kWh, a four-fold reduction. Through the gradual introduction of new models, the averate unit consumption of refrigerators in California fell between 1972 and 1980 by close to 24 per cent - a trend which is still continuing as more and more old models are replaced. The reduction in Californian air conditioning over the same period has been more modest at 7 per cent, but this is partly because there is a slow turn-over of the equipment stock.

These figures illustrate the enormous potential for providing the same energy service with less energy - or absorbing large increases in demand without any increase in fuel consumption. However, these super-efficient technologies often - but not always - involve certain penalties. With lighting, for example, the initial cost may be substantially higher than the standard equipment, thus deterring all but the wealthiest consumers even though payback times may be less than a year. With automobiles, in developing countries long life, ruggedness over poor roads and low maintenance requirements may over-ride concerns about fuel consumption.

Arising from this, two main questions for planning are (a) whether these 'trade-offs' are acknowledged, and (b) whether in the light of them the 'best' technologies are being imported or manufactured domestically. A third question is whether the planning system has adequate 'look out' capacities for answering these two questions. These issues apply to many of the topics raised below.

4. Redesign of the end-use system

In this case one is considering not a single device such as a car, but a series of energy processes as shown in Figure 1. Co-generation and heat recycling are good examples for the industrial sector. Energy savings can be very large, but usually carry penalties not only of cost but greater complexity and maintenance needs. The choice of which technology is best can rarely be generalised since it is almost invariably site specific. This means that successful, widespread implementation depends on having a system of detailed energy audits by outsiders or on a good level of energy expertise within industry.

5. Controllability and matching energy quality

Theoretically, any demand for heat should be met by the lowest grade of heat supply that can do the job. This is a basic principle of least energy use. Thus $50^{\circ}C$ hot water for washing in a factory should ideally be met by a waste heat stream at, say, $70-100^{\circ}C$ (the difference to allow for losses in heat exchangers, etc.) rather than by electricity which is produced at low thermal efficiency from a heat source of $600^{\circ}C$ or more in a power station. However, the latter source can more easily and cheaply be made to deliver heat at a much higher rate and is more controllable. It can therefore be switched on and off to match demand, saving energy.

An important example of this principle is in cooking, where heat is normally needed only in relatively short bursts. Controllable high grade fuels such as electricity, kerosene, biogas or bottled gas have, for this reason, much higher technical efficiencies than 'permanently-burning' wood, charcoal or coal fires. While there are normally many other reasons in developing countries why the former fuels are not appropriate, scarce biomass sources now used for cooking can be converted into high grade, controllable fuels. Bearing in mind other 'trade-offs' such as the warmth and light given out by cookers, the most appropriate solutions are not at all obvious.

6. Can other fuels be used?

This is a large question that has been touched on earlier. One of the most important demand-related priorities for the urban/industrial sector of developing countries is to survey the potential for switching from oil (and in some industries, wood) to coal or clean-burning solid fuels made from coal. Coal can now be imported at roughly half the equivalent price of oil, world reserves are enormous, and the producers have not (yet) formed a cartel. If these potentials seem attractive, further assessments would have to be made of the costs of setting up a 'coal chain', including port facilities, transport, final distribution, and storage and handling facilities for final consumers.

7. Saving energy or saving capacity?

Many technologies for saving energy over the year do little or nothing to save the maximum supply capacity that must be provided to meet the peak demands on the energy system. The classic example is solar water heating with electricity as a back-up for times when the sun is not shining. The electricity system must provide the same capital-intensive peak capacity as before, but loses sales and revenue to the solar heaters over the year. Conversely, if renewable sources such as wind and wave power have their highest outputs during periods of highest demand - as they do in Europe, during the winter season - then they save both capacity and fuel.

This extremely complex issue is familiar to electricity planners. Technologies that save high cost peak demands are generally attractive for the electricity system. So are those that save most of the year round. Technologies that reduce demand mostly in base load periods may be counter-productive for the supply industry, if not for the user. However, the supply industry usually has the last word and can alter the tariff structure to suit itself, unless governments intervene to impose a tariff structure that provides the best national solution.

8. What kind of energy is being saved?

This point is closely related to the previous topic. Again solar water heating serves as a good example. In many developing countries solar heating is being promoted for offices, schools, hotels, hospitals, etc., with little information on whether it will replace oil or electricity. (An intensive study by Schipper on just this question in Kenya, where solar water heating has been doubling in volume for each of the past six years, was unable to provide any but the crudest and most approximate answers). If electricity is being replaced, planners also need to know whether the savings will be in base load hydropower plant with almost zero marginal cost, or in oil- and perhaps coal-fired plant with large marginal (ie, fuel) costs.

9. Intermittency, storage and back-ups

It is well-known that many renewable energy sources - especially solar and wind - give intermittent energy outputs. Small-scale hydropower is a firmer source except when streams fail after prolonged dry conditions. Biomass, in contrast, is a consistent energy source because it can either be cropped as needed or stored after harvest for later use (although even here drought can decimate supplies).

One of the most difficult set of energy planning problems is how to balance these issues in relatively small-scale and self-sufficient communities, such as the remote developing country village. To what extent should energy intermittency be smoothed out by energy storage, be backed up by other 'firm' sources, or accepted by users (which means that they have to do without the source when it fails)? If these issues are not resolved properly the technology may simply be unacceptable.

Storage may be cheap, as in the case of water pumping for irrigation or drinking. The product of energy use - the water - can be impounded in a pond or tank. It may be extremely expensive, as with batteries in solar electric systems. Back-up systems such as diesel generator sets may be relatively cheap if they are used frequently (high load factors) or very expensive if not, and if problems of reliability and maintenance mean that they have to have their own back-up; ie, a spare machine.

These considerations make it virtually impossible to generalise about the appropriateness of village-scale renewable energies even on purely technical grounds. Successful choice and implementation of technologies can in most cases come only through a thorough understanding of all village activities, preferably initiated by the villagers themselves. This important aspect of energy planning is now increasingly recognised by governments and aid agencies, if not by some of the hopeful suppliers of renewable energy devices. And, of course, it has brought this lecture once again, full circle, into the realms of the fine structure of energy demand.

WHERE DOES ENERGY PLANNING STAND?

This lecture may have left the impression that energy planning is not possible without complete information on every consumer and his circumstances. Of course this is not so. Consumers are the actors who respond to circumstances partly created by planning. The role of planning is to create the right circumstances and opportunities. Nevertheless, what I hope the lecture has shown is that planners must move further down the energy flows towards consumers and their energy uses.

Most energy planning has been and is 'from the top'. If one wishes, for example, to promote solar water heating to replace oil, many kinds of signal can be sent down the energy system. One might tax water heating fuels. Or remove taxes on solar equipment. Or provide subsidies for installations and otherwise support a solar industry. In countries where resources for planning and data collection are scarce, this must remain the dominant approach.

But it is not ideal. Which are the water heating fuels to tax? Are solar water heaters acceptable alternatives? How many might install them at given costs and fuel prices? Would a solar industry have a sufficiently large market to survive? Are there better uses of scarce resources than to encourage one? The heavily subsidised German solar water heating industry has recently come near to collapse for lack of forward information of this kind. This type of information, as I trust I have demonstrated, can come only from a much better understanding than now normally exists of the intricacies of energy demand.

Developed countries have got this message and are trying to improve this understanding. But they have large resources of skills to call on. An important question for energy policy in developing countries is how far the planning system can and should move in the same direction. I hope this question will worry you during the remainder of this seminar.

ENERGY BALANCES: THE BASIC TOOL FOR ENERGY PLANNING

Pierre VERNET

IREP-Développement - GRENOBLE (France)

The energy balance is a simple and reliable tool for energy planning especially during the first steps of this process. The advantages and limits of the use of balances are presented. The OLADE methodology for energy planning has been made to fit the characteristics of energy systems in Latin America : it appears as a consolidated balance sheet in which the main energy information appears in an easily understandable way. It is emphasised that the process of energy balance elaboration can be mastered by national technicians and contribute to more independent planning in developing countries.

This text is partly based on the document : OLADE Methodology for the Elaboration of Energy Balance, OLADE Document Series N° 8, Quito 1980, in the elaboration of which we participated actively.

The basic need concept as a strategy of development is very questionable. It has been chosen as a title here because of some statements that are worth quoting :

* An energy planning effort is of absolute necessity in each country either developing or industrialized.

* This effort should be worthless if national institutions and personnel do not manage and realize the necessary studies.

* Even where there exists a national capacity to develop

H. Neu and D. Bain (eds.), National Energy Planning and Management in Developing Countries, 195–212.

complex econometric instruments, energy problems need to be presented in a reliable and didactic form (at least, to be understood by the political decision-makers).

There exist few countries, either D.C.'s or L.D.C.'s, which have a global energy policy. In many cases, however, there are good sector plans. Electricity generation may be the energy sector the most studied in the world. In the library of the World Bank there are probably the same number of "electrification plans" as of member countries. Hydrocarbon production, transport and refining are also widely studied ; but this sector and the relevant studies are generally controlled by the oil-companies. During recent years rural energy has been the new theme of many meetings, experts' missions, and publications.

Nevertheless, a country which has all types of sectorial energy studies, still needs comprehensive and global analysis of the entire energy system (1). The allocation of finite ressources, either financial or human, is the main problem of the economic decision-maker whatever the ideological orientation of his government. The global energy comprehension is important in order to take such decisions. Some day, it might be useful to establish the priority between a reforestation programm and a hydroelectric project. The global energy assesment should indicate that the part of the involved population -in terms of energy (firewood and domestic electricity)- is more important in the first case than in the second one... Apart from such extreme examples it is worth insisting on the fact that it is always useful to consider energy problems as a whole and also in relationship with the rest of economy.

The efficient tool that is generally recommended to be used for a global analysis is the energy model. Some experiments indicate that begining a work of energy planning with a modelling effort is not always the most appropriate way to start this process. Some desirable characteristics of the first phase of an energy planning effort are :

- to be made with the lightest technical assistance,
- to be made without sophisticated computers,
- to be made in a short period of time,
- to give results in a clear and simple presentation.

Making and analysing energy balances (2) gives the possibility to reach these targets. This is the reason why this work should be recommended in any case. An energy balance is not a target in itself : it should be considered from two points of view.

Firstly, in the training process of national personnel, it appears from experience that the construction of energy balances is a good didactic process.

Secondly, with the help of some socio-economic indicators, energy balances permit good analyses.

In that sense, energy balances are at the beginning of autonomas planning :

- self-made recollection of information,

- self-made matrix, construction,

- self-made analyses of the results.

The last part is obviously the most important.

In a Latin-American country and with the help of accurate electricity and oil statistics, it has been possible to make a ten-year series of energy balances and the respective comments within one year and a half. The team in charge of this work was composed of only four persons.

- 1) DEFINITION, INTEREST AND LIMITATIONS OF THE ENERGY BALANCE.

Every form of energy can be reduced to a common physical unit due to the specific heat measured in laboratories. This characteristic is unique in the field of the economy of intermediate and final consumption goods (3) and contributes to the specificity of the economy of energy that "crosses" all the socio-economic sectors of a country.

1.1.- Definition of energy balance.

The energy balance (4) is a group of equilibrium relationships that numerically express the physical flow when energy is produced, exchanged with the outside, transformed (5) and consumed, all calculated in a common unit, within a given country or region and for a given time period, generally one year. Taken in isolation, the balance gives an image of the physical relationship of the energy system in a particular historical period. It also gives the opportunity to consolidate the forecasts made in the field of energy (see below).

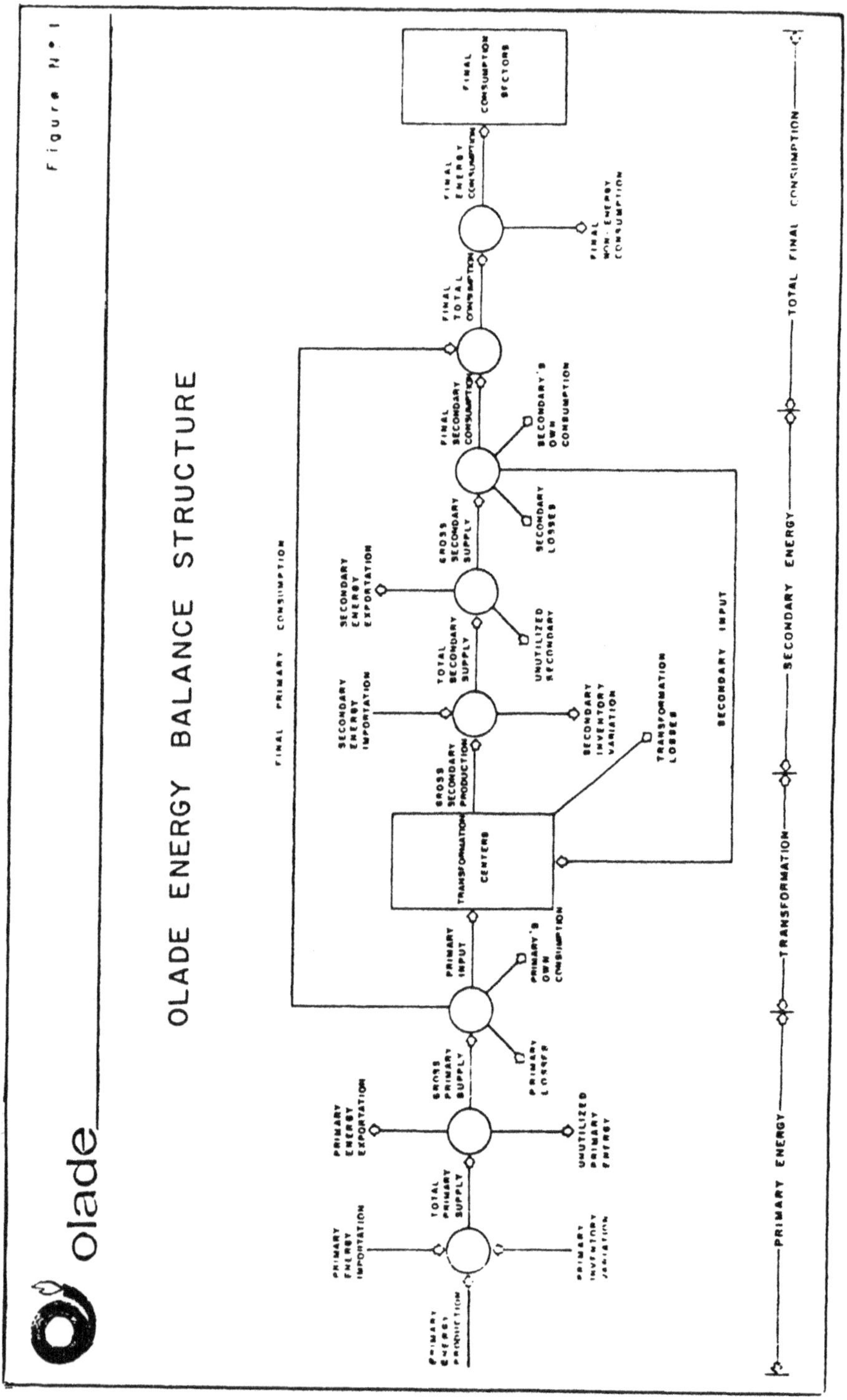

Chart 1: OLADE Energy Balance Structure

The balance tells how energy is produced, exported or imported, transformed and consumed by each economic sector. It permits the calculation of certain efficiency relationships and the formation of a diagnosis of the energy situation for a given region, country or continent. Nevertheless, it is by means of this relationship that the balance becomes an instrument for planning. In this sense, an energy balance is a pre-condition for the energy planning process. A balance fulfills a role in the energy sector analogous to that of the input-product matrices in the economic sector.

The balance sheet shows (see chart n° 6) :

- In the columns the different energy products that are to be taken in account in the specified country sub-totals and a general total which are worthwile for analyses.

- In the rows the origins and uses of energy, broken down by sector.

(obviously columns and rows could be interchanged).

1.2.- A simple and powerful tool for energy planning.

One of the greatest advantages of the energy balance can be characterized in a paradoxical way. Being simple and easily intelligible it should not be the subject of a myth. Like any other economic decisions, energy policy finally depends on a political decision process. The technicians working in a national energy group have to present organized information. They do not have to hide the political aspects behind a sophisticated presentation of the information. Apart from this, it should be emphasized that the energy balance has six fundamental objectives :

* To evaluate the dynamics of the energy system, in accordance with the economy of each country and determining the principal economy-energy relationships among the different sectors of the national economy.

* To serve as an instrument for energy planning.

* To establish the structure of the national energy sector in detail.

* To determine the competitive and non-competitive uses of each energy source, which allow for the stimulation of the substitution process whenever possible.

* To create the appropriate bases which carry with

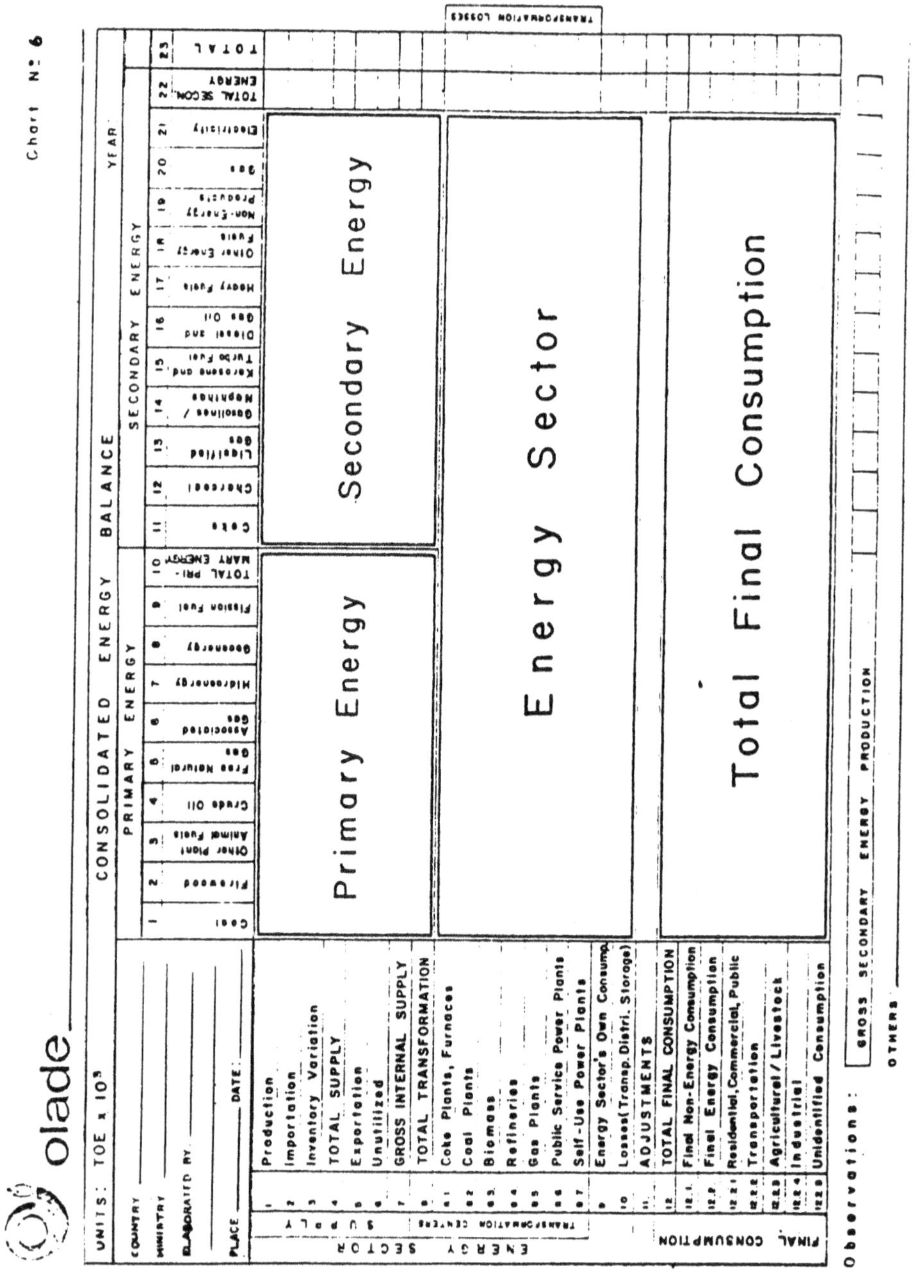

Chart 6: OLADE Energy Balance Sheet

Chart N.° 7

EXAMPLES OF CONVERSION FACTORS

PRODUCT	CONVERSION FACTOR (1)	CONVERSION FACTOR (2)
1.- COAL	0.700 TOE/TON	——
2.- FIREWOOD	0.360 TOE/TON	——
3.- CRUDE OIL	0.139 TOE/BI	1.020 TOE/TON
4.- FREE NATURAL GAS	0.0235 TOE/1000 FT^3	0.830 TOE/1000 M^3
5.- ASSOCIATED GAS	0.0252 TOE/1000 FT^3	0.890 TOE/1000 M^3
6.- COKE	0.680 TOE/TON	——
7.- CHARCOAL	0.690 TOE/TON	——
8.- LIQUIFIED GAS	0.093 TOE/BI	1.08 TOE/TON
9.- AUTOMOTIVE GASOLINE	0.124 TOE/BI	1.05 TOE/TON
10.-AVIATION GASOLINE	0.119 TOE/BI	1.05 TOE/TON
11.- NATURAL GASOLINE	0.106 TOE/BI	1.06 TOE/TON
12.-KEROSENE	0.133 TOE/BI	1.03 TOE/TON
13.-TURBO FUELS	0.136 TOE/BI	1.04 TOE/TON
14.-DIESEL OIL	0.139 TOE/BI	1.02 TOE/TON
15.-HEAVY FUELS	0.143 TOE/BI	1.00 TOE/TON
16.-FURNACE GAS	0.090 TOE/1000 M^3	——
17.-COKE GAS	0.420 TOE/1000 M^3	——
18.-DISTRIBUTED GAS	0.0113 TOE/1000 FT^3	0.400 TOE/1000 M^3
19.-REFINERY GAS	0.0311 TOE/1000 FT^3	1.100 TOE/1000 M^3
20.-ETHYL ALCOHOL	0.083 TOE/BI	0.52 TOE/ M^3
21.-NON-ENERGY DERIVATIVES	0.99 TOE/TON	0.139 TOE/BI
22.-ELECTRICITY	0.086 TOE/MWh	——

Chart 7: Energy Conversion Factors

them improvement and systematization of the energy information.

* To be used in order to permit energy projections and its short-term, medium-term and long-term prospects.

These objectives show how to use the instrument energy balance which can never be an objective in itself.

1.3.- Limitations of the tool.

It is important to keep in mind that the energy balance- like every tool- has its limitations.

First of all, the physical approach to the energy systems leaves apart not only the human and social dimension - like any economic or monetary instrument of analysis - but also the monetary and financial part of the analysis. So, with a single balance, nothing of the economic dimension of the operation of an energy system is said. The introduction of socio-economic elements of analysis is indispensable before it is attempted to use the balance for planning purposes. This is especially important as far as the problem of substitutions between energies is concerned. A physical optimization may be very far from the socio-economic one.

Even in its own field-aphysical approach to the energy sector- the balance has some limitations. It does not evaluate either energy reserves, on the one hand, or useful energy on the other hand.

Knowing the respective importance of the reserves of each form of exhaustible energy and the resources of each form of renewable energy is nevertheless of vital importance. This evaluation (6) has to be done in any case.

The useful energy is the energy actually used in a final energy process, due to the fact that not all the energy that enters a consumer system is utilized ; the part of useful energy in consumed energy depends on the efficiency of the consuming apparatus. Although it is impossible to know the efficiency of all the consuming apparatus of any country, the mean efficiences of each concerned form of energy involved in a substitution process have to be estimated.

Keeping in mind these limitations will help the use of the energy balance as a tool in a whole which -even simplified- is complex.

- 2) EXAMPLE OF SELF-MADE METHODOLOGY.

Latin-American and Carribean countries, through the Latin-American Energy Organization (OLADE) have decided to elaborate their own methodology despite the numerous existing ones. Why ?

2.1.- OLADE and its energy balance programm.

It is an international organization, with its headquarters in Quito (Ecuador). The council of Ministers of Energy (or equivalent) of the member countries is the leading board. During recent years, OLADE has sponsored a lot of meetings, courses, etc. in the field of new and renewable energies. In the field of energy policy, it has avoided political problems -that would have blocked any regional effort- by using a technical approach : the implementation of a general energy balance programme.

Two objectives were sought in this programme:

1) to give the opportunity to member countries which already have energy balances to reproduce them in a common way in order to permit international comparisons and regional consolidations.

2) to offer a regional technical assistance to the countries that did not have any national energy balance programme, in order to give them the opportunity to elaborate their own energy balances.

These objectives have been fulfilled.

2.2.- The proposed balance methodology.

The methodology had to be very flexible in order to fit the energy characteristics of countries as different as Brazil and a little Caribbean island. Each country can -in a fixed scheme- adapt the proposed methodology to its specific characteristics and needs.

It is not worth presenting here the complete methodology of OLADE energy balance (7). Nevertheless, some characteristics of it are worth underlining.

The general presentation of the balance-sheet is based on the O.E.C.D.-type balance that is now also recommended by the U.N.'s organizations. The reason for this choice is that it is largely known all around the world. Moreover, this presentation makes possible a clear presentation of transformation losses within one sheet (8) (see charts 2, 3, 4, 5 and 8).

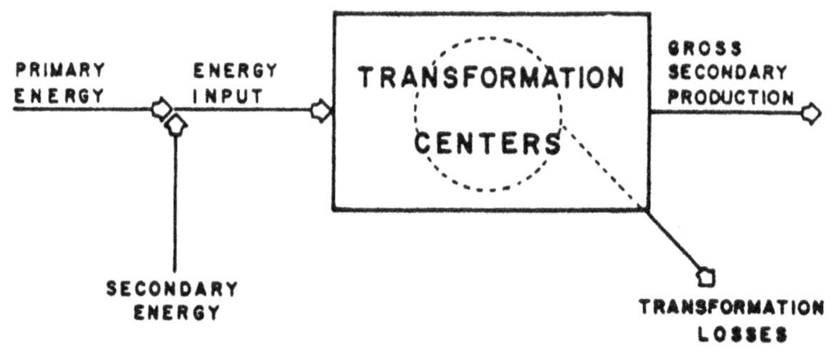

Figure Nº 2

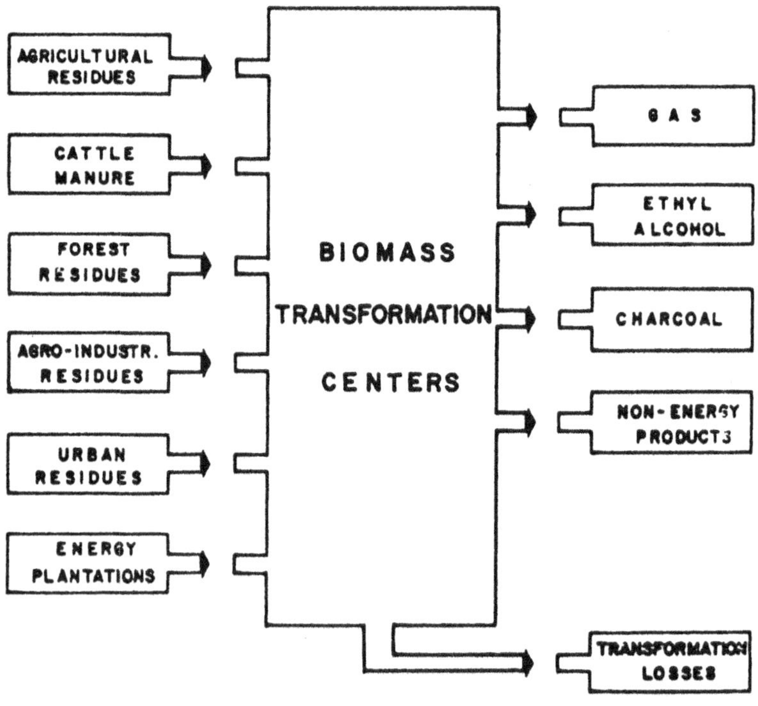

Figure Nº 3

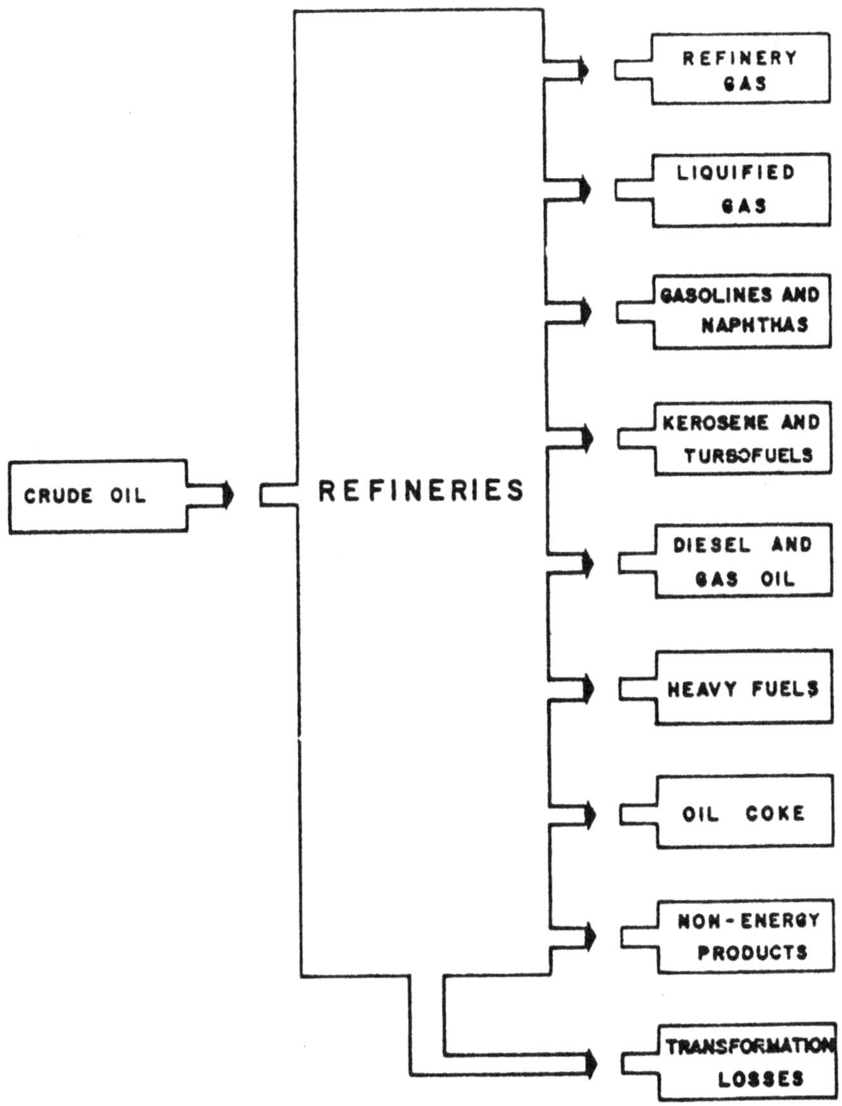

CRUDE OIL
REFINERIES
REFINERY GAS
LIQUIFIED GAS
GASOLINES AND NAPHTHAS
KEROSENE AND TURBOFUELS
DIESEL AND GAS OIL
HEAVY FUELS
OIL COKE
NON-ENERGY PRODUCTS
TRANSFORMATION LOSSES

Figure Nº 4

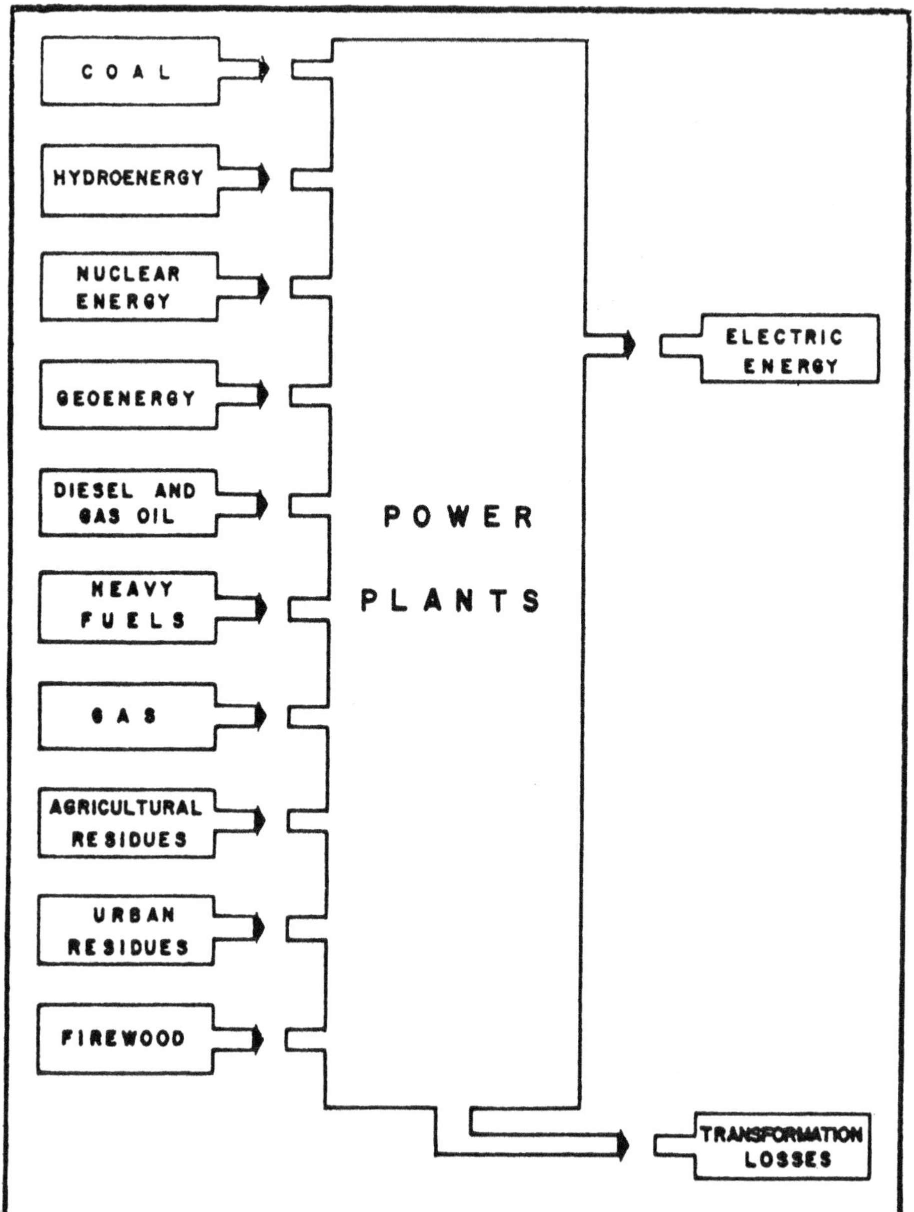
COAL
HYDROENERGY
NUCLEAR ENERGY
GEOENERGY
DIESEL AND GAS OIL
HEAVY FUELS
GAS
AGRICULTURAL RESIDUES
URBAN RESIDUES
FIREWOOD
POWER PLANTS
ELECTRIC ENERGY
TRANSFORMATION LOSSES

Figure Nº 5

Within the energy products taken in account, it should be noted the importance given to the so-called "non commercial energies" firewood and charcoal. These products are generally "forgotten" in other balance-sheets though they may often be the most important ones in final energy consumption. As far as the relationship between energy and standards of living is concerned, the importance given to non commercial enérgies in the balance is a good means to call for the attention of the political executives. Biomass Transformation Center (see chart n° 3) as Other Energy Fuel are introduced because of the Brazilian efforts in the field of ethanol fuel.

The conversion factors for each form of energy into the common unit are based on specific heat. This choice avoids all the misinterpretations due to the conventional conversion factors - especially with electricity. It appears that the problem of end use efficiences has to be treated apart from the process of balance elaboration, for instance with an approximate end use matrix in term of useful energy.

The group which prepared this methodology did not pretend to make the perfect balance. But the objectives of flexibility and simpleness led to a compromise which is perhaps the best model of balance not adapted to but conceived for the L.D.C.'s conditions (the Latin-American and Caribbean ones in this specific case).

- 3) USING THE ENERGY BALANCE AS A PLANNING TOOL.

Once a historical series of balances has been realized and the concrete results presented, the next and most important step is to analyse the elaborated information. This analysis must confront the study of the national energy system and its evolution in absolute and relative terms.

It should not be forgotten that all the information necessary for the analysis are not presented in the balance. The information that should be also presented concerns :

- the country's energy reserves and resources and the respective production capacities.

- the non-energy determinants of the organization of the energy sector : politics, international relationships, technologies, and so on.

- the general production of goods and services and their organization.

In effect, the demand for energy is determined either by the requirements for production (industrial, agricultural, services) by means of the technical options in the production system, or by the requirements for final consumption through the social behaviour of the consumers. So the energy system can only be fully defined to the extend that its place within the socio-economic system is fully known.

For the management of the energy system, the balance should provide the possibility of establishing some criteria with respect to the "efficiency" of this system either in the physical sense or in a more general sense of economic efficiency.

3. 1. - Analysis.

The analyses can observe successively the energy supply, the energy sector, the final consumption and the relationships between energy and social and economic growth.

The evolution of each primary energy production and the total of it should be observed, calculating the annual growth rates. Likewise the relationship between national production and importation will give an idea of the degree of self-sufficiency. The exports will illustrate the importance of this activity especially for oil producing countries. It will be important to observe the evolution of unutilized energies such as : volumes of spilled crude oil, flared natural gas and refinery gas burned in the air or spilled volumes of water in hydroelectric intallations. It is particularly interesting to consider the ratio in which hydrocarbons participate in the energy supply.

In the analysis of the energy sector, it deserves special interest to observe the modifications with regard to the inputs of the transformation centers, their production structure and the corresponding transformation losses. Each transformation center, should be studied separately. Moreover, with respect to the power plants the comparison between public service ones and privately-run ones is of great interest, giving the opportunity to draw pertinent conclusions within the framework of each nation's particular situation. Another important point is to measure the contribution of each type of power plant in the national mix of electricity generation : in the programmes, this framework will aid in elaborating a policy for substituting the more abundant energy-sources for those less available.

Apart from the transformation process and its efficiency it is worth analysing through their losses the production, transportation, storage and distribution processes. The percentage of national production vs importation in the total secondary energy demand is another interesting indicator.

The final consumption, which is detailed in the last part of the balance, has to be cross-checked ; this means a double analysis : product by product sector by sector. The apparent dynamics of a product and/or a sector and its determining causes should be presented.

An awareness of the energy sector's evolution and composition permits the making of a sort of correlation with the macro-economic variables ; in this way, it is possible to analyse this sector's behaviour and that of the economy as a whole.

The following are some interesting indicators among others :

- elasticity of total energy consumption vs gross internal product.

- elasticity of electricity consumption vs gross internal product.

- total energy consumption per unit of the gross internal product.

- relationship between commercial and non-commercial energy.

- energy consumption per inhabitant.

- proportion of the population having access to each form of energy (supplied population by electricity for instance).

- part of energy imports and/or exports in the commercial balance.

- and so on...

The diagnosis of the energy problems may now be made ; this is an important step in a planning process which has been possible utilizing energy balances. The balance can also be useful in the following steps of the planning process.

3.2.- Forecast and optimization.

Being by definition equilibrated, the energy balance,

when used in forecasts has the advantage to assure the coherency of these.

In general, within a short term perspective, the balance permits knowledge of the different energy alternatives (substitution process) according to the final use in the large consumer categories. Within a long-term perspective, using an energy balance permits either classical forecast methods (prevision of the demand and then of the necessary production and imports) or the more comprehensive ones (planning the desirable capacities of production leads eventual modification of the volume and structure of forecast demand of energy, in a double feed-back process).

Most countries have sectoral plans in the field of energy. In a first step, establishing previsional balances based on these plans will assure the coherence between them. For example, a rapid work of elaboration of balances showed, in a latin-american country, that while the oil and electricity companies agreed on the total volume of fuels that will be necessary for thermo-electricity generation in the next five years they had not planned the same structure (respective parts of gas-oil and fuel-oil). Coherence can also be checked between sector plans for energy and the global economic objectives of development. Consolidation of the previsions through the balance makes it possible.

All type of forecast or prospective methods are used in energy planning. Consolidating the results of any of them will at least assure the internal coherence of these results if never their intrinsic worthiness.

A simple optimization process is also feasible when alternative energy balances are elaborated for each one of the groups of hypotheses made ; the impact on the key-variables being calculated from the consolidated data that appear in the balance.

Thus, it appears that the energy balance can be the most important tool used in analysis of historical and present evolution of the energy system. In the forecast and optimization steps of the planning process its importance diminishes. Nevertheless it still may be a useful tool combined with others.

CONCLUSION

Neither the energy balance nor any economic or econometric instrument can be the perfect instrument for energy planning. It has been insisted here on this tool because of some characteristics that are worth repeating.

- it is a good instrument, in the sense that it can support comprehensive studies.

- it is an "easy to master" instrument and eases the development of a national capacity in the field of planning.

- it does not shut any access to the more sophisticated instruments of planning such as models.

This will be the conclusion. The more sophisticated instruments for energy planning may be mastered by the same national group that elaborated the energy balance (plus eventually a programmer). If the instrument is either indigenous or imported, after an experiment of elaboration of energy balances the need for foreign assistance is reduced and also from this point of view, the time spent in elaboration of energy balances will not have been lost.

(1) See contribution by H. Neu.

(2) Flow diagrams offer the same general characteristics as balance sheets. With a high level of aggregation, the diagrams appear more didactical than the balance. On the other hand, with many sectors and energy products they become too complex.

(3) In a classical economic approach energy is either an intermediate good for industry or a final consumption good for individual transportation, cooking and space heating.

(4) The same definition should be used for a flow diagram (see chart n° 1).

(5) One of the most important transformation processes is, for every country, electricity production.

(6) It is necessary to insist on the fact that reserves and resources are not determined once and for ever. They depend on a lot of changeable factors : exploration, state of technologies, ruling economic conditions.

(7) See OLADE Methodology for the Elaboration of Energy Balance, OLADE Document Series n° 8, Quito 1980, also available in Spanish (original) and the appendix.

(8) For more details on the discussions about energy balances, see P. Ramain. Réflexions Critiques sur les Bilans Energétiques, C.N.R.S., Collection Energie et Société, 1978.

INFORMATION GATHERING FOR ENERGY PLANNING IN DEVELOPING COUNTRIES

Pierre VERNET

IREP-Développement - GRENOBLE (France)

Information gathering is an important step in any planning project. Existing information is often important but is not always systematic or consistent. Three criteria for testing the consistency of information are presented. The non-existing information has to be created : inquiries. The main sectors of energy data are those of : non-commercial energy, hydrocarbons, electricity. Non-energy information is necessary for analysis purposes

As a first consideration, it must remembered that the problems of information gathering depend on the type of energy planning that is desired : the more sophisticated the necessary planning instruments, the more complete the necessary informati Based on a personal and limited experience, this paper refers to the problems of information gathering in the case of a first planning effort such as, for instance, the elaboration of an historical series of energy balances.

It should be noted -and this is a second consideration- that information gathering problems are more important during the first steps than later on, especially if information circuits have been set up.

Each country has its own characteristics in the organization of the production of goods and services. Gathering information should largely vary from one country to another depending on their "specificities" : this is a third consideration. Experience shows that the socio-political organization is not

H. Neu and D. Bain (eds.), National Energy Planning and Management in Developing Countries, 213–221.

always determinant : it is not easier to get information in a centrally planned economy than in a market one ! The number of economic agents involved in energy, and the interest they may have in energy planning are the important conditions, whatever may be the juridical statuses -public or private ones-

The need for information can be observed at three different levels. Two types of information have to be collected for the analysis-diagnosis part of a planning study : the energy and the non-energy relevant parts. For the proper planning part, all the existing forecasts and any sectorial plans -either energy-related or not- has to be known.

- 1) GENERAL PROBLEMS OF INFORMATION GATHERING (1).

The problems met in gathering information vary from one energy sector to another. Nevertheless, a general guide for the diagnosis of information may be useful. In the establishment of historical series, the situation of information status may vary as indicated in the following scheme :

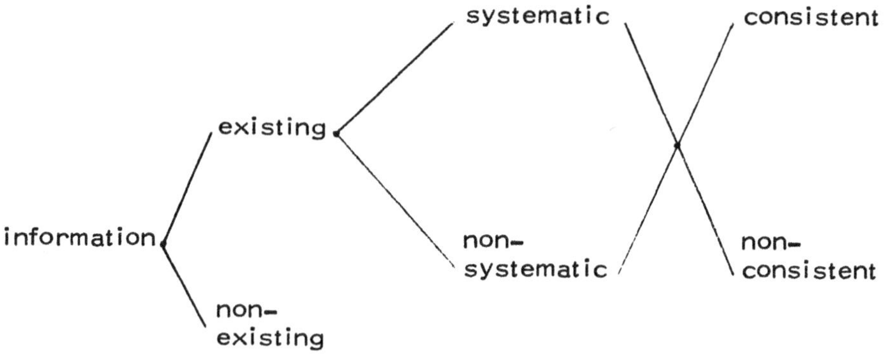

The ideal case is to get information that is : existing, systematic and consistent.

1.1.- The consistency of information.

This characteristic of information is very important ; it has to be checked every time. For energy information there are three criteria that are to be used.

The equilibrium between supply and demand is a good means of controlling the information. Knowing that the demand for an energy product is known from one part, and the figures of the production and imports from another part, it is then easy to get cross checked information.

The need for an explanation of any historical discontinuity constitutes the second way to control the quality of information. This permits the resolution of the cases in which there have been unspecified changes in the compilation criteria of the information.

The third criteria of consistency is more specific to energy statistics. It concerns the average efficiency of a transformation center which has to be within the technical limits of the process. For example, thermal power plants cannot normally appear with efficiencies higher than 0.35.

1.2.- Missing information.

The diagram presented before mentions the possibility of non-systematic information, that is information published in isolated reports and not in regular editions. If non-systematic information makes leave only a few missing points, they should be estimated through interpolation, survey or inquiry.

Finally, the missing information should be established through statistical treatment of the results of inquiries. The means of elaboration of this information are detailed later in this paper, through some examples.

- 2) ENERGY INFORMATION NEEDS.

As indicated before, the problems vary from one energy sector to another one.

2.1.- Non-commercial energy.

Firewood and other non-commercial energies (2) are in many countries one of the most important energy products, at least for household consumption. In the Ecuadorian case, in 1978, they represented 62 % of household consumption and 22 % of the total final energy consumption.

The values that are met in the literature are not be trusted as far as any particular country is concerned. Any national planning group has to look for reliable data and this is possible only through a national statistical inquiry.

The way it has been made in Ecuador may serve as an example (3). Firstly, the existence of two different geographical regions within the country was taken into account : the coast and the "Sierra". The design of the sample survey had to be different. Then a questionnaire was designed after some concrete tests (see chart). People in charge of the inquiry were carefully selected. Agricultural officers in charge of villages were chosen : they know the rural people and are able to understand the interest of the job.

A statistical processing of the data collected allowed the calculation of a national average daily consumption : 2,1 kg of firewood per person per day, with a lot of information about regional variations, percentages of penetration of commercial energies (kerosene, l. p. g.) and about the historical processes of substitution.

In this inquiry, it was necessary to be careful about the units in which the inhabitants express their consumption. These units vary from one region to another. The calorific value of the fuel used in each region was also measured.

Multiplying the average consumption by the firewood using population (4) and the average calorific value of the firewood, it is possible to obtain the national consumption with a reasonable accuracy.

2.2.- Hydrocarbon sector.

In this sector, the internal organization of the oil sector determines the availability and quality of the information. Precise information always exists.

The main problem of the information gathering depends on relationship between the energy planning group and the oil companies. Even in a centrally planned country, with the planning group belonging to a ministry and the existence of only one state oil-company, there is no evidence of the willingness of "oilmen" to give the information. This will generally be eased through personal contacts much more than through the normal administrative processes.

2.3.- Electricity sector.

On the contrary, electricity people are generally "information minded". There exists good statistics on :

- installed capacity with the type of each plant,
- production, losses in transport and distribution,
- fuel consumption,
- quantities sold and at least ideas on the structure of consumption with the different tariffs (residential, commercial, industrial...)
- some financial information.

The existence of such transparency of the utilities comes from the fact that they always need public financing for their investments -either national or international- so they are not reluctant to give information.

Some difficulties are encountered with the problem of self-producers -either industrial or commercial. Information on electricity generation may be obtained either from public utilities statistics -they make their own inquiries for obvious reasons- or by making a short inquiry at least for the main ones. But the fuel consumption for electricity generation is not broken down- not separated from the other fuel consumption of the industrial plant- and generally has to be estimated as a function of the types of the privately-run power plants in activity.

There exists utilities that give the sectoral break-down of industrial and commercial consumption. In the majority of the cases they do not, and this useful information will have to be estimated through existing industrial surveys.

2.1.- Other fuels.

The three energy products mentioned above generally represent the most important part of the energy sector in any given country. Depending on the specific characteristics of it, it is worth estimating what the other important fuels are, such as for instance : coal (mineral), charcoal, firewood for industry, heat (i.e. geothermal energy directly used), ethanol, and so on. For each one of these forms of energy it would be necessary to look for information through inquiries based on the detailed knowledge of the economy that is always necessary in an energy planning group.

- 3) NON-ENERGY HISTORICAL INFORMATION NEEDS.

The analyses of the situation of the energy sector cannot be completed without some more general information than that directly relevant to energy. This can be classified into two groups presented below.

3.1.- Organization of the energy system.

A good knowledge of the legal and concrete terms of the organization of the energy system is indispensable. For instance, it has to be realised that the influence of a national planning group on the investment decisions of a public utility is tiny. It is always a long range process to impose the idea that electricity is only a part of a global energy problem and not so important in itself. In some rural regions, the most important energy problem is more getting cheap and reliable cooking than the availability of electricity.

So, the organization of the energy system determines the decision process. Though it should not be translated into figures or even not written in an official way it has to be known.

The best energy plan is completely useless if not applicable because of not inserting itself in the existing "scheme". There are a lot of examples of this type of unsuccessful job.

3.2.- Economic indicators.

Nearly all existing socio-economic information is useful for energy planning : the energy sector through the production, transport and utilization processes, "crosses" all the sectors of economic activities. Energy for the households, and all demographic, habitational, "standard of living" statistics are needed ; energy for transportation and agriculture, and all transport and agricultural statistics are needed ; energy imported or exported, and all the harbour facilities should be known...

The socio-economic information has to be broken down into different sectors : the best scheme would be that where the energy and economic sectors are the same. This is not always possible ; in the socio-economic statistics, energy production appears within mining activities either energy or non-energy ones ; energy transformation like refining appears within the chemical industrial sector, while electricity production and water pumping belong to the same sector. Utilization of these statistics for energy planning purposes implies then that some precautions should be taken.

In this field, the problem is generally to select the information rather than to look for it. Nevertheless the optimal information -in the sense that it should be used directly for energy planning purposes- is not always available. It should be created through the cross-checking of existing socio-economic information, or with the means of limited inquiries.

- 4) PREVISION AND PROSPECTIVE INFORMATION NEEDS.

It should be possible to an energy planning group to elaborate its previsions in an independent way. This is a hazardous method. On the contrary, the group is better to look for all the existing previsions in different directions.

4.1.- Existing sector energy planning or forecast.

As mentioned above, in the majority of the cases, there exist some electrification plans that correspond to an effort of sector planning based on prevision. Any other sector planning should also be looked for. The internal coherency of these works has to be checked ; but the first objective of a national energy planning group should be to study the possibility of consolidation of such sector plans into a global plan. It has to be known that such a consolidation leads to modified sector plans and has to be accepted by the energy institutions.

4.2.- Resources and reserves inventory.

A global energy policy would favour the utilization of less expensive and more abundant forms of energy. This supposes that a good evaluation of the relative abundancy of each energy product is available. This is a hard task. On the one hand, oil reserves are considered by anyone who has a bit of information as strategic : the information is not "easy to obtain". Moreover there are two other factors that transform any inventory into a headache.

Firstly, the comparison between finite resources and renewable ones is not really feasible. Nevertheless it has to be done. Giving a conventional finite duration to a renewable energy is a means of solving this contradiction ; it remains hazardous.

Secondly, the reserves (and resources of renewable energies) are anything but fixed data. They vary largely depending on many factors, such as :

- the degree of exploration,
- the costs of operation,
- the political strategy of a government or an oil-company,
- the criteria of compilation,
- and so on.

A planning scheme based on the fact that one energy form is scarce, could be made redundant the next year if it appears opportune to someone to say there exist important reserves of this energy form.

This part of the information gathering process appears, from experience, to be the most difficult one.

4.3.- Global and detailed economic projections.

The sector planning is checked as mentioned above through its energy coherency. It also has to be through the consistency it has with the general economic projections that exist in the country. The information has to be looked for at two levels. The global economic projections allow the establishment of general energy indicators such as : total demand, total volume of investment, and so on. Sectoral projections permit a more detailed work in the energy field. If physical objectives are known in some sectors -one of the most important in that sense is that of energy intensive industries- the use of technical relationship factors permits the elaboration of detailed energy previsions.

The relevant information for such work is not readily available. The energy planning group would be better to insist on some economic institutions elaborating this information. If this is not possible, it should be done by the energy group itself.

There exists one general danger regarding utilization of such information. The energy projections that are obtained from socio-economic projections cannot be good ones if the base is not good. It is important to avoid the "confidence vicious circle", that is to say that one forecast reinforces the preceding one and gives a false feeling that both are valid.

CONCLUSION

Information gathering is an important stage of the energy planning process. Experience shows that in every country there exists much more information than was thought a priori. Nevertheless, access to it is never evident and has to be looked for through pragmatic and flexible approaches.

The precision of the information has to be considered. Energy informations which claim an accuracy of 1 % are useless if all demographic and socio-economic informations are known with a 10 % precision. In the case of Latin American countries, it was assumed that a 5 % accuracy for energy information was a reasonable objective. It is not always easy to reach, and supposes a lot of effort where double checking of data, verification of inquiries -and common sense- are useful.

(1) This paragraph is largely based on : OLADE Methodology For the Elaboration of Energy Balances, OLADE Document Series n° 8 (Chapter 5) which me helped write.

(2) The term non-commercial, though it is useful, does not mean that firewood, charcoal and so on are not subject in some cases to commercial process.

(3) See : Jaime Andrade, Eduardo Moran, Estudio sobre el consumo energético de leña en el Area Rural del Ecuador, Enero de 1981, Estudio INE n° 6, Instituto nacional de Energia, Quito, Ecuador (in Spanish only).

(4) Taking into account the total or partial penetration of commercial fuels in the concerned population.

THE IAEA ENERGY AND ECONOMIC DATA BANK (EEDB)

Jean-Pierre Charpentier

Economic Studies Section
Division of Nuclear Power
International Atomic Energy Agency
P.O. Box 100
A-1400 Vienna, Austria.

ABSTRACT In 1977, the Agency initiated a computerized Energy and Economic Data Bank (EEDB) to provide in a unified and systematic way energy and related economic data needed for analyzing the role of nuclear energy within the national energy plans of its Member States. The available data on different sorts of energy production and consumption in most Agency Member States are rapidly accessible and permit the production of a variety of tables, graphs and reports. For the time being, the EEDB is restricted to in-house use and an appropriate announcement will be issued as soon as external availability has been decided upon.

INTRODUCTION

In order to ascertain the role which could be played by nuclear power in developing regions of the world, the International Atomic Energy Agency in the past has carried out market surveys for nuclear power in a number of developing countries. In 1977, the Agency initiated a computerized Energy and Economic Data Bank (EEDB) not only on nuclear energy but on other forms of energy as well. The purpose of the data bank is to provide in a unified and systematic way energy and related economic data needed for long-term energy planning.

The Agency has now gathered the available data on energy production and consumption in most of its Member States and basic information on the world's energy situation and related economic parameters.

H. Neu and D. Bain (eds.), National Energy Planning and Management in Developing Countries, 223–230.

Data required for analysis of the role of nuclear power in Member States, for official Agency reports, or for national, regional or world surveys can now be obtained rapidly and automatically. Flexible computer programs permit the production of a variety of up-to-date tables and graphs. Aggregation of results on a regional or world basis, which previously called for time-consuming individual calculations, are now easily and quickly obtained.

There were three main developments that led to the setting up of the computerized data bank:

a. The first was a considerable increase in the volume of information that was being gathered. For a number of years, a variety of energy and economic data have been gathered and classified by the Agency, and management of the continuously growing data base was becoming more and more difficult.

b. The second, and probably the most important, development was the need for faster selection and retrieval of the basic information. With the computerized data bank, it is now possible to provide a quick answer to many questions about energy production, consumption and resources, and about related economic aspects.

c. Thirdly, there was a need for data for making forecasts of energy consumption and production through trend extrapolations, cross-comparisons between countries as well as through more sophisticated models. Nevertheless, it should be clear that the data bank will not be a new mathematical energy model but an easily accessible system of energy data that may be used in any model as well as for more general purposes.

ORGANIZATION OF THE DATA BANK

The setting up of a computerized data bank requires essentially four things:

1. Input data;

2. A computer and a linked magnetic storage system for these data (the hardware support);

3. A computer system (the software) which allows organization and classification of the raw input data and easy retrieval of the output data from the magnetic storage system;

4. A series of additional computer sub-programs to process the raw output data into tables, graphs, statistical analyses, etc.

Points 1 and 3, the gathering of data and adaptation of a computer program for the classification and retrieval of raw input data, were the keypoints for the development of the data base. The computer hardware support was already available at the Agency (an IBM 3032 operating under MVS), and the development of sub-programs related to the final uses of data was a "relatively" easy task.

Data gathering: As shown in Figure 1, the data sources of the IAEA energy data bank are twofold: either data are sent to the Agency directly on magnetic tape by other international organizations, or the Agency develops its own files.

Concerning the data received from outside the Agency, good co-operation has been established, including loans of computerized tapes and data up-dating procedures, with the United Nations Statistical Office in New York, the Statistical Division of the OECD International Energy Agency (IEA) in Paris, the World Bank in Washington, D.C., and the Statistical Energy Division of EEC in Luxembourg.

Contacts have also been initiated with other national and international bodies such as the World Energy Conference in order to set up a similar procedure. These external contacts are oriented toward the gathering of non-nuclear information. Data related to the nuclear field are on the whole gathered directly by the Agency through specific organizations, through perusal of literature from meetings and by staff members on mission.

Figure 2 shows the structure of the content of the data bank. The thick lines on the figure refer to data presently gathered and available, and the thin lines indicate the type of data that are still being collected.

The Computer System: The magnitude of basic information in the data bank is presently close to four million or 40 megabytes. The derived data such as growth-rate or per capita consumption, etc. which are calculated by sub-programs, are not included in this amount. The processing of such a large amount of data, even computerized, is not an easy task. Originally, the data bank used a program that the Agency had developed for library purposes

Figure 1: Data sources of the IAEA Energy Data Bank

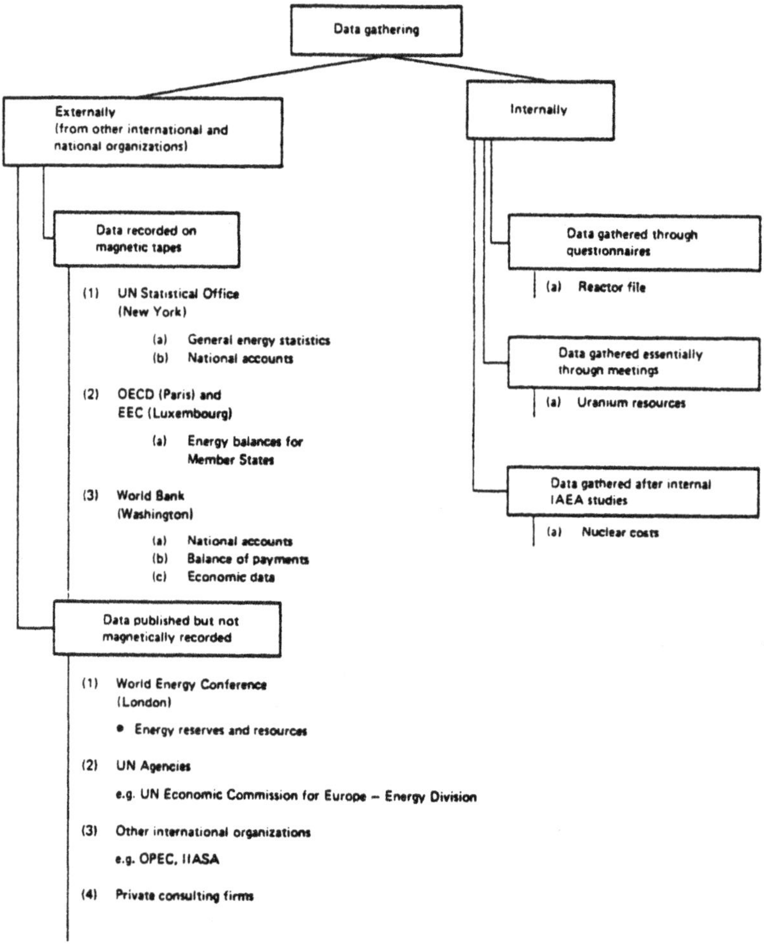

Figure 2: Structure of the Content of the Data Bank

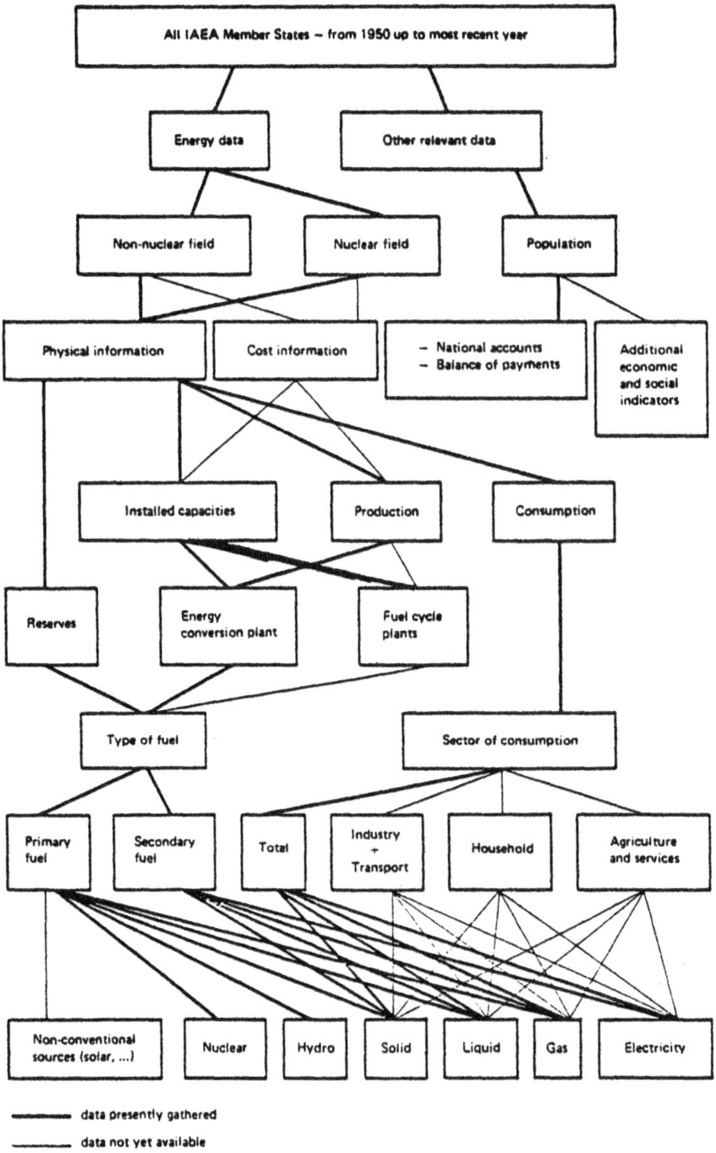

(code LIPSY). But even with modifications, it was clear that this program was very difficult to adapt with sufficient flexibility to the management of information which is essentially numerical.

After examination of several possibilities, it was decided to make use of the general data management system called ADABAS. This system for computer programs is specially written for the management of data. It organizes input data in such a way that they can easily and quickly be retrieved without the user knowing the exact location of the data in the data base. There is a simple language which allows one to make simple commands and retrieve data from very large and complicated data bases.

The data are grouped by homogeneous topic and separately stored on magnetic support which is called file. Eight different files are presently in use or under development.

1. General energy information on production, consumption, trade and plant capacities.

2. Energy resources.

3. Nuclear reactors.

4. Fuel cycle facilities.

5. Cost information on investment and operation for energy facilities.

6. National accounts.

7. Forecasts.

8. A summary file gathering the data most often requested in order to allow easy access to the EEDB for non-specialized persons.

ADABAS contains in its internal structure features which allow the easy coupling of files containing at least one common keyword. For example, since every data in each of the files always refers to the name of the country, a country's name could be used as the keyword. With such features it is also possible to couple the general energy file with the national account file in order to find the energy consumption of all countries where the Gross National Product (GNP) per capita is a certain value.

SERVICES SUPPLIED BY THE DATA BANK

The data bank has been structured so that it can fulfill three main services:

1. The "day to day" service, which consists of being able to answer quickly any questions on energy consumption, production, trade, resources, and their related economic aspects.

In the Agency an on-line video-terminal is also available so that users of the data bank are able to make simple queries by themselves and the answers are displayed directly on the screen.

A test period for direct access to the EEDB by outside users is in progress, but no decision has been taken yet with regard to further development of this new facility offered by the Agency to its Member States.

Example of a simple request: "Find all countries which, in 1975, had an average load factor of their national plants greater than 50%".

The computer query was formulated as follows: "FIND ALL RECORDS IN WORLD-ENERGY-STATISTICS WITH YEAR EQUAL 1975, IF ELEC-CAP NOT EQUAL 0 THEN COMPUTE ≠ UTIL = ELEC-PROD * 114/ ELEC-CAP ELSE SET ≠ UTIL = 0. END IF ≠ UTIL LT 500 THEN REJECT RECORD END DISPLAY COUNTRY ELEC-PROD ELEC-CAPA ≠ UTIL 10 'LOAD/ FACTOR/%'".

A list of countries and output data appeared on the screen from which the following were selected in order of decreasing electric energy production.

Country	Elec-Prod Gigawatt-Hours	Elec-Capacities Megawatt	Load Factor %
USSR	1 038 625	217 484	54.4
Canada	272 624	59 886	51.8
Poland	97 168	20 057	55.2
German Dem. Republic	84 505	16 929	56.9
Norway	77 564	17 090	51.7
South Africa	74 914	13 990	61.0
Romania	53 720	11 577	52.8
Yuogslavia	40 040	9 043	50.4
Venezuela	21 179	4 705	51.3
Hungary	20 465	4 291	54.3

2. The regular supply of statistical bulletins and booklets summarizing in standardized tables and graphs the past and present energy situation in the different IAEA Member States. All information is given in a unified unit: Joule.

3. Support for energy forecasts. Several mathematical models have been developed and systematic series forecasts have been carried out through a computerized link between the data and the models.

CONCLUSION

A world-wide energy data bank has now been created. It is hoped that the service offered by the data bank will assist individual States and international organizations in planning future power systems. However, there remains much to be done. It goes without saying that the development of an energy data bank is a long-range undertaking requiring constant additions and improvements.

BASIC TOOLS AND TECHNIQUES FOR ENERGY PLANNING

Jean-Pierre Charpentier

Economic Studies Section
Division of Nuclear Power
International Atomic Energy Agency
P.O. Box 100, A-1400 Vienna, Austria.

ABSTRACT This paper first draws attention to various difficulties encountered in energy planning, and then describes some basic mathematical tools and techniques used today in the domain of energy forecasting. As the recent events have given man a harsh reminder that history is the study of events which never repeat themselves, this paper tries to critically review the explicit or implicit simplifications on which most methods used for energy analysis are based and concludes that only a balanced combination of judgement for the choice of realistic hypotheses and the power of the computer models could permit better investigation of the complexity of the present and future energy situation.

INTRODUCTION

Whatever word is chosen to describe the events of 1973-74 in the oil world - crisis, reappraisal or adjustment - it is clear that for energy analysis they involved if not a total re-examination then at least a very critical revision of the method of approach employed.

One of the compensating advantages of economic or political crises is that they often bring about a series of revisions of concepts and methods which seemed to be firmly established but were frequently nothing more than simplifications which had been adapted to a certain economic context and too quickly raised to the status of immutable theories.

The analysis of past changes can lead to a better understanding of the present but does it provide a sufficient basis for pre-

H. Neu and D. Bain (eds.), National Energy Planning and Management in Developing Countries, 231–251.

dicting the future? This is the question which every honest energy analyst must face.

What, in fact, do we expect from a model? Is it an exact prediction of what is to happen or just a general idea of the problems which may confront us in the more or less distant future? Unfortunately, an analyst who seeks a positive answer on the first point is likely to be disappointed sooner or later. In order to understand why, it is worthwhile, before looking at the tools and techniques, to briefly recall various hypotheses and explicit or implicit simplifications on which most of the energy planning activities are based.

1. DIFFICULTIES RELATED TO ENERGY PLANNING

The main objective of energy planning is to predict in due time the investment which will be required to satisfy the demand in the various sectors of the economy. This aim is generally achieved by two successive stages:

(a) The first stage is to forecast intrinsically what the future demand for different froms of energy will be in various sectors of the economy;

(b) The second stage is a "confrontation" between this "demand" and the physical and economical possibilities of the country studied.

A brief review of the ambiguities to be found in the definitions used at each of these stages provides at least some idea of the uncertainty of measurement and of the difficulties encountered when assessing each of these aspects.

1.1. Field of Energy Demand

Let us start by recalling the definition of demand in theoretical economics. Demand for a given product or service stands in relation, on the one hand, to the amounts of the product or service that the consumer wishes to receive, and, on the other hand, to the price of the product or service, the prices of all other products, the consumer's income and the time involved. Thus, for every product, demand is a function of a number of variables.

A first simplification consists of reducing the variables to the price of the product itself and its substitutes and to the consumer's income. A second step can be taken by assuming that the impact of fluctuation in the prices of substitutes is negligibe. A third approximation consists of assuming that for practical purposes demand does not depend on the price of the product itself, but only on income and time or, to use economic jargon, that it is completely inelastic in relation to price.

A last step is taken when income is given in advance as a function of time and demand is reduced to a function of time alone. Despite the radical simplifications that these approximations involve, their use in forecasting the demand for certain forms of energy has been producing results of surprising accuracy for several decades, as for example in the case of electricity demand predicted in industrialized countries between 1950 and 1970. It is, however, essential to bear the neglected variables in mind if, for instance, one wishes to analyse the effect of sudden jumps in prices, such as we are seeing at the moment.

So what does one mean when one says, for example, that world demand for energy will reach 550 or 650 Exa-Joules (10^{18} J) by the year 2000? What series of simplifying hypotheses and what set of aggregations can be used to replace a large number of functions of several variables relating to different services by a single value expressed in terms of a single unit?

In order to answer these questions it is necessary to briefly trace, from the initial producer to the final consumer, the route taken by energy and to note in passing the definitions used for the different forms of energy demand, the methods used for reducing these different forms to a common denominator and the way in which all independent variables with the exception of time are excluded.

(a) Problems of Definition. It should always be borne in mind that the final consumer requires not the energy itself, but a service of which energy is an essential ingredient. Examples of such services are the maintenance of living space at a given temperature, the illumination of a given surface area at a certain luminous intensity, private or public transport in given conditions of comfort and speed over a given distance and the performance of a given chemical reaction. Such a service can be obtained by means of a series of processes available at a given level of technological development. For example, in the case of heating, use can be made of a coal fire, a wood stove, an individual heating system, gas- or oil-fired central heating, or electrical heating of the resistance type or using a heat pump. The amounts of energy required for providing the services in question are therefore highly variable as a function of:

- the quality of the service required (continuity, flexibility, ease of operation and comfort);
- the conditions under which the service is provided (good or bad insulation of living space, etc.);

- the technology chosen for obtaining the service required.

Here we see that even at the level of the consumer, energy can be expressed in terms of different forms and amounts, for example, the amount of energy required to provide the service ("useful energy") or the amount of energy required at the input to the equipemnt ("final energy").

In most cases, final energy reaches the consumer in a form that is not found in nature. This is known as "secondary energy", as it is liable to losses during its transmission and distribution from the point of production. In addition, its production requires a certain amount of primary energy which, in the case of a thermal power station, can be greater by a factor of approximately three.

One can go even further. Since the primary energy required is measured at the point of input to the power station and not at its production point, we require the addition of a further transport penalty in order to find the gross primary energy requirements. In addition, all the components of the two systems require energy not only for their operation but also for their construction materials and asssembly. There is, therefore, a whole set of energy accounting operations to be performed, although they do not affect the force of the conclusion reached: Any energy service can be expressed in terms of many different <u>useful</u>, <u>final</u>, <u>secondary</u> and <u>primary</u> energy requirements, depending on the technologies chosen and the conditions in which they are used. **Examples chosen are far from unique, and** one could just as easily take individual passenger transport or an industrial process. It is in any case clear that every forecast of energy demand summarized in a few figures involves the conscious or unconscious assumption of a vast number of explicit or implicit hypotheses regarding the technological and economic options on which it is based.

(b) <u>Problems of Aggregation</u>. The addition of figures for different froms of energy in order to obtain global figures also presents a problem.

The first difficulty is of a statistical order: It is necessary to ensure that all energy sources have actually been taken into account. In practice, however, only commercial forms of energy are reflected in national accounting systems.

The second difficulty is of a methodological nature: If it is assumed that figures for fuels have been added up on the basis of their energy content - which already involves the assumption of simple substitutions - how does one account for electricity from hydroelectric and nuclear power? On the basis of consumption and in terms of their thermal equivalents? Or on the basis of production and in terms of the amount of heat which it would be necessary to expand in a thermal power station of given efficiency in order to produce the amount of electrical energy involved? In other words, should one take the kWh of electricity from hydroelectric or nuclear power as being equivalent to 860 kcal or to between 2200 and 2600 kcal?

It is clear that, depending on the system chosen, not only do the totals differ (and, therefore, also the relative proportions of the different fuels), but also the growth rates during the period considered.

In fact, one could go even further and give thermal energy unit coefficients reflecting their value as substitutes in different uses; this, however, goes beyond the scope of the present introduction, the purpose of which is merely to demonstrate that the horizontal integration of different energy sources presents the statistician and the analyst with perhaps just as many problems as the vertical integration discussed in the previous section, and these are problems which are equally difficult to solve.

1.2. Field of Energy Supply, Reserves and Resources

(a) Problems of Definition. In this case, as previously, it is essential to bear in mind that energy supply, reserves and even resources (with the possible exception of resources known as "ultimate") are, at least in theory, functions of price and cost and are not fixed values. This group of functions has some rather unpleasant factors.

First, the number of functions to be taken into consideration is uncertain as soon as the period covered by the forecast extends beyond a few decades. Who would have included uranium as a source of energy forty years ago? And should one include thorium or lithium today?

Secondly, analytical, graphical or tabular representations of these functions are possible only in certain cases and for very short ranges of the cost and price variables. At present we have only rather uncertain information on the quantities of oil than can be extracted at production costs double or triple the current price, and even more uncertain data on uranium deposits, the mining of which would become economically viable if present prices rose substantially.

In the absence of reliable numerical data on which to base forecasts, use is made of a large number of definitions and adjectives aimed at providing, instead of a quantitative analysis (which is impossible), a qualitative classification, with which it is hoped to clarify ideas if not to solve the actual problems.

It would not be appropriate here to discuss the different definitions of resources and reserves and the extraordinary number of adjectives qualifying them.

(b) Problems of Aggregation. The aggregation of different energy reserves and resources for global balances presents methodological problems as difficult as those presented by demand and we shall only mention two of these. First, there is the problem of adding "stock" (fossil fuels, uranium, etc.) and "flows" (hydroelectric power, geothermal and solar energy). Secondly, it is impossible to assign an energy value to certain resources without previously defining these systems of production by means of which this resource is utilized. The most striking example is that of uranium, since the same quantity of natural uranium can produce amounts of energy differing by a factor of 100, depending on whether it is used in a thermal reactor without recycling or in a fast breeder reactor. With the first option, demonstrated uranium reserves would, at present, be smaller than those of oil and could only provide a transitional source of energy. With the second option, they would be able to meet the world's energy needs for centuries.

These considerations, although presented in a fragmentary fashion, are sufficient to demonstrate the provisional and uncertain nature of estimates of reserves and, even more so, of resources. Even the figures put forward for demonstrated reserves depend on the availability and cost of mining technology, which is in the course of rapid development. This can be seen, for example, with secondary and tertiary recovery of oil from well-defined deposits which have already been exploited.

As in the domain of energy demand, the notion of costs and prices plays a major role in the definition of useable reserves, therefore, we shall now consider these two factors.

1.3. Notion of Costs and Prices

The gulf between theoretical concepts and their practical application, which the previous sections have hinted at, becomes apparent when one attempts to quantify the costs and prices of different forms of energy and their future trends.

Present production costs for certain fuels are, if not unknown, at least often inaccessible to those outside the producing companies. This is the case with uranium and, to a certain extent, also with oil and gas.

Estimates of future production costs lie within such wide ranges that their analytical or practical value is singularly limited. Thus, in seven years the estimated cost of producing oil from bituminous schists has been multiplied by a factor of three or four and that of uranium recovery from sea water by a factor of more than ten.

For certain particularly important energy sources production costs are now out of all proportions to the prices charged. This is the case with oil, for example, where a barrel of crude from one of the larger deposits in the Middle East, costs 30 or 40 cents to produce, while the products derived from it can be sold for sums of the order of US $ 35 - 40 in Western Europe ($ 15 or $ 16 going to the producer countries and approximately the same amount to the tax authorities of the consumer countries). Thus, over three quarters of the final price, a figure more than one hundred times the cost of production, is determined by political decisions. Analysis of natural gas prices would give the same sort of results, and a similar situation is beginning to emerge in relation to uranium.

Finally, costs and prices are quoted on the markets in current money, but the analyst wishing to perform comparisons or discern trends will have to transform these values into constant money in such a way as to make allowance for inflation, which affects all prices and distorts the actual trend for each product. This is frequently a delicate operation which adds a further element of complexity to the difficulties already mentioned.

1.4. Some Conclusions on Difficulties Related to Energy Planning

In a simplified and fragmentary review of complex and interconnected problems it would be highly presumptuous to speak of conclusions. However, certain impressions can be gained from this review.

1. A certain degree of scepticism about energy supply and demand forecasts is inevitable. It is sufficient to look at the forecasts made before 1973 for this to be obvious.

2. The extent of this scepticism should depend, among other things, on the following:
   a. the period to which the forecast relates;
   b. the degree of aggregation of the different variables;

c. the method used for linking energy to the economic and political environment in which it is to be used.

This last point is particularly important since, despite the sophistication of certain energy prediction models, they tend not to take into account the uncertainties affecting the input data. Thus, much time can be wasted in discussing the divergences found in results obtained using different theoretical refinements, the order of magnitude of which may be much smaller than the divergence which would have been produced by the adoption of different but equally plausible assumptions regarding input parameters.

3. This scepticism should not take the form of criticism alone, if only out of sympathy with the energy analyst who already has enough difficulty obtaining much essential information relating to the past and the present, not to mention the future. Nor can one or should one ask him to avoid using simplistic hypotheses and aggregations of questionable validity, since without these techniques no provisional estimates, and consequently no information for public consumption, would be possible. One can, however, ask him to explain such techniques clearly and not to ignore them, leave them implicit or disguise them.

Finally (and perhaps this is the most important impression gained), it seems clear that the energy economy is more than anything else a question of _political economy_.

2. BASIC TOOLS AND TECHNIQUES USED IN ENERGY PLANNING

2.1. Energy Demand Forecast

At present there are two types of opposing "philosophies". One is to use a study of past needs to draw conclusions about the future. We shall call this the _prospective_ in contrast to the _normative_ approach, which consists of defining "_a priori_" the objectives to be achieved and then looking for possible ways of reconciling these with the present situation.

Prospective Methods Used to Forecast Energy Demand. Again, it is possible to distinguish between:

a. Purely statistical methods based on regression and extrapolation, in which the only explicative parameters for the energy consumption are time, space (i.e., a comparison between countries) or both of these together;

b. Econometric methods where the explicative variables are price and income.

In both these cases, a very careful analysis of the data referring to past and present consumption should be made before the exploratory study of the future is undertaken.

a. Statistical Methods Involving Correlation-Regression Followed by Extrapolation. If we assume, as was often done before the 1973 oil crisis, that economic mechanisms are highly stable and that the future is nothing but a continuation of the past then we can use purely statistical methods based on the various procedures for the extrapolation of trends. Although these methods are very much open to criticism they have the advantage that they enable very rapid tests to be carried out on what might happen if "all other things remained unchanged".

Although they are based on slightly different philosophical ideas, we shall group together methods which use time and/or space as explicative parameters because they rely on identical mathematical formulations.

The use of time as a unique explicative parameter implies that this variable combines the possible effects of both technical progress and economic change. The methodology is aimed at finding the best correlation function between energy consumption and time and at extrapolating the resultant "law" into the future. This extrapolation only makes sense, of course, if it is accompanied by some sensitivity study of the quality of the correlation obtained from past statistical data. The statisticians speak of a confidence interval, which can be imagined as a zone surrounding the extrapolated mean value in which the future demand can be located with a certain degree of possibility. Without going into statistical details, we can easily see that this confidence interval increases with the period covered by the forecast.

The choice of space (or, more precisely, the analogy of situations among various countries) as the explicative variable implies that there exists an "ordained" path along which every country must eventually pass during its development. For example, the global distribution of energy consumption growth rates of all countries of the world is often regarded as a function of the absolute per capita consumption. Such "law" is used to estimate the possible future evolution of a given country.

It has to be borne in mind, however, that these methods are normally used to study only one fuel at a time - electricity, fuel-oil, coal, and so on. These approaches were in common use before the first "oil crisis". They are nowadays heavily criticized but their very simplicity will continue to ensure that they survive.

The various mathematical functions used in these correlation/regression studies can be classified as follows:

We shall denote by x the quantity being studied. This may be the per capita consumption of primary energy, electricity, fuel-oil or any other component of the energy system. Then, $z = (dx/x)(dt/1) = (1/x)(dx/dt)$ represents the (instantaneous) growth rate. We shall distinguish between functions where z (the growth rate) is constant, a function of time or a function of x itself (i.e., of the existing level of consumption).

(i) z = k: constant rate of growth

If $z = (1/x)(dx/dt) = k$, we can easily carry out the integration and we then find that x is an exponential function of time:

$$x = \exp\ (kt + h)$$

The constants of integration k and h are determined from a statistical regression study on past data.

(ii) z = f(t): growth rate as function of time alone

The differential equation can now be written $(dx/x) = f(t).dt$. This type of function can usually be integrated quite easily and gives families of exponential growth equations:

$$x = A \exp\ \left(\int_{t_0}^{t} f(t)dt\right)$$

(iii) z = g(x): growth rate as function of the current level of consumption

A qualitative study of evolution of the consumption growth rates for any type of energy shows that, whichever country is considered, the growth rates first increase and then decrease when a certain level of per capita consumption has been attained.

Qualitatively, we say that the energy consumption growth rate is related to the per capital consumption by a "bell-shaped" function.

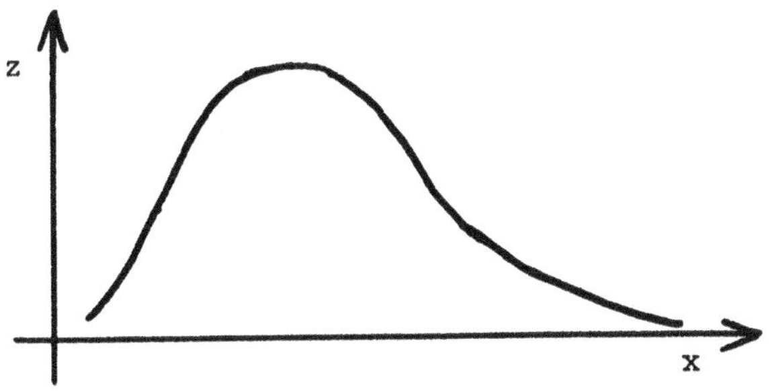

In the literature different mathematical functions are proposed to link the growth rate to the per capita level of consumption.
The major functions are as follows:

(a) Linear function: z and x are generally related by two straight line segments

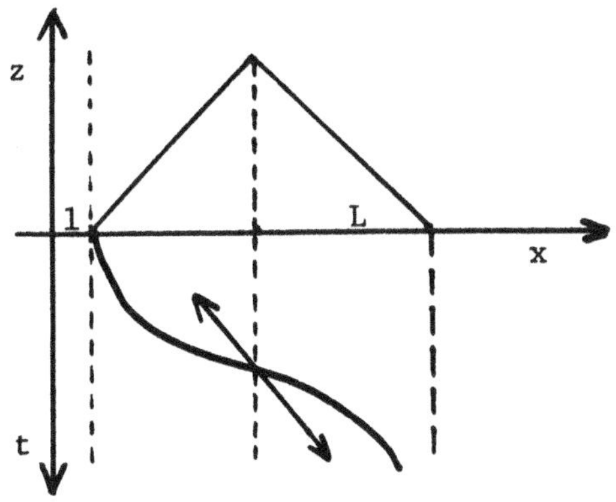

$z_1 = (1/x)(dx/dt) = a\,(L - x)$
$z_2 = (1/x)(dx/dt) = a'(x - \ell)$
The separate integration of each of the two linear paths leads to two saturation functions (also known as logistic functions) of the form $x = L/(1 + e^{-(at+c)})$ where L, a and c are constants obtained from the regression analysis.

(b) Logarithmic function: $z = (1/x)(dx/dt) = g(\log x)$. For this case it is found in practice that three principal types of function can be fitted to the basic data relating z and log x.

First case: two straight line segments

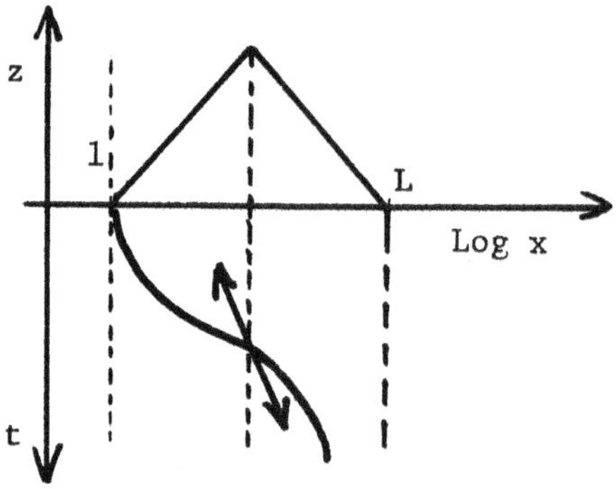

It is sufficient to analyse each straight line segment separately:

$z_1 = k(\log x - \ell)$

$z_2 = K(L - \mathrm{Log}\ x)$

After the change of the variable $X = \log x$, the integration of the two differential equations leads to two exponential branches with opposite curvatures which are matched to have the same value of log x at the intersection of the two straight line segments:

$x_1 = \mathrm{Exp}\ (\ell + \exp\ (kt + h))$

$x_2 = \mathrm{Exp}\ (L - \exp\ (kt + H))$

Second case: a parabolic function

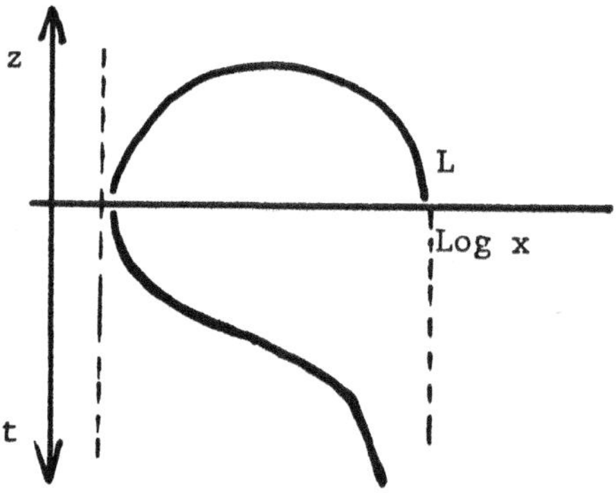

$z = (1/x)(dx/dt) = A \log x\ (L - \mathrm{Log}\ x)$

After the change of variable ($X = \log x$) we obtain a saturation function in X:

$$X = (L/(1 + e^{-L(At+B)}))$$

or in x:

$$x = \exp\ (L/(1 + e^{-L(At+B)}))$$

Third case: a function of the "gamma" group

Although this type of representation often cannot be integrated directly and computer simulation must be used, it is the one that is most frequently found in practice. It reflects a rapid growth for low values of consumption (case of developing countries) followed by a slow increase which never quite becomes zero though it tends asymptotically towards it (case of countries which are already highly industrialized). We have noted in practice that the most suitable function is

$$z = a \log x \cdot \exp(-b(\log x)^n)$$

where a, b and n are three parameters obtained by regression analysis on the basic data.

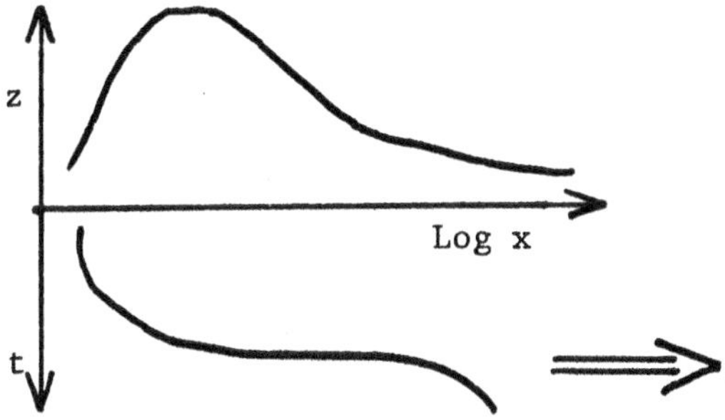

b. Econometric methods. With econometric methods it becomes necessary to take into account the conditional effect of parameters such as the price of energy and/or the level of income (the latter often being approximated by the gross per capita national product in the case of studies made at country level).

Generally, the production function is of the Cobb-Douglas type, e.g.,

$$Q = AL^{\alpha} K^{\beta} E^{\gamma}$$

where Q is the quantity of goods produced,
and L,K,E represent a simplification of production factors namely, labour, capital and energy. α,β,γ are constant coefficients (also called elasticity coefficients).

This equation clearly reflects the type of simplification mentioned above and that energy analysts should make in order to be able to mathematically describe the evolution of the economy.

Similarly, a simple consumption function is used to describe the quantity of goods consumed (C) as a function of the price system (P) and the income (Y):

$$C = AP^{a} \cdot Y^{b} \quad .$$

By assuming that the production-consumption system can reach the optimal situation and after elimination of certain variables, the demand for a given type of energy may be written as follows:

$$E_i = A\, p_i^{\eta_i} \prod_{j \neq i} p_j^{r_j} \cdot y^{\varepsilon} \qquad (1)$$

where p_i is the price of energy i; p_j are the prices of the competing forms of energy; ($\pi_j\, p_j$ is the product of the prices), y is the consumer's income; A is a constant; η_i, r_j and ε are the elasticities of E_i to the price and income.

It is relevant here to make a number of comments about the elasticity:

(i) We recall first that the elasticity η_i of the consumption E_i to the price p_i represents the variation in the consumption E_i when p_i changes by 1%:

$$\eta_i = \frac{\partial \mathrm{Log}\, E_i}{\partial \mathrm{Log}\, p_i} = \frac{\dfrac{\Delta E_i}{E_i}}{\dfrac{\Delta p_i}{p_i}}$$

For example, if the consumption of petrol decreases by 3% the day after a 10% increase in the price, we would say that the elasticity to the price is -0.30;

(ii) In Eq. (1) E_i is the optimum consumption at time t, i. e., the theoretical or desired value. Many authors consider that the sensitivity of the variation in actual consumption which occurs at time t is, in fact, the result of a linear combination of the sensitivity at an earlier time (t-1) and the theoretical consumption at t:

$$\frac{dE_t}{E_t} = \lambda \frac{dE_{t-1}}{E_{t-1}} + (1 - \lambda) \frac{dE^*_t}{E^*_t} \qquad (2)$$

where E_t and E_{t-1} are the real consumptions at times t and (t-1) and E^*_t is the theoretical "desired" consumption required to satisfy Eq. (1). Eq. (2) can also be written:

$$E_t = E_{t-1}^{\lambda} \cdot E^{*\,1-\lambda}_t$$

or:

$$E^*_t = \left(\frac{E_t}{E^{\lambda}_{(t-1)}}\right)^{\frac{1}{1-\lambda}} .$$

Substituting this into Eq. (1), we get:

$$\text{Log } E_{i,t} = B + \lambda \text{ Log } E_{i,t-1} + \eta_i (1-\lambda) \text{ Log } p_{i,t} + (1-\lambda) \sum_j \nu_j \text{ Log } p_{j,t} + (1-\lambda)\varepsilon \text{ Log } y_t \quad , \qquad (3)$$

where $E_{i,t}$ is the amount of energy i consumed at time t; $p_{i,t}$ is the price of energy i at t; $p_{j,t}$ is the price of the competing energies at t; y_t is the income at time t, as measured by the Gross Domestic Product (GDP); η_i, ν_j, ε are known as the long-term elasticities; and $\eta_i(1-\lambda)$, $\nu_j(1-\lambda)$, $\varepsilon\ (1-\lambda)$ as the short-term elasticities.

The short-term elasticities measure the immediate variations in consumption following a perturbation in prices, incomes or both together. If, after a certain time, these variations produce more profound changes in the economic structures (for example, replacement of individual by collective means of transport, or a change in the methods of heating, etc.) and thereby modify the consumption of energy, the resultant variations are measured by the long-term elasticities.

The measurement of elasticities is a question which divides many economists. Some experts believe that the demand for energy is relatively inelastic to prices and they explain this by the limited possibilities for short-term substitution and the very lond time required to develop new sources.

Methods involving normative scenarios. Practically all these methods are still in a state of development. They appeared as a result of the oil crisis which greatly reduced the credibility of purely statistical and econometric methods.

In general, all these approaches involve three stages of reasoning:

a. The breakdown of an economy into sub-systems, such as urban or inter-city transport, accommodation, etc. This breakdown is clearly rather subjective and may give rise to controversy.

b. Certain key parameters are sought to describe each sub-system.

c. A quantitative estimate is made of the useful and final energy which will be used for each of these economic sub-systems. A considerable amount of data is required at this stage.

Based on a collaboration between the Institute for Economic and Legal Aspects of Energy (IEJE Grenoble, France) and the International Institute for Applied Systems Analysis (IIASA Laxenburg, Austria), the International Atomic Energy Agency (IAEA Vienna, Austria) has developed a model based on the normative scenarios approach and simulation. This model, called MAED (Model for Analysis of Energy Demand) is, in fact, a follow-up of the MEDEE model that was already presented to you. MAED is a little bit simpler than MEDEE because it is oriented toward studies of developing countries where the data base is more limited, and more to the details of the electrical system. For example MAED supplies the future annual electrical load curves, that is to say, the hourly power needed by the country over the whole year.

It is not the place here to describe MAED (or MEDEE) in detail; but it should be borne in mind that in order to analyse energy demand of a given country, MAED (or MEDEE) provide a breakdown of the energy needs of each economic sector (household, transport, industry, service) into various elementary needs of final and useful uses (needs for space heating, cooking, furnaces, inter-city transport, etc.).

The useful and final energy requirement is drawn by two types of parameters: one linked to the technical considerations (such as the efficiency of different appliances, etc.) and the other linked to life-style considerations (e.g., average distance covered by car during a year or surface of dwelling, etc.).

MAED (or MEDEE) is a simulation model for evaluating the energy demand implications of the economic and life-style scenarios describing the long-term evolution of a country.

2.2. Supply models and supply-demand confrontation

Once the demand has been intrinsically determined or at least "prospected", it is necessary to ensure that its level is compatible with the constraints relating to both the existing energy system (infrastructure, energy reserves, etc.) and the rest of the economy. This comparison between what is desirable and what can be achieved can only be studied by means of models which describe the impact of each sector of the economy on the others.

In many countries, the fairly systematic use of models began around 1970 and underwent an increase after the 1973-74 oil crisis. Before that date, only a few large enterprises (mainly those involved in the production of electricity) were using sectorial models and these were essentially oriented towards the choice of investments. More recently, numerous other types of models have appeared. We can try to classify these by distinguishing:

a. <u>their sectors of application:</u>
   - sectorial enterprise models;
   - models for studying national energy systems;
   - macromodels of economic growth in which energy forms only a small part of the field of study;

b. <u>the mathematical models used:</u>
   - optimization models (linar, non-linear or dynamic);
   - simulation models.

We shall not go into the different possible classification schemes but simply refer those interested to the various studies published in this subject:

<u>Examples:</u>
- "A Review of Energy Models", by J.P. Charpentier, IIASA-4 Issues: RR-76-10, RR-75-35, RR-76-18, RR-78-12.
- "Global and International Energy Models: A Survey", by J.M. Beaujean, J.P. Charpentier and S. Nakicenovic, Annual Review of Energy, 2 (1977), 253-270.
- "Overview on Techniques and Models used in Energy Field", by J.P. Charpentier, IIASA RM-75-8.
- "Energy System Modelling and Forecasting", by K. Hoffmann and D. Wood, Annual Review of Energy, 1 (1976), 423-453.

<u>Supply Models.</u> The WASP model (<u>W</u>ien <u>A</u>utomatic <u>S</u>ystem <u>P</u>lanning package) developed by the International Atomic Energy Agency could be taken as example of a supply model. The WASP model is designed to find the optimum power system expansion plan for a given country.

This model uses the dynamic programming technique. Its objective function is the present-worth discounted value of all operating costs plus all investment costs of the different plants taken into consideration.

In addition to the data describing the potential power plant system and the related economic information, the major input into the model is the seasonal electrical load duration curves. (This last information is supplied by the MAED model previously mentioned).

The WASP model is, in fact, composed of five modular programs:
- A program to describe the existing power system and all firmly scheduled additions;

- A program to describe the alternative plants which could be used to expand the power system;
- A program to generate alternative expansion configurations;
- A program to determine whether a particular configuration has been simulated and, if not, to simulate operation with that configuration. Using a probabilistic simulation model, energy generation by each plan and the corresponding operation cost are calculated. The reliability of the generating system and the probable amount of unsatisfied demand are estimated.
- Using the data files created by the other modules together with economic inputs and reliability criteria, the last program chooses the lowest cost expansion schedule for adding new units to the system over the period of interest. The objective function of this (dynamic programming) optimization is the present-worth discounted value of: all operating costs plus all capital investment costs of the variable-system less a salvage value credit at the horizon for the remaining economic life of its plants.

Supply-demand confrontation.
General structure. The structure of the energy system and its relation to the rest of the economy can be represented diagrammatically as follows:

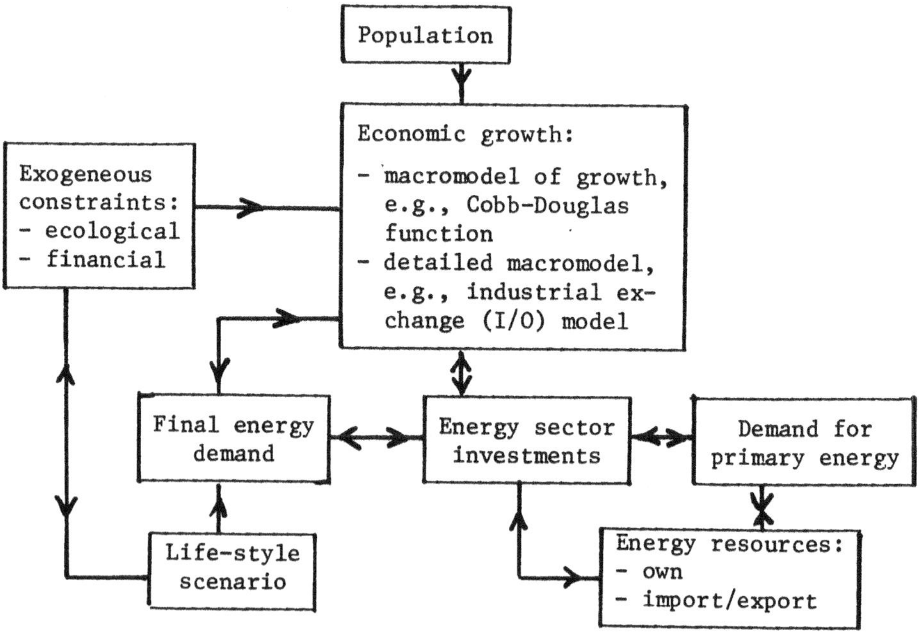

Each box or combination of boxes in this diagram can, in fact, represent the subject for a model study. If, for example, we retain only the "final energy demand" box we have, in general, just the simple correlation-extrapolation approaches mentioned above.

By adding the "life-style scenario", we get the simulation models relating to the study of the links between the energy domain and the way of life - for example, the use of a particular amount of steel per head of population, the benefit associated with a particular form of housing or means of transport, and so on - that we talked about previously in connection with scenarios.

If we consider just the link between the final energy demand and the investments, we come back to the sectorial models of the electricity producers.

Inclusion of the whole set of processes involved in economic growth leads to the most highly developed studies involving inter-industrial (Input/Output) exchange matrices used to investigate the reciprocal impact between the energy sector and the other sectors of economy.

Mathematical techniques used. There are two major categories of modes: simulation models and optimization models.

(a) Simulation models, as their name indicates, simulate situations which are described in mathematical language and represent:
   (i) The results of a statistical analysis. In this case correlation and regression are the most commonly used basic techniques. For example, consumption of one type of energy can be related to the level of the GNP, the price of that energy or the price of competing forms. Regression and correlation techniques are used on past data to deduce "laws" for simulating the future.
   (ii) The description of physical relations, such as the law governing the energy required to heat a building under particular climatic conditions, or again the models describing the nuclear fuel requirements for different development policies.
   (iii) Hypotheses which it seems "reasonable" to formulate in a particular way. The world model constructed by Dennis Meadows for the Club of Rome, which served as a basis for the well-known study on "The Limits of Growth", involves numerous relations of this type. Although heavily criticized by a number of university economists, these models can, if used with care, be of much more value in fore-

casting than statistical extrapolation in which the future only appears as an extension of the past.

(b) Optimization models essentially rely on linear, non-linear and dynamic programming. Game theory, which would seem to be well adapted to the competition existing between the various forms of energy, as as far as we know hardly ever been used.

We shall not discuss linear (or dynamic) programming here since this could in itself form the subject of an entire lecture. (A summary of these various techniques can be found, for example, in "Overview of techniques and models used in the energy field", by J.P. Charpentier, IIASA, 1975). However, the objective function envisaged for these models is worth a little attention. The majority of the models use the minimum total present-value of expenditures (investment plus operation and management costs). The more sophisticated models introduce probabilities (often subjective) and then minimize the mathematical present expected costs.

It is quite clear that this minimization of outlay, however important, represents just one of numerous criteria used by investors. We must therefore be aware that whatever the quality of a model, the presentation of it in an optimizable form will restrict the reliability of its results to the value accorded to the single optimization criterion employed.

No general methods for multi-criterion analysis are yet available. Some "heuristic" projects have been developed but their field of application remains restricted to very particular situations. It seems, therefore, that optimization is very often only used for "the self-satisfaction of academic analysts" who enjoy playing God and who forget that a model is not in any way a decision-making tool but simply one means among others designed to allow them to test the consequences of some particular decision or scenario. As M. Lesourne, ex-Director of the Inter-Future Centre of OECD, said recently in a lecture "...we must bear in mind the fact that many system analysts, despite what they claim, are only interested in the study of the mathematical reactions of a system which will allow them to play the decision-making game. In practice, however, they probably under-estimate the fundamental problems in this decision making ...".

CONCLUSION

Among the energy forecasting models briefly described above, which seems the best adapted to the needs of the present situation?

It is clear that simple models based on extrapolations of past tendencies or on historical analogies cannot deal with the discontinuous changes characteristic of our time. Recent events have given man a harsh reminder that history is the study of events which never repeat themselves and it is probable that the regularities in growth rates observed during the past thirty years will have been only a transitory phenomenon.

Optimization models offer a series of important tools for the analysis of well-defined sectors of the energy economy but their reliability decreases very rapidly with extension of their domains in space and time. Moreover, their methodological complexity leads sometimes to an irrational allocation of effort between processing of the data and critical evaluation of their validity.

It is probably the simulation models based on hypothetical scenarios which offer the best chance of success. The combination of judgement in the choice of realistic hypotheses and the power of the computer in the study of their consequences sometimes makes it possible to escape the unpleasant dilemma so well expressed by the philosopher: "All that is simple is wrong, but all that is complicated is useless".

USE AND ABUSE OF ENERGY MODELS

D. Finon

Institut Economique et Juridique de l'Energie
Université des Sciences Sociales de Grenoble
France

ABSTRACT. The use of energy models in a planning process must be based on a good communication between modeler and decision-maker. A model is only a tool intended to improve understanding in order to assist decision-making. Its use necessitates a sound understanding of the underlying mechanisms, of the hypothesis of the methodology and of the validity of data. The principles of a cautious use of sectorial optimisation model are detailed in the second part of the analysis.

As a result of what has come to be called the energy crisis, energy consultancy to large government departments and energy firms has developed into a new source of profit in Western countries in the last ten years or so. A great number of consultancy firms, university institutes, international institutes, and former nuclear research teams have gradually made a place for themselves in the various national markets. The most open of these have proved to be the American and German markets; unlike in France or the U.K., the government departments and energy firms are accustomed to making frequent use of expert opinion from outside. Expertise has in some cases been based on the use of various types of models (simulation, optimization, macroeconomic formalization), the application of which cannot always be said to have been appropriate.

It was inevitable that these Western experts should, sooner or later, turn their attention to the developing countries, some of which were faced with particularly critical energy situations, amplifying the contradictions in their economies. The action of the World Bank in this field was decisive, for from 1977 onward

H. Neu and D. Bain (eds.), National Energy Planning and Management in Developing Countries, 253–269.

it made a minimum of energy planning one prerequisite for granting financial aid to energy projects. A swarm of consultant groups, armed with variously well-devised models, buzzed excitedly about this tempting new cake. Their models were, in most cases, merely adaptations of models designed for the study of energy problems in the developed countries: macroeconomic energy-economy models, extrapolation models of demand, optimization models and so on. Even in the developed countries, these models are hampered by intrinsic limitations and can be inserted into the decision support process only with great difficulty, because the communication between modeler and decision-maker is far from perfect. No wonder, therefore, that importing a foreign expertise, based on modes of reasoning which take no account of the underlying mechanics of the technological and economic systems in developing countries, has often proved useless or even harmful, or that the experts' reports have frequently ended up as decoration in ministers' offices.

As a first stage, we shall attempt to establish some general principles which should govern the use of models, in order to prevent their misuse; we shall then look more specifically at optimization models, which seem often to be proposed as part of consultancy programmes.

1. HOW CAN THE ABUSE OF MODELLING BE AVOIDED?

1.1 Misuse of Models

A model is based on a representation of reality, but must never in any way be confused with it. The map is not the territory. This implies that there are necessarily limitations on its uses, due both to the model's underlying assumptions and to the quality of the data used. The results can have no significance in themselves except in relation to those assumptions and those data.

The design of a model will depend on its function: the distinction is made here between cognitive models (models meant to improve a knowledge of reality) and decisional models (models embodied in a decision process in order to help a person or an institution to make the best possible decision).

In the energy sector, the great rigidities in the structures of production and consumption, and the long periods required for creating new plant, make it necessary to plan for very distant horizons. This implies that the bases of models must be suited to this precise temporal specification; in this respect, methods of extrapolation from the past provide very scant information and are incapable of integrating new information in situations of sudden change or total breakaway.

These methods blindly reproduce a given movement, as we could anticipate the way taken by somebody in the snow by the footprints. They are more suitable for cognitive interpretation of past developments, or for short-term forecasting, than for investigation of the long-run future.

It is clear that the first function of an energy model is as aid to the planning activity in different ways:

- Analysis of system behaviour: impact analysis of different endogenous and exogenous events ('what happens if?')

- Analysis of system controlability: limits of the manoeuvrability of the system, considering the actual set of decision variables available to a decision-maker ('what can we do?')

- Analysis of policy alternatives: to generate different policy alternatives with respect to generally formulated goals concerning the future state of the energy system ('what shall we do?')

Models are misused whenever the attempt is made to overshoot this function, i.e. whenever the application involves ignoring a model's methodological limitations and going beyond its function as simply an aid to decision-making.

It may seem rather obvious to point out that an institution (government department, firm, etc.) or a social group uses a model in order to increase its influence on collective choices by bestowing a pseudo-scientific legitimacy on certain images of the future which conform with the interests of that organization or group. Schematically, we can say that the decisions that will be taken subsequently may be foretold in part from the forecast, from the manner in which the future is represented. And using a model gives a pseudo-objective content to a reasoning that is implicitly loaded with values, in particular with the values of an idealized Science and Technology; thus, a sociologist would observe that the energy models actually used in the decision-making process are those constructed by engineers; the models produced by economists, which are based, essentially, on econometrics, are most often still confined to the academic games of the young international community of energy modelers who endlessly express their disagreement concerning price-elasticities, functions of production with energy factors, and so on; there is plenty to keep them busy for many long years, especially if the engineers get involved.

The models we are dealing with here necessarily incorporate

engineers' representations of socio-economic reality. Let us take the example of sectoral optimization models, which we will discuss at greater length later. These are merely an extension of the programming models used for the management of complex technical systems, such as refineries or electricity production and transmission systems; for this type of application, they take into account a considerable number of parameters representing the complexity involved in managing these processes. But already they embody a highly integrated, highly centralized, highly rationalized representation of the production activity concerned. This is especially true of electricity models in which the whole stock of production plant is concentrated at one specific location, and the choice of transmission equipment is divorced from the choice of production equipment; necessarily, therefore, no allowance is made for the spatial heterogeneity of the electrical economy, as regards consumer locations, hydropower resources that could be developed because of their proximity to potential consumers, etc. If this shortcoming is an obstacle to energy policy design in the developed countries where space is now relatively homogeneous, what can be said when these electricity optimization models are used in developing countries? The Western promoters of such models may have the best of intentions; the fact remains that their action ultimately favours integrated electricity production by large plants, which means that it encourages developing countries to consider building nuclear power stations.[1]

The practice of applying optimization models to the energy sector is even more significant. It is an impoverishing influence, because it reduces the socio-economic system embracing energy supply and demand to its mere technico-economic reality. It ignores the diversity of local and regional conditions. It ignores the effects of domination by which large firms influence the choices. It refuses, as indeed does the engineer, to recognize uncertainties that cannot be probabilized; it rejects ignorance of the future, which means having to spread the risk, make provision for reversibility, etc. A sectoral optimization model reproduces industrial integration and the structural rigidities that accompany it. It relates to a purely operational outlook: an objective to be attained, means to be specified.

At a different level, the approaches that are less heavily influenced by the hidden values of the engineer, like the needs forecasting models of the MEDEE type (an engineering approach combined with scenarios), are still just as impregnated with the dominant representations of reality. They postulate a number of technological developments or concurrent policies (transport, etc) within a range of what are considered reasonable assumptions; they are based on extrapolations of strong trends with no major breaks, particularly in the area of social values.

Pointing out this social role and these implicit values will do nothing to alter the fact that models exist, for, with the large-scale development of computerized means of calculation, it has become an exciting prospect for futurological engineers to use models to study the phenomena in complex interdependence within the energy system. If, however, there were wider awareness of this social role, there would certainly be fewer absurd and useless models, and less mystification. In the worst cases of the social abuse of models, the intention to mislead higher officials, politicians or public opinion is so patent that the model-maker can no longer be defended on the grounds of naiveté.

This appears to be true of certain long-run optimization models in which there are severe constraints on the development of all energy technologies except breeder reactors. It is equally true of models in which the cost and performance data for the proposed equipment reflect the same intention to subvert the options while ignoring the huge uncertainties involved, of which past events have provided ample evidence.

Let us now move on to quite a different problem: the model as a market commodity. To come back to our initial postulate: a model is a tool intended to improve understanding in order to assist decision-making. But the marketing of models will be based first of all on symbolical applications, on the wish to justify rather than the wish to understand. The customer purchases off-the-peg know-how which is not always relevant to the situation for analysis.

Competition aggravates this weakness, for, with it, content becomes totally subordinated to the appearance of being scientific. The engineer's naiveté is now combined with the pursuit of monetary or institutional profit. But pseudo-scientific appearances are kept up, for the market functions on the sole basis of reputations acquired within the small community of model-makers who willingly present one another with certificates of scientificness.[2]

It is not my purpose here to give a type-classification of the "products" sold in the market. I shall dwell only on the caricature: the "black-box" model. Apparently there are firms of consultants which draw up pseudo-plans of development; after collecting some data in the field they adjust their models, which have almost always been designed for the study of energy structures in developed countries,[3] and supply the contracting State with a ready-made plan. Nothing could do more to foster scepticism and distrust. True, the international consultant is sometimes intelligent enough to adapt the application of the model to the specific characteristics of the local situation, especially where it is a matter of substituting commercial forms of energy for non-commerical forms. But the fact remains that such practices will prevent the emergence of local expertise in the field of energy planning.

Using "black-box" models is thus a denial of planning. It goes against any real planning process, which implies a political and socio-economic vision of the energy supply structure based on an analysis of development mechanisms, and which must involve the recipients of the plan (government departments, national corporations, subsidiaries of multinational firms) in its preparation. Models are used only to provide a better understanding of the areas of leeway open to the planner. True planning is thus the contrary of any "black box" process, for the decision-makers have to be in a position to understand the contents of the model and to compare it with their analysis of local problems. The ideal situation would be for models to be developed by local experts, possibly with assistance from foreign consultants; one advantage of a model is that its preparation forces planners to analyse and comprehend reality, by trying to locate the most important variables.

Another disadvantage of the current practices is that they may encourage local officials to blame the computer or the models for any failures in the energy programmes thus prepared.

Dwelling on such practices will not give a very favourable view of the prospects of applying energy models to developing countries. However, models do often prove most useful in improving the understanding of phenomena. We have therefore attempted to detail how the abuse of models can be avoided.

1.2 Principles for the Correct Use of Models

In the following tables we have specified the principles by which model-makers and their clients should be guided.

Models are misused:

1) when the role of the model is to provide scientific justification for decisions that imply veiled political choices (this situation is inevitable but it sometimes reaches very great proportions),

2) when the model is used as a mysterious black box, with no real dialogue between the model-makers and the decision-makers,

3) when the model replaces the planner, whereas it should exist only to improve his understanding of the phenomena,

4) when simpler and less obscure methods could be used. Models must not be thought of as the only suitable means for investigating the future.

Table 1

How can the abuse of models be avoided?

For there to be no abuse of models:

1) the implicit theoretical hypotheses must be made clear,

2) the level of detail with which the phenomena are represented must match the availability of data,

3) the fundamentally uncertain aspects of the parameters on which the decision is based must not be concealed,

4) the model used must be appropriate for the problem investigated, i.e. for the question which the decision-maker has to answer,

5) the model must be relatively simple and remain comprehensible to users,

6) in a word, fruitful communication must be established between model-makers and model-users.

Table 2

The prerequisites for constructing a model

1) Statement of the right questions (as a general rule models have always been specified before the questions to be answered)

   Sample questions:

   - How much leeway does the planner have with respect to energy consumption and production?

   - What might be the energy system's reaction if such-and-such an event should occur in the environment?

   - What would be the effects of this or that policy (reduced dependence, energy saving, transport, national development)?

   - What possible bottlenecks or contradictions might arise?

2) A sound understanding of the underlying mechanisms (development mechanisms, formation of energy demand, influence of supply structures, etc.),

3) Identification and assesment of the uncertainties of all kinds,

4) An understanding of the bases for the proposed method (simulation, extrapolation of econometric relations, optimization, etc.),

5) The existence of data and specifications at the right level of aggregation for the time horizon involved (for example, the more distant the horizon in a macro-economic model, the lower the level of disaggregation required for the parameters).

Table 3

The conditions of good communication between the modeler and the decision-maker

1) Ideal situation:
the decision maker is associated to the definition of the model

2) Ideal conditions:
the modelers must have an accurate understanding of the perceptions of the problem to be questioned by the decision-makers and their objectives. The client's hope is that the modeler's analysis will add to his own understanding concerning the problem area or specific question under study

the model users must understand the underpinning of the model and the limitations of the results. The model must be conceived in such a way that their understanding is not too much time-consuming.

3) The implicit hypothesis (behavioural hypothesis, theoretical hypothesis) must be clear
the data must be clearly visualized by the users-clients

the forgotten aspects of the model must be made precise.

4) An effort must be made in the explanation of the results. (relations with hypothesis, with data, with constraints, etc.).

Table 4

Notes concerning table 4

1) The intricacy of models often leads to resulting estimates which are very opaque to the user, considering the basis of the estimates. Scepticism is generally the consequence of eomplexity. An arbitration must be found between too much simplicity and the necessity to have a pertinent representation of the phenomena.

2) The normativity of implicit behavioural hypothesis is often the origin of great scepticism from decision-makers. There is from them a fear to see their modicum of power reduced or to find themselves obliged to modify their ways of thinking and choosing.

3) The reason for success of engineering approaches combined with scenarios to ensure an empirical consistency of exogenous parameters (MEDEE-type approach) is certainly the clearance of hypothesis and data for model users.

2. FOR A CONTROLLED USE OF SECTORIAL OPTIMIZATION MODELS

2.1 The Limits of Sectorial Optimization Models

2.1.1 <u>Simplified representation of energy process</u> (quality dimension, time dimension, space dimension ...). The representation of energetic chains must be the result of an attention carried on the necessity of not exceeding reasonable dimensions. Because of this, many aspects of the problem of energy supply, which are perhaps decisive in the establishment of future optimum supply, are neglected.

However, we could include energy uses in order to embrace the processes of consumption and the possibilities of substitution of energy carriers by the medium of the substitution of processes. But on the other hand, it seems difficult to consider

- the load curve of certain uses
- the thermodynamic quality of the energy resources or needs
- the geographical location of production and consumption activities - which are three key factors of the possible

organization of national energy systems. For instance, in order to include the sources of carriers of "low power density", it is necessary to assimilate each of the local problems to a typical problem determined from an empirical mean of technico-economic characteristics (mean period of sunshine, mean distance of transport, mean residential density, mean cost of distribution ...). Now the space and time dimension of the exploitation of energy resources and the supply to different types of area ranging from districts with sparse population to the industrial and urban megalopolis should be the key factor of an energy optimization approach.

a) a space dimension, since the local characteristics of the sources and needs will be the deciding factors of the local supply structure;

b) a time dimension, since the demand of certain economic agents is irregular and the ability to stockpile local sources/carriers is not always ideal.

Furthermore, many new energy chains can develop only through the intermediary or rigid infrastructures (electricity, gas, hot water supply systems) the economics of which can very often be appreciated only in relation to local conditions (proximity of the source, existence of a supply network, etc.).

In developing countries, the spatial dimension is crucial, because of the heterogeneity of the space. The best choice could not be defined by a unique centralized approach. We need regional or local studies where arbitration between national networks supply and local sources (hydroelectricity, biomass, small coal deposits ...) is preliminarily done on the basis of costs of centralized energy supplies.

Faced with these limitations, different options could be defined:

a) to "regionalize" the model, i.e. to interlink several regional submodels but, at the level of the region, the problem could be the same as at the level of the national system; moreover, the problem of data and of energy needs forecasting could be very important;

b) to obtain a crude global model which could be broken down, by a zoom effect, at the level of study of precise processes or of a given region whose interactions with the rest of the energy system would then be considered;

c) to obtain a succession of very detailed submodels demanding more sophisticated calculation tools and studying chains or particular processes, it being possible to link the submodels in a formal or informal manner to form a global model.

2.1.2 Simplified representation of the optimum of a national energy system. The decision making hypothesis: the optimization programme could be interpreted as the quest for the equilibrium of the energy market in a situation or more or less pure and perfect competition. But we prefer a dual interpretation: the model would be the formalized expression of the problem confronting a body responsible for drafting and implementing energy policies. And one of the conditions of optimality in the two cases is the alignment of selling prices with the marginal costs.

Reality is often far from this scheme; the strategy of enterprises, whether they be public or private, is not always compatible with what could be the public interest in the energy field. Their capability of expertise is often greater than the public administration's capability, and, it is not seldom that public interest is sometimes assimilated to the interest of certain firms which dominate the energy sector in various countries. (Development of energy systems took place around a concentration of the major activities in the hands of a small number of firms on either a national or international basis. In some oil countries, the state oil company is generally predominant). The results of the model could demonstrate the interest of development of new energy sources or carriers; but large enterprises would not wish to face the risk of reduction of their autonomy or of their potential development. The inertia of the structures has to be analyzed just as much in relation to the organizational forms created around the structures (large public or private groups, controlling administration, framework of regulations, etc.) as in relation to existing plants, to future technologies and to the skills and knowledge developed around these structures.

So in the model it appears crucial to define certain constraints of development applied upon certain technologies in order to keep the results credible. But this problem gave evidence that energy policy is not only the result of a technico-economic optimization, but also the result of confrontation of varying interest.

The oversimplification of the objective function of the planner in a context of deep incertitude: generally speaking, the objective function of an optimization model is the cost of energy supply which has to be minimized in order to determine the sectorial optimum for the whole.

It would be naive to think that optimization of the national energy system is perfectly separable from optimization of the remainder of the economy. Energy programmes are tied to much more general political decisions, such as economic and industrial policies, national independence. Some elements could be more or less faithfully reintroduced into the model either in the form of constraints or in the form of social costs in order to complement or moderate the simplistic objective of minimization of the discounted cost of long-term supply: limitation of oil dependance, outflow of currency, possibilities of financing of investments, environmental impacts, etc: for a country in a situation of dependance. On another ground, the essence of the key factors of an energy policy rests, in fact, on a profound uncertainty. The level of future needs, the fossil and fissile fuel resources/reserves, the relations between countries importing and exporting energy, the role of large international firms, the energy policy of the United States, the feasibility and cost of new technologies (nuclear reactors, coal gasification, solar power, hydrogen, etc...), the transfer of technologies are all subject to deep incertitude. The fact that the model rests on a hypothesis of perfect forecasting prevents it from providing a satisfactory reply to the adaptability problem, although it enables a reference energy scheme to be described for well-defined and well-forecast situations of its environment. But, it anticipates future events a long time in advance, which enables it to set up an already appropriate structure when these events supervene. Many events are for the most part impossible to effect with an objective probability and the state of the system can only be inappropriate at the date of the eventual ruptures and its response can only be very slow, in view of the enormously long periods of development, construction and life of energy plants for production and consumption. The decision trees method would not allow major improvements to face this problem, because it needs to attach subjective probabilities to each event.[4]

The model is then incapable of ascribing an objective advantage to solutions which would improve adaptability or would prevent some irreversibilities. Energy systems which are too specialized -and for this reason too rigid - are condemned to be extremely vulnerable to sudden changes in their environment; the consequences of the rises in world oil prices of 1973-74, for dependent countries, illustrate this problem very well. An energy strategy based on the reproduction of a centralized and specialized energy system with two or three dominant vectors or chains which depend on either imported resources (coal, gas, petroleum) or on future processes based on technical gambles (gasification, fast breeder reactors) presents unsuspected disadvantages to weigh against the advantages of an alternative based on a diversification of energy sources and the use of carriers adaptable to different sources. This alternative would by itself give rise to a de facto stability because it would lead to a distribution of the risks and limit the vulnerability of the whole.

Our purpose is not to inititate here a comparison of the advantages and drawbacks of the one or the other process, but simply to emphasize that economic calculations in general and optimization models in particular are blind to this type of preoccupation, since they do not know how to treat satisfactorily the problem of choice in a future of complete ignorance. Moreover, they never assess the socio-political advantages of more neutral technologies that are more easily politically controlled, or that are more adapted to the technical capability of the country.

2.2 Use of the Model for the Study of System Behaviour and Controlability

These various theoretical and practical limitations of an optimization model ought never to lead users to frame energy policy proposals on the basis of the model's results. It would be naive to think that a model of this type would allow one to decide technocratically and once and for all:

a) the investments to be made at different periods in each

energy chain;

b) the distribution of the market between the various chains in the long-term;

c) reference prices of each of the products (with the help of the dual programme).

It would be a very rationalistic reaction to wish to determine a single optimal solution to such a complex problem in such an uncertain environment. But is it necessary to throw the baby out with the bath water?

The model must be used as the outcome of a global and detailed study of the political, economic and technological aspects of the energy supply, a study which is made by trial and error and by successive approximations between the whole and the parts, between a global approach and partial studies at the margin of the system or at the regional level. To do this, several interactive stages could be followed in succession:

1) Establishment of the geopolitical, economic, technological, social and institutional framework by a study of interactions of the national energy system with its environment (for example, a study of the world energy system, of the world resources, of the role of the USA, of the role of large companies, of the evolution of under-industrialized economies, of the national industrial system, etc...). It is at this level of study that the possible evolutions of the national economic system and their influence on the energy requirements would be investigated with the help of MEDEE-type models.

2) Description of the whole of the energy chains and partial studies of some of them. To do this, an evaluation of the resources and uses by regions and districts would have to be performed at the same time as the possibilities of development of new technologies were analyzed. We could proceed at this level to an initial selection of technologies to consider. This second stage should already enable us to outline the frontiers of the technical field of possible energy systems.

3) Placing the different energy chains in competition, doing so within each energy use with the help of a simple model within the framework of different consistent scenarios established at the first stage, and with reference (among other things) to geopolitical relations, economic development, the extent of institutional innovation and the predictable state of the new technologies.

An optimization model could be fitted into such an investigation process. The model thus has, as its principal contribution, the automatic and global formalization of the competition of the various energy chains: firstly, within an inter-time framework enabling time constants to be taken account of; secondly, within the coherent framework of the neo-classic theory of the partial equilibrium at the level of the energy market. The primary purpose of the sectorial problems is their interaction and their globality and to integrate the rigidities of the system since the specification cover plants (capacity, age, lifetime), and eventual political or sociological inertia. The fact of placing its use at the last stage of a process of studies which would enable the whole of the data and the economic, technological and political parameters to be established reliably leads to the running of "what if?" and "what can we do?" tests rather than to the normative search of an optimal energy structure ("what shall we do?" tests).

The elaboration of sets of contrasting hypotheses on the international price of petroleum, the cost of energy equipment, energy needs, innovations both technological and institutional depend on extensive studies preceding the use of optimization models. Each test of a variant may involve the variations of one or more sets of parameters of constraints. It would thus be possible to test different pairs of hypotheses on the price of petroleum and the cost of electrical plants or "local" technologies ensuring that there is consistency between the level of future prices, that of the needs and the institutional policies adopted to help in the development of local systems, all of which means setting fairly loose limits on the development of the latter.

The model also enables us to appreciate the global cost of adaptations necessitated by taking this or that decision: limiting the dependence on petroleum or energy at a given level of consumption for an importing country; developing a given equipment produced by national industry but uncompetitive with other processes implying importation from advanced countries; developing local systems by comparing the objective value of the base case with that of the variant considered. Should this value prove too high, it is always possible to pass to a new variant where the demands of the policy expressed by the constraint considered would be reduced[5]

The "what if ..." use of the model does, however, have some normative features, because the underlying theoretical foundations of the optimization remains normative. As a consequence, if the conjoint optimization of all the energy chains leads to results that diverge appreciably from the partial studies of the second stage or from the conclusions and recommendations in the matter of official energy policy,[6] it would be wise to ask the reasons for this divergence:

a) Might the official programmes result from partial optimizations of the agents of the different sub-systems considered that were not compatible with the general economic optimum?

b) Might the economic information implicit in the bases of the model (shadow prices) be too remote from the information really exchanged between the agents? Is it not necessary to adapt price and tariff structures?

c) Would the partial ignorance of the structural, institutional, or industrial constraints in the model be sufficient to explain the divergence between the results and the official plan?

d) Could not the simplification of the different energy chains and, in particular, of the transport and distribution activities explain in part the divergences between the model results and the official forecasts? In this case might it not be necessary to improve the studies upstream of the model?

The results from the model would give those who decide on optimal energy choices cause for thought, by making them assess the margin of manoeuvre they have in different structural and institutional contexts, by making them appreciate the significance of the rigidities of all orders and the commitment to measures aimed at reducing their impact in the event of the divergence between the optimum of the model and the official plan proving too serious.

In summary, sectoral optimization models must keep reasonable dimensions. They have to be used in a process of studies of global energy systems only following more specific studies of the environment of the system, of the different technical process and of the regional aspects of energy supply and demand. Partial studies of this type should permit the establishment of the data and technical parameters to be considered. Work on sectorial models is always in advance of these studies. It can only breed mistrust and scepticism on the part of the decision-makers which could be aggravated if the programme seems too complex.

Notes:

1. Have a guess: which of the international organizations has been the most active in promoting electricity optimization models in developing countries? The International Atomic Energy Agency.

2. It would be interesting to analyse the interaction between recognized consultants (private firms, foundations) and the academic or scientific world in the USA or Europe: the movement of former academics or nuclear research engineers into private consultancy; joint ventures; model copying etc.

3. I myself had the following surprising experience. In 1976-1977 I prepared for the EEC Commission a prototype for a very detailed optimization model extending as far as the industrial processes for the use of energy.

   At that time I had thought it advantageous to have one model covering the whole system of energy production and consumption. Later, a multinational firm of consultants assumed the role of prime contractor to the Commission for the model, co-ordinating its application to the European countries. Then I heard by chance about a consultant's report prepared by a joint venture, led by a very respectable American institute, for a country in South East Asia. Imagine my surprise when I found it to be my own monstrous model, slightly modified and used with its 64 energy uses. It even optimized the transportation sector.

4. The approach based on decision trees offers the advantage of being able to determine a sequential strategy which is a function of the achievements of the various uncertain parameters. But it scarcely takes into consideration the advantages of flexible and adaptable structures.

5. It is important to stress also that the test may apply to the study of the adaptation of the system to an objective to be attained in a given year. For example, oil imports may not exceed a certain amount: 50% of the total primary energy in 1990. Because of the rigidities of the structures, these have to be adapted in advance to the imposed reduction of imports of crude oil, the model determining the route as a function of the target so defined.

6. These results would, of course, be obtained by keeping the same hypotheses concerning the economic and political parameters.

THE MEDEE APPROACH AND ITS APPLICATION TO THE DEVELOPING COUNTRIES

B. Lapillonne

Institut Economique et Juridique de l'Energie
Université des Sciences Sociales de Grenoble

1. THE MEDEE MODELS

The MEDEE Approach

A critical assessment of the limits of traditional econometric methods for long term forecasting led the IEJE to develop in 1974 a new methodology for the forecasting of the energy demand called the MEDEE approach[1]. The general principles of this approach have already been explained in detail in several documents (1, 2, 3, 4); we shall therefore only outline it briefly in order to concentrate on a description of the MEDEE 2 and MEDEE 3 models.

Whereas traditional econometric methods merely explain the development of the energy demand in terms of macro-economic aggregates, the MEDEE approach takes into account quite explicitly the direct and indirect "determinants" of this demand at a disaggregate level: social, economic or technical determinants mostly measured in physical units (for example, the number of journeys per capita, steel production, household equipment in various durable goods). This approach, which is typical of the technico-economic forecasting methods, revolves around the following general pattern, which may be applied to all energy uses and consumer sectors: (Fig. 1)

From the basis of this pattern the energy demand is examined at the level of sub-groups or "modules", quite homogeneous in terms of the development of their energy needs. The development of the energy demand is then deduced from the development of the technical, social and economic factors or, more precisely, from the development of needs and activities as well as from technical

H. Neu and D. Bain (eds.), National Energy Planning and Management in Developing Countries, 271–298.

choices. A model is needed in order to render these detailed calculations more systematic and coherent.

Level of Satisfaction of Need or Economic Activity Level → Useful Energy Need (Reference) → Final Energy Demand

Physical and Technological Context ↑ (Useful Energy Need (Reference))

Substitution Coefficient among Energy Forms[2] ↑ (Final Energy Demand)

Figure 1.

A scenario is used to coherently define the development of certain important variables, the scenario variables, which are exogenous to the model. The structure of the scenario and its mode of construction are closely linked to the model used. Two sets of "model-scenarios" have already been developed: MEDEE 2 and MEDEE 3.

The MEDEE 2 model (5) was essentially conceived as a very fast and very flexible calculation tool; it focuses more on the calculation of the energy demand for a given year than on the development mechanisms of the energy demand determinants which have in general been included in the scenario. Thus, MEDEE 2, is first and foremost, an accounting model allowing for the automatic calculation of detailed energy balances per sector, per end use and per energy form from the projection of social, economic and technical indicators. It is simple, transparent and of great pedagogic value; its limitations come from the great number of exogenous variables there are to handle.

The MEDEE 3 model (6, 7, 8) at first answered much more ambitious goals. In this model, the energy demand is broken down as far as possible, taking into account existing information and the relevance of the disaggregation. The simulation of the development of energy demand determinants in the future is systematically included as often as this neither proves meaningless or over-simplifies reality. Thus, MEDEE 3 has been given a systematic method of scenario construction which reduces the arbitrary aspect and the risks of incoherence. Because it is largely based on a systemic analysis of the energy demand (it is, in fact, its formalisation) this model may be considered as a "systemic model".

If the detailed conception of a model is very much dependent upon the type of country for which it was originally designed, it nevertheless is true that the way in which the energy demand is grasped and analysed and the scenario method used are sufficiently general in their range to enable the study of the future energy demand of any country, especially that of semi-industrialised or developing countries. Furthermore, the general philosophy of the MEDEE approach has inspired prospective studies of the energy demand in various countries or regions.

The MEDEE 3 Model

MEDEE 3 was conceived to help in the elaboration of energy policies; it aims, in the long term, to associate useful and final energy profiles to major hypotheses concerning the economic and social development. In effect, it was meant to fit into the model system of the European Community Commission in between a macro-economic model (S.L.T.) and a supply model (E.F.O.M.)[3] However, the great specificity of the MEDEE 3 approach, both on the theoretical and conceptual level, at least in relation to the SLT and EFOM models, meant that in fact MEDEE 3 was designed and built in view of a mostly autonomous usage.

The disaggregation of the model is designed in order to study energy consumption categories (or more exactly units of energy needs/equipment) which are relatively homogeneous in terms of the behaviour of the actors and the dynamics of the development in energy needs and technical choices (means of transport, of heating) The choice of the level of disaggregation is a central phase in the construction of the model. It results from a systemic analysis of the energy demand which allows one to identify the consumer sectors and sub-sectors which are most relevant in relation to the goals at stake and the country in question. The level of disaggregation is obviously limited by the need to represent relevant groups, in terms of energy consumption, and about which the present scarcity of information may be remedied in the future. For example, in the MEDEE 3 models developed for the DG XII for EEC countries, a central role was accorded to the urban system and a more secondary role to the rural system (Table 1) unlike those models developed for Spain and Portugal.

As regards each homogeneous energy consumption category, or "module", the development of the energy demand is deduced on one hand from the development of socio-economic factors (individual mobility according to trip purposes, steel production level, for instance) and on the other hand from technological factors (production distribution according to processes, the importance of the various means of heating ...). These developments are intergrated within the model in quite a different way from one module to another, according to the relevance of the available information and the mathematical formalisations likely to be used: from the basis of econometric relations (for instance steel or cement consumption), of procedures simulating the choice of economic agents (for modes of heating and passenger transport), of discontinuous relations ("tabular functions"). The choice of explicative variables, in other words, the factors taken into account in each module to explain the dynamics of energy demand, is also the result of a systemic analysis; beginning with the needs, this analysis has allowed for the identification of all the factors likely to significantly influence the demand, by focusing on the factors reflecting the intervention of public authorities and other major decision-makers having direct or indirect bearing upon the demand (energy and transport companies, councils, public and private real estate promoters ...). The calculation of the development of the energy demand at the level of each module involves four main kinds of relations:

- relations calculating the energy demand for a given year, which appear as accounting identities linking, on one hand, the demand for energy products to useful energy needs and to the efficiency of the appliances and on the other hand, linking the useful energy needs to the levels of economic activity (for example steel production) or to the level of satisfaction of social needs (for instance the indoor household temperature);

- relations of simulation which allow for the development of the levels of economic activity or of social needs from one year to another, on a five-year basis (these relations are essentially econometric); furthermore, these relations will allow the development of specific energy needs from one year to the next. These relations are mostly linear;

- relations simulating energy equipment choices which rely upon the explicit representation of choice behaviours, involving two criteria, one which is purely economic and applied to the cost of the equipment and the other, which is social, applied to its externalities;

- and finally, relations of control, the function of which is to ensure that the data entered into the model is coherent and that the developments simulated by the model are relevant and compatible with the original hypotheses. In the case of an unsatisfactory control, these relations either bring about an interruption in the execution of the model, or a visual impression of warning messages.

Any technico-economic model which is disaggregate like MEDEE 3 calls upon a relatively significant amount of exogenous variables, technical and economic; upon which it is necessary to build hypotheses.

URBAN SYSTEM (5 classes of cities)

1 - Households (three classes)
- heating (crossing dwelling type/heating mode)

dwelling type	heating type
-multifamily/single family dwellings	-district heating
-dwelling age	-central heating (fuel, gas, coal)
	-electric heating (direct, heat pump)
	-solar heating
	-room heating (fuel, gas, coal, electricity)

- water heating
- cooking (gas, electricity ...)
- electrical household appliances
- urban transport (crossing household class/model/purpose)

mode	trip
-automobile	-home/work
-public transport	-others
-motor fuel	
-electricity	

2 - Tertiary sector (crossing sector/use)

sector	use
-education	-space heating
-commerce	-water heating
-offices	-electricial appliances
-others	

RURAL SYSTEM (crossing use/household class/dwelling type)
(same as above)

TABLE 1 - Breakdown into modules of the socio-economic system in MEDEE 3*

* each module results from the crossing of several criteria; version developed for EEC countries.

PRODUCTIVE SYSTEM

1 - Energy Intensive Industries (crossing sector/process/use)

sector	process	energy use
-steel	-process no.1	-low temperature
-aluminium	-process no.2	-medium temperature
-cement	-etc...	-high temperature
-pulp and paper		-specific electricity
-glass		
-ammonia		-non energy uses
-chlorine		
-ethylene		

2 - Other Industries

sector	uses
-food industries	-low temperature
-textile industries	-medium temperature
-other chemical industries	-high temperature
-etc...	-specific electricity

3 - Agriculture (three uses: heat, specific electricity, motor fuel)

4 - Construction (two uses: tars/motor fuels)

TRANSPORT SYSTEM

1 - Freight Transport
- short distance (<50km)
- long distance (crossing transport mode/goods transported)

mode	goods
-trains	-food and agricultural products
-barges	-building materials
-trucks	-other manufactured products
	-energy products

TABLE 1 (continued)

2 - Passenger Transport (crossing household class/mode/purpose)

mode	purpose
-car	-private short distance trips
-bus	-private long distance trips
-trains	-business trips
-planes	

TABLE 1 (continued)

Amongst these exogenous variables, some have a range of variation which is relatively restricted within the time span under consideration: either because they are characterised by a great inertia (demography, for instance) or because they correspond to trends over very long periods; such variables are considered as being purely exogenous, upon which one builds a central hypothesis.

The other exogenous variables, having a range of variation which is relatively open within the time-span under consideration (20 to 30 years) and the development of which will ultimately depend upon decisions taken by the main actors (public authorities, companies ...) are considered as scenario variables. The quality and reliability of the results depend of course upon the coherence and the plausibility of the scenario hypotheses. In order to attain this goal of coherence and plausibility, one bases the elaboration of these hypotheses on a real method of scenario construction, which is characterised by the creation of a hierarchical structure between all the scenario variables (on the basis of the relations evidenced during the systemic analysis of the energy demand). This hierarchical structure reflects the order of dependence between events, decisions and hypotheses on the relevant variables.

When building hypotheses, one begins at the higher hierarchical level by defining, first of all, the international setting chosen, by means of quantitative variables (concerning world production, for instance) and of "qualitative descriptors" identifying great events (the trend in the price of oil, for example); once this setting has been established, and within the limits determined by this very setting, one defines the national political, economic, social and technological setting, by using once again quantitative variables (essentially macro-economic and sectorial variables) and qualitative descriptors reflecting the major political choices or decisions. Up to this stage, the coherence and plausibility may not be formally guaranteed by any method: at most one can back them up with macro-economic model or ad-hoc techniques.

The next and highly specific phase of this method consists in the introduction of discontinuous functions or "tabular functions" relating the quantitative descriptors mentioned above to the variables that they may influence. These functions associate to each state or option of one or several descriptors (which specifies a particular decision or policy) one and only one value for the "explained" variable.

Having established the tabular functions for a particular country or region, the scenario writing will follow the following steps.

- direct formulation of hypotheses for the indicators which are not described by tabular functions
- choice of options selected for each descriptor (for example: option 2 for the energy policy, 1 for the industrial policy ...); the model will automatically identify the corresponding values for all the scenario variables defined by a tabular function.

The implementation scheme of MEDEE 3 is summarised in Figure 2. This model, in our opinion, stands out quite clearly from most of the technico-economic models, generally of the "accounting" kind developed since 1973-74.

- based on a systemic analysis of the energy demand, the model necessarily integrates the important relations linking the modules, relations of a technical, economic or decision-making nature; so that the structure of the model does not consist in the juxtaposition of largely independent parts, as is the case in most analytical models, but truly consists in an integrated whole of sub groups.
- the decision-making dimension, absent from traditional analytical models (at least in any explicit form), is one of the principal components of the MEDEE 3 Model, both as regards the representation of the actors' behaviour and of the relations among the actors.
- the scenario relies upon a structure which takes into account the interdependence of all these indicators and because of this goes well beyond a mere juxtaposition of hypotheses on exogenous variables which is often the case in the other technico-economic models.

The MEDEE 2 Model

The MEDEE 2 approach is quite clearly a simplification of the MEDEE 3 approach: it was essentially developed with three goals in mind[4].

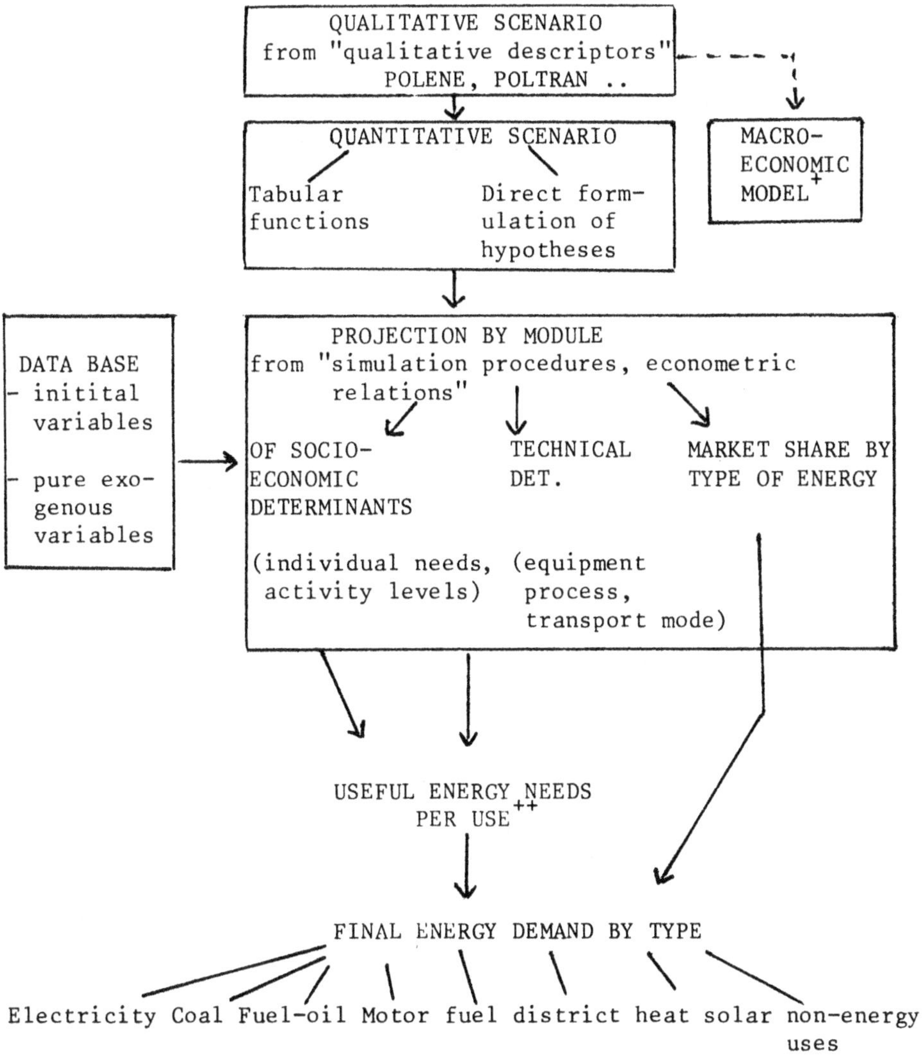

Figure 2. Scheme of implementation of MEDEE 3.

+ facultative use
++ for period of five years: 1980, 1985, 1990 ...

TRANSPORT	1. Passenger transport - urban/interurban -car -urban public modes, planes, trains (distinction diesel/electricity) 2. Freight transport - short distance transport: trucks - long distance transport: trucks, trains, barges 3. International transport: bunkers
HOUSEHOLD AND SERVICES	1. Households - space heating -dwellings built before and after initial year -single family/multi family dwellings with central heating, dwellings without central heating - water heating - cooking - air-conditioning - lighting - electrical household appliances 2. Services - space heating and water heating - air-conditioning - electrical appliances
INDUSTRY, AGRICULTURE	1. Branches+ - steel - chemicals - other intermediate goods industry - equipment and durable goods industries - consumer goods - construction - mines agriculture 2. End-uses+ - low-temperature (drying, space heating ...) - medium temperature - high temperature - specific uses of electricity - motor fuels - coke

TABLE 2. Breakdown into modules of the socio-economic system in the MEDEE 2 Model++

+ crossing branches/end-uses
++ standard version for industrialised countries

- to supply an accounting framework for the easy and rapid transcription of a growth scenario in terms of useful energy needs and the final energy demand per product; this scenario being defined in a normative fashion from a certain number of social, economic and technical factors; the model having to be capable of estimating the impact upon the energy demand of changes in the life style, in the economic and industrial structures, in the modes of transport and in the energy appliances and installations.

- to allow the evaluation of the potential of penetration, per use and also globally, of new energy techniques and sources: heat pumps, solar energy ...

- to represent a sufficiently general framework allowing for the application of its basic structure to any country.

The general structure of the model appears in Figure 3. The model is driven by the scenario; which consists of a certain number of macro-economic and energy indicators. The scenario variables are selected on the same basis as for MEDEE 3. The scenario is written by directly defining the values of these indicators for certain years of reference, 1985-2000, for example. There is no systematic procedure as in MEDEE 3.

The level of disaggregation adopted at the sub-model stage (sector/use) was selected in terms of its ability to answer, with a maximum of relevance, the questions asked. Table 2 indicates, as an illustration, the different modules considered in the "standard version" of MEDEE 2, developed for industrialised countries. Obviously it is quite easy to adapt to the conditions of each country by resorting to a more aggregate level; or by adding new uses ... Thus, in its adaptation to developing countries, the heating section was withdrawn or simplified, other uses were added such as pumping, irrigation, traditional lighting and a distribution between urban and rural population introduced[5].

As regards each particular end-use, the energy demand is estimated in two stages:

- the calculation of the useful energy need from the basis of a defined level of activity within the scenario or from an econometric relation (goods traffic in ton per km, agricultural activity ...) or from a level of social needs satisfaction, generally defined in the scenario (cooking, lighting, hot water, or individual mobility needs ...)

- the calculation of a final energy demand per type of energy (electricity, biomass ...) associated to each need from the basis both of the substitution coefficients

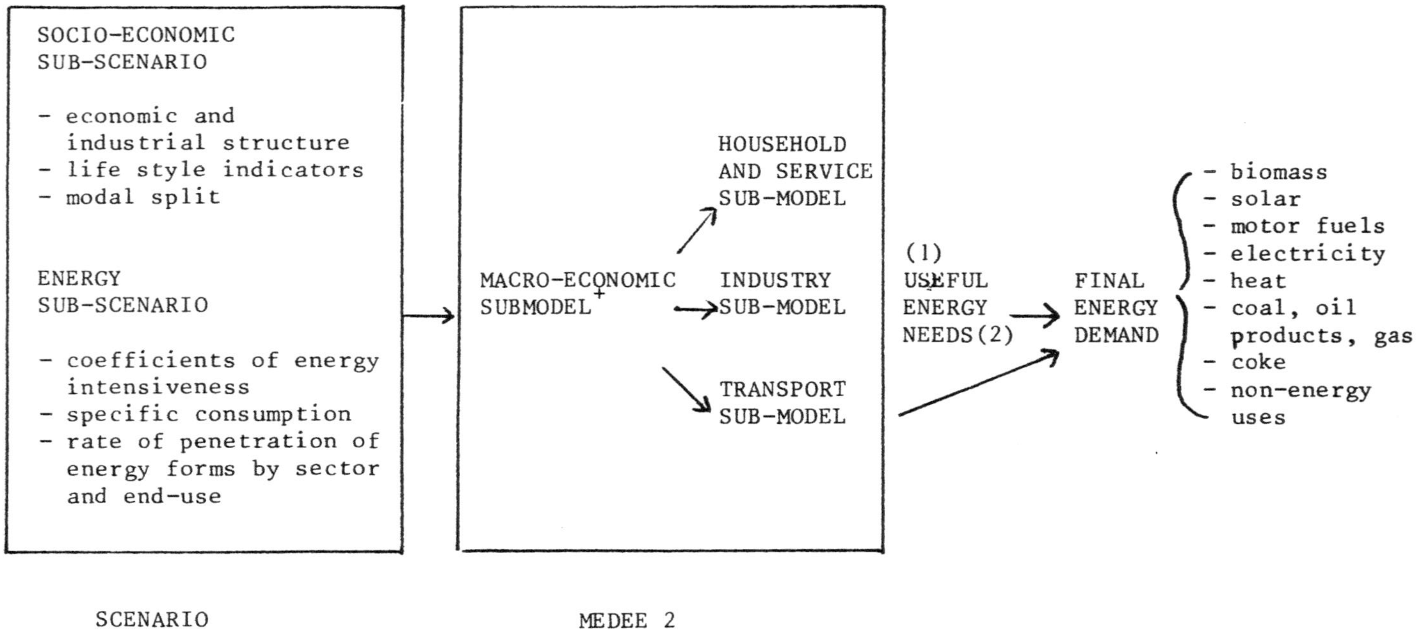

Figure 3. General Structure of MEDEE 2, of its Inputs and Outputs

+ facultative use

(1) substitutable uses
(2) specific uses

between energy forms, defined for each kind of equipment, and of the penetration rates of each energy form defined in the scenario as characteristic of the chosen energy policy (percentage of needs satisfied by each energy form); the concept of substitution coefficient, broader than that of end-use efficiency, which is purely technical, includes the behavioural changes linked to the shift from one energy form to another (for example, from wood stoves to LPG cookers).

The implementation of the MEDEE Models

We shall refer here to the numerous applications and adaptations of the MEDEE approach (see bibliography). As a conclusion we shall merely recall the different phases involved in the application of the MEDEE models, from the composition of the data bases to the construction of scenarios, emphasising the major difficulties likely to be encountered.

There are three kinds of data necessary for the application of MEDEE 2: firstly, socio-economic data, characterising levels of economic activity or levels of need satisfaction, followed by "technical coefficients" (specific consumption of equipment, of modes of transport ...) and finally data concerning equipment stocks per energy form (percentage of the market per energy form for cooking, drying ...). The technical coefficients are probably the most difficult data to assess but present the advantage of being relatively stable from one country to another, if only because much equipment or many of the techniques used are the same in a great number of countries and in a first stage transposition among countries may thus be envisaged; the other two kinds of data are more specific to the countries or regions studied. Socio-economic data involves a number of traditional indicators used by the U.N. or the World Bank, to characterise the level of development of countries, and generally may thus be quite clearly identified. Data concerning market shares is certainly more difficult to collect and requires the use of market surveys or specific studies (often undertaken by energy producers).

In addition to the above mentioned data, the application of MEDEE requires more detailed information - which is thus harder to find - and a greater knowledge of the development mechanisms of the determinants of the energy demand - thus this necessarily entails specific studies for each kind of country.

Second aspect of the implementation of the MEDEE models: the relations. Within MEDEE 2 most relations are accounting identities thus directly adaptable to any country: the other relations, of the econometric kind must be suitably adapted in each case.

The application of MEDEE 3 is, on the contrary, much less simple. Indeed, as we have seen, this model involves several kinds of relations, a certain number of which, simulation relations in particular, must be adapted to each individual context: for example, relations simulating energy substitutions are quite specific to the countries for which they were designed.

And, last but not least, the scenario. No systematic procedure exists in MEDEE 2 and the construction of the scenario requires the definition of the development of each scenario variable, thus involving the risk of incoherence. The results being directly linked to the scenario, and obviously to the quality of the data -and being meaningful only in terms of the scenario -it is important to work with relevant and coherent scenarios ; in order to do this, the scenario writer will, as far as possible, have to include in his scenarios the conclusions of other prospective studies on the habitat, the modes of transport as well as the existing plans and sectorial programmes. In any case, the use of MEDEE 2 presupposes that the user is well acquainted with the social, economic and technical problems related to energy: its use cannot and must not be of the "press button" kind.

Though more complex in its adaptation, MEDEE 3 has the advantage of being easier to use and of requiring less competence - the competence being especially needed for its adaptation.

2. APPLICATION OF THE MEDEE APPROACH TO THE DEVELOPING COUNTRIES

The MEDEE approach has aroused a certain amount of interest in the semi-industrialised countries, either because the general philosophy behind the approach has been adapted to specific countries or regions (Mexico (24), Algeria (25), Catalonia (26), Third World regions (27)) or because the MEDEE 2 model has been applied in an adapted form to generate energy demand forecasts (Brazil (22), Algeria (25), Portugal (15, 16), World Regions (23, 27)).

This success which is due more to the type of approach chosen, the technico-economic approach, rather than to the MEDEE approach itself, may be explained by the inability of the methods traditionally used for energy demand forecasts to take into account a certain number of crucial problems and moreover to provide answers to the questions which arise. The prospective tools of the technico-economic kind, however, by the disaggregate nature of their approach per use and the introduction of scenarios to describe the future enable (28):

- the visualisation of the existing margins in the energy/

GNP relationship, by allowing the impact of choice upon the demand to be taken into account in the transport, dwelling or industrial sectors; in terms both of the development trends or the energy savings;

- the identification of the demand in the modern sector, group of activities, regions, or social classes reproducing the western consumer modes, and the demand of the "traditional" sector within which the mere satisfaction of basic needs is predominant;

- the assessment of the development potential of each energy form by means of a precise evaluation of the possibilities of adaptation of the supply/demand (by paying particular attention to the quality of the energy required) in order to assess the range of possibilities for and by the valorisation of local energy forms, whether they are renewable or not;

- the acknowledgement of the relationship between the penetration rate of classical commercial energy forms in the traditional sector and the growth rate of the needs in this sector.

The MEDEE models, in their present state and such as we have described them previously, are ill-adapted to the reality of Third World countries. The MEDEE approach does strike us, however, as quite relevant, and on the basis of the above-mentioned studies, in particular those of the CIRED, we shall thus explain how a new model based on the MEDEE approach could be designed for Third World countries (29).

The implementation of the MEDEE approach involves three main steps:

- partitioning of the system i.e. disaggregation of the total energy demand into relevant homogenous end-use categories or modules (e.g. cooking, transport);

- analysis of the energy demand determinants i.e. systemic analysis of the social, economic and technological factors determining directly or indirectly the long-term energy demand evolution within each module, identification of their relationships, and finally organisation of all determinants into a hierarchical structure from the macro to the micro level, showing how the "macro determinants" (e.g. land use planning, agricultural change, industrial growth or energy policy) affect each end-use category;

- construction of a simulation model and of an associated

scenario structure.

The partition of the economic system into homogeneous modules will depend upon the objectives pursued by the modeller, the data availability and the characteristics of the country under study. The level of disaggregation should in particular make it possible to take account of the effect on energy demand of the pattern of development adopted by the country and to resituate the dynamics of energy demand around certain fundamental axes related to the characteristics of underdevelopment:

- the evolution of the rural sector: role of agriculture in economic growth, relative weight of food and export crops, agroindustrialisation and modernisation of techniques, replacing of human labour, monetarisation of trade, development of poor regions, extent of rural exodus ...

- growth of the industrial sector: type of insertion in the world economy, choice of international specialisation and structure of foreign trade, development of heavy industries (steel, basic chemicals, non-ferrous metals, pulp ...), link with consumption model, type of technologies adopted...

- urbanisation process: development of conurbations, location of industry, statute of city suburbs, housing policies ...

- developments in the transport sector which will themselves be predetermined by the orientations of industrial and town and country planning, by the capacity to set up infrastructures and by the role played by Western modes of consumption.

The MEDEE method opts for a breakdown into systems: industrial system, urban system (urban households) and transport system. This breakdown provides a better representation of the urbanisation phenomenon than the classical breakdown into sectors. The modern/traditional distinction cuts transversely across this breakdown. We shall explain below how the disaggregation into sub-systems and homogenous modules could be carried out [6].

Agriculture (and fishing)

The main uses of energy to be considered in this sector are mechanical energy for irrigation (wind energy, diesel engines, animal energy, solar pump ...) and traction (e.g. fuel for tractors, animal energy), propulsion energy for fishing boats (fuel,wind energy ...) and perhaps, in certain countries, the energy used for the conservation and drying of agricultural products (wood, fuel, solar energy ...). Because of the comparatively small share of this sector in total commercial energy consumption (generally less than 5%) a fairly rough breakdown should be sufficient. The development of energy requirements in agriculture (excluding fishing) will be closely linked to the development of farming technology (mechanisation, irrigation, use of waste ...) and thus it would appear expedient to distinguish initially between two sub-sectors: traditional subsistence farming and modern intensive farming. This breakdown will also be relevant for evaluating chemical fertilizer requirements, and therefore the intermediate consumption of energy associated with their production in the country under consideration. Moreover, the identification of sectors with homogeneous trends in agricultural production could lead to a three-fold distinction in modern agriculture between three sub-sectors whose growth dynamics are quite specific: export crops (sugar, coffee ...), energy crops and other crops. In certain countries, it may be necessary to distinguish between regions in order to take account of climatic and geographical disparities, as well as differences in the availability of energy resources, whether it be biomass or conventional energy. We will return to this problem of regionalisation, since, of course, it is not particular to agriculture.

Industry

The breakdown of industrial activities should be operated on the basis of the following principles:

Identification of the main energy intensive industrial consumers (steel production, construction materials, ammonia and nitrogenous fertilizers, non-ferrous metal; paper...); analysis in terms of processes for this type of production in order to grasp in more detail the trends in specific energy consumption and the type of energy required (e.g. steel production on a gas, coke, electricity or charcoal basis), as well as the possibilities for recycling waste (scrap iron); grouping of the other industrial sectors in a few major categories, which make it possible to characterise accurately the nature of industrial development (e.g. agricultural and food industries, capital goods industries..) The choice of the sectors to be examined will depend on the country, its level of development and the availability of agricultural and mineral resources. This choice should make it possible in particular to simulate alternative scenarios on the international division of labour, the strategies of valorisation of national resources and the extent to which the national economies depend on the international market (import substitutions, choice between export and home market oriented industries, international subcontracting): in short, scenarios on the various patterns of economic and industrial development. In order to obtain a more precise idea on the possibilities of energy substitution, and their impact on overall demand because of differences in efficiencies it would seem necessary to introduce a breakdown according to major uses: specific uses of electricity (electrolysis, electronics, small engines), other mechanical uses, thermal uses according to the different temperature levels, in order to analyse more accurately the possibilities of developing solar energy and cogeneration (low, medium and high temperature).

Households

Most of the energy at present used by households in the developing countries is for cooking. The rest of the energy used is for lighting, and, in certain temperate or continental zones, for heating. Distinctions in this sector should also take into account the other energy uses which are likely to become more widespread if average incomes rise (domestic hot water, electrical household appliances, heating of premises, air-conditioning). Different social groups should be distinguished in order to take into account the social heterogeneity which may be observed in life styles and in the levels and modes of satisfaction of household needs. This breakdown into social groups makes it also possible to integrate the effect of changes in social structure on energy needs. These groups could be set upoon the basis of income or living standards (e.g. households living in the poor suburbs, high-middle and low-income households integrated into the market sector ...)

In large countries with substantial regional disparities which result in different degrees of access to commercial and non-commercial forms of energy (electrification or not in certain zones, difficulty of access to the distribution network of petroleum products, availability of the biomass), or with differences in climate which involve major distortions in heating or air-conditioning needs, a breakdown on the basis of major regions is necessary. Thus the disaggregation should be operated in several ways: end-uses, social groups (including urban/rural distinction), regions ...

Tertiary sector

This sector is never easy to define as it groups together a multitude of often heterogeneous activities. However, it is possible to distinguish between two major uses (specific electricity and air-conditioning). In countries where tertiary activities account for a substantial share of overall energy consumption, a breakdown of major types of activity could be operated (offices, hotels, commercial establishments and artisan premises ...). Concerning commercial establishments and artisan premises, an urban-rural distinction could be made, if it is necessary to take into account the conditions of access to energy resources (e.g. electrification or not in the zones under consideration). For those countries where air-conditioning may become important, an approach based on climatic zones should be considered.

Transport

The distinction, traditionally used in the industrialised countries, between the transport of goods and people remains relevant although some of these activities are covered by vehicles which may be used for both. For the transport of people, an approach based on the mode of transport, purpose and household category, makes it possible to characterise trends in individual mobility, in substitution between modes (foot/bike/car...)and in the mode/mobility relationship (access to such and such a mode of transport generating new trips). A distinction between trips within towns and trips between towns or from town to country, whenever possible, is necessary in that it is then possible to deal with homogeneous areas of competition between modes of transport.

For goods transport, it will be possible to make distinctions on a mode and product basis. Long- and short-distance transport can also be distinguished (human or animal conveyance); and long-distance trips can be examined on the basis of major axes, in order to take into account existing infrastructures and the possibilities of substitution between modes of transport. On this level, the elements of heterogeneity between different regions will be dealt with in the scenario in order to determine future requirements of transport between towns or regions. These estimates will take into account the present and future location of industrial activities and of the population. The transport requirements for agricultural products, or minerals will thus be related to the hypotheses concerning export.

Table 3 summarises the various modules which are arrived at in the most general case.

The analysis of the energy demand growth is carried out at the level of each module in order to identify the main factors acting upon the energy demand pattern and the interactions existing between these factors: factors affecting changes in individual needs or economic activities underlying energy demand; factors influencing changes in the mode of satisfaction of these needs or in technologies; finally factors determining changes in end-use equipment (e.g. cooking, appliances, vehicles). The analysis of these determinants should in particular focus on the role of energy prices on individual needs and technical choices, on the possibilities of technological changes as well as on the constraints to these changes, and finally on the influence of the main decision makers' policies on the factors acting upon energy demand (government, municipalities, large industrial firms, energy utilities ...). This system analysis is a central element in the model design. It should make use of the various case studies devoted to specific aspects of energy depand. With regard to geographical zones, to economic sectors or to types of uses which

Agriculture	- Fishing - Traditional agriculture - Intensive agriculture . export crops . energy crops . others	Propulsion energy for boats (fuel, wind ..) Mechanical traction energy (fuel, anim.en.) Mechanical energy for irrigation Conservation, drying
Industry	E.I.I. by product and process (steel...) Other industries (capital goods..)	Specific electricity Other mechanical uses Low-temperature thermal uses Medium temperature thermal uses High temperature thermal uses
Households	Crossing of three dimensions: Socio-economic classes, urban/rural, climatic zone or region	Heating Hot water Cooking Lighting Household electrical appliances Air conditioning
Tertiary	Offices, hotels, shops Artisan workshops (distinction between urban and rural)	Specific electricity Air conditioning
Transport	Transport of people: Crossing mode/purpose/socio-economic class/urban-rural Transport of goods: Crossing mode/product/distance	Fuel Electricity Coal etc...

TABLE 3. Breakdown of the energy demand of developing countries into homogeneous modules

are highly dependent on non-commercial energy, this analysis should mainly focus on the manner in which energy demand is modified by the system of supply and its development (choices of equipment and technological processes, move from non-commercial to commercial energies). For some sectors or countries, it may be possible to transpose the analysis made on industrialised countries (only if it is meaningful, of course). As an illustration of what should be such a system analysis, Figure 4 indicates the relation between the main energy determinants in the case of cooking needs for poor rural households.

The construction of a simulation model and of an associated scenario structure is carried out by simplifying the structure of the socio-economic system and grouping the determinants into those the evolution of which can be formalised through mathematical equations (endogenous determinants or variables) and those the evolution of which cannot be modelled, the exogenous variables on the one hand, the scenario variables on the other hand.

In its simplest form, the model may be reduced, in the same manner as in MEDEE 2, to an automated accounting system, carrying out additions and multiplications, and making it possible to transcribe a scenario automatically in terms of useful energy needs and finally energy demand by energy form: the simulation of the evolution of the economy and of the structures or choices of installations being integrated in the scenario. In a more elaborate form, the model will be able to incorporate a certain number of endogenous simulation procedures (behaviour of decision-makers, choices of equipment ...). (MEDEE 3 type model).

The scenario indicators should be organised in a hierarchical structure, based on the overall structure of the system, as well as on a decisional structure, characteristic of the country under study, which accounts for institutional and power relations between all decision makers. The scenario structure could be organised on three levels as outlined in Figure 5. In all cases, the scenario will play an important part; the relevance and significance of the energy forecasts will depend very largely on whether it is coherent or not. Thus the use of a structured scenario method which integrates the interdependences between indicators becomes of paramount importance. It is, for example, absolutely necessary to make sure that the assumptions concerning the determinants of the future energy demand of the rural populations are not self-contradictory: population growth, rural exodus, availability of non-commercial fuels, rate of electrification, type of agricultural development... In addition, the scenario should include the feedback effects of energy development on economic activity (e.g. capital or currency requirements); this may imply iterations between the scenario and the forecast results.

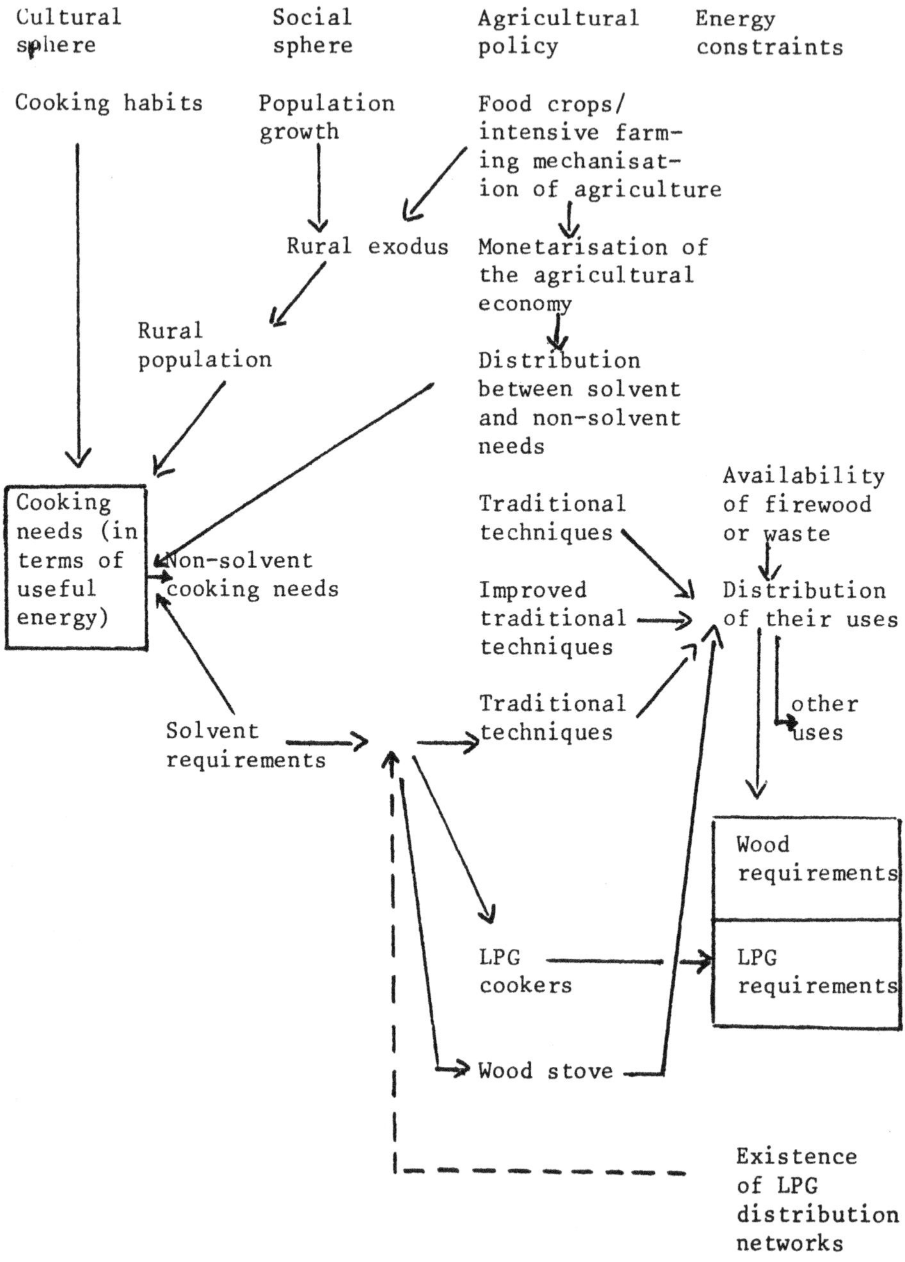

Figure 4. Cooking requirements in rural zones

First Level: Description of the Environment of the country

Assumptions about the power relation between developing countries and industrialised countries, about the strategy of multinational companies, the cooperation between developing countries, about the world energy supply/price of primary energy sources, availability and cost of nuclear technologies, about the world food production.

This level is intended to set up roughly the physionomy of the world economy in the characteristics which may have consequences on the energy demand.

Second Level: Description of the Economic Development Pattern of the country

a) Definition of the general policy goals: for instance limitation of the dependence on the outside, development of technologies adapted to developing countries or on the opposite, imitation of the development pattern of industrialised countries.

b) Specification of the content of the development pattern:

- demography — agricultural policy
- economic growth — foreign aids
  dependence on imports
  industrial policy
  GNP structure
  employment distribution
- distribution of the population into several classes of incomes: poor/others in rural areas
  rich/middle-class/poor
- urbanisation — rate of urbanisation
  distribution of the population into urban and rural areas
  distribution of the urban population between megalopolis, towns of more than 50,000 to 100,000 inhabitants and others
- transportation policy objectives
  degree of dependence on imports

The assumptions defined at this level are partly qualitative, partly quantitative.

Their formulation is made according to the assumptions of the first level

Figure 5. Scenario Structure of the MEDEE Approach for Developing Countries

NOTES

1. MEDEE: Model of the Energy DEmand Evaluation. This approach was developed at the IEJE from 1974 to 1975 due to the financial help of three energy companies: first of all the "Société Nationale Elf Aquitaine (SNEA)", later joined by "Electricité de France (EDF)" and the "Commissariat à l'Energie Atomique (CEA)". (1)

2. This coefficient includes both the differences in the technical efficiency of utilisation, and the behavioural changes (as regards the needs) brought about by the transfer from one energy form to another; this coefficient is defined in terms of an energy of reference, the corresponding need of which serves as reference (one generally adopts the most popular energy form in the use under consideration)

3. The MEDEE 3 Model was developed at the IEJE from 1977 to 1981 by B. Chateau, B. Lapillonne and I. Renzetti due to the financial help of the European Community Commission, Section Science Research and Development (D.G.XII): today it represents one of the basic elements of the system of energy models of this Directorate General. The EFOM Model (Energy Flow Optimisation Model) is an optimisation model of the energy system. The SLT Model is a long-term macro-economic model.

4. The MEDEE 2 approach was developed from 1976 to 1978 by B. Lapillonne at IIASA (International Institute for Applied System Analysis) within the framework of the Energy Programme. This programme, directed by W. Haefele aimed to outline world energy prospects, according to major world regions and in a far distant time-horizon. Within this context, the model served to project the final energy demand for each region. These applications carried out by A. Holzl, A. A. Khan and J. di Primio are presented in detail in (23).

5. A version of MEDEE 2 adapted to developing countries was developed for the IAEA (International Atomic Energy Agency) under the name of MAED: this model has already been applied to Algeria.

6. The Instituto De Economia Energetica (Bariloche, Argentina) and the COPPE (University of Rio de Janeiro) are proposing a similar breakdown to analyse the energy demand in developing countries; the philosophy is the same even if the detailed modules considered are not strictly identical; this work is carried out in the framework of a "cooperative program of studies on energy in the developing countries", sponsored by the European Commission (Energy directorate).

BIBLIOGRAPHY ON THE MEDEE MODELS AND THEIR APPLICATIONS

A. MEDEE APPROACH

(1) B. CHATEAU: The Methodology of Long Term Forecasting: Limitation of Traditional Methods and Proposals, in "Energy Systems Forecasting, Planning and Pricing", Madison (Wisconsin, USA), Institute for Environmental Studies, University of Wisconsin, February 1975, pp. 37-49.

(2) B. CHATEAU, B. LAPILLONNE: La prévision à long terme de la demande d'énergie: propositions méthodologiques, Paris, Editions du CNRS, 1977, p. 225, (Collection "Energie et Société)

(3) B. CHATEAU, B. LAPILLONNE: Long Term Energy Demand Forecasting: a new approach, Energy Policy, vol. 6, No. 2, June 1978, pp. 140-157.

(4) B. CHATEAU, B. LAPILLONNE: The MEDEE Models for Long-Term Energy Demand Forecasting, Socio-Economic Planning Sciences, 15, 1981, pp. 53-58.

B. MEDEE 2 MODEL

(5) B. LAPILLONNE: MEDEE 2: A Model for Long Term Demand Evaluation. Laxenburg, Austria, International Institute for Applied Systems Analysis, 1978, (RR 78-17).

C. MEDEE 3 MODEL

(6) B. CHATEAU, B. LAPILLONNE: Long Term Energy Demand Simulation in "Energy Model for the European Community", Guildford, Surrey, U.K., IPC, 1979, pp. 120-128.

(7) B. CHATEAU B. LAPILLONNE: L'Approche MEDEE 3, Grenoble IEJE, La Lettre de MEDEE No. 2, 1982.

(8) B. CHATEAU, B. LAPILLONNE: Long Term Energy Demand: Method of Constructing Scenario Linked with MEDEE 3, Brussels, CEE-DG 12, January 1979.

D. APPLICATIONS OF THE MEDEE APPROACH

(9) B. CHATEAU, B. LAPILLONNE: Les acquis de la méthode MEDEE dans sa formalisation et dans ses applications, Grenoble IEJE, La Lettre de MEDEE, No. 1, 1981.

(10) B. CHATEAU, B. LAPILLONNE: La Méthode MEDEE et ses applications Revue de l'Energie, avril 1982.

FRANCE

(11) B. CHATEAU, B. LAPILLONNE, M. ALINHAC, B. LAPONCHE, G. MOURY La demande d'énergie finale de la France à l'horizon 2000, Grenoble, IEJE, mars 1979, p. 77

(12) D.A.T.A.R.: Energie et régions: production et consommations; perspectives 1985-2000, Paris, La Documentation Française, (Collections Travaux et Recherches de Prospective), 1980.

(13) J. M. CHASSERIAUX, B. CHATEAU, B. LAPILLONNE: Un scénario de croissance sobre en énergie pour la France, Paris, Ministère de l'Industrie, (Collection "Dossier Energie et Matières Premières No. 21), 1980.

(14) B. LAPONCHE, P. VALETTE: Le modèle énergétique des Communautés Européennes: exemple d'application des modèles MEDEE et EFOM avec des données françaises.

PORTUGAL

(15) S. PAES, A. MARTINS: L'évolution probable de la demande d'énergie au Portugal, in Proceedings Conférence Mondiale de l'Energie, 8-12 septembre 1980, Munich, vol. 11, p. 179.

(16) C. NETO, J. CRAVO, A. de SOUSA, H. BAGUENIER, A. MARTINS, S. RIBEIRO, L. MATHIAS: A procura de energia en Portugal (cenarios alternativos), Lisbonne, Direction Générale de l'Energie, 1980, p. 208.

UNITED STATES

(17) B. LAPILLONNE: Long Term Perspectives of the US Energy Demand: Application of the MEDEE 2 Model to the US, Energy, 5, pp. 231-257, 1980.

EEC COUNTRIES

(18) B. CHATEAU, B. LAPILLONNE: Projections à l'horizon 2000 de la demande d'énergie finale de l'Europe des Neuf à l'aide du modèle MEDEE 2, Grenoble, IEJE, septembre 1978, p. 50.

UNITED STATES, USSR, EEC COUNTRIES, EASTERN EUROPEAN COUNTRIES

(19) Conseil Economique pour l'Europe: L'économie énergétique de l'Europe et de l'Amérique du Nord: perspectives pour 1990, Bulletin Economique pour l'Europe, vol. 33, No. 2.

(20) F. ROMIG: An Efficient Energy Future for the ECE Region - Symposium on Energy Conservation and International Cooperation, Bucharest, Romania, March 1981.

SWEDEN
(21) G. SAROS: Energy Demand Analysis: present structures and future changes in the Swedish households, service and transportation sectors, Stockholm, Swedish Council for Building and Research, 1979 (D. 12: 1979)

BRAZIL
(22) L.Tadeo Siqueira PRADO: A utilizaçâo do modelo MEDEE na availiaçâo da demande energia no Brasil, Estudos Economicos; vol. 11, numero especial, 1981.

WORLD (by regions)
(23) W. HAEFELE: Energy in a finite world, Cambridge, Mass. Ballinger Publishing Company, 1981.

OTHER COUNTRIES (publications in progress for the EEC countries, Spain and Portugal these studies are carried out with the Commission of European Communities)
ALGERIA (AIEA, SONELGAZ), QUEBEC (Ministère de l'Energie et des Ressources), SPAIN (Ministère de l'Industrie et de l'Energie), BELGIUM (Bureau du Plan), IRELAND (National Board of Science and Technology), UNITED KINGDOM (Science Policy Research Unit, University of Sussex), DENMARK (RISO National Lab.), ITALY (CNEN), GERMANY (IKE, University of Stuttgart).

E. <u>ADAPTATIONS OF THE GENERAL MEDEE PHILOSOPHY</u>
(24) CIDE, Groupe de Economica de Energeticos - La Evaluacion de Projectos y Metodologias de Demanda en el Sector Energetico Nacional, Mexico, I.D.F., October 1976.

(25) SONATRACH: Le Gaz naturel et les GPL dans la consommation d'énergie à long terme en Algérie, Centre Arabe d'Etudes Pétrolières, août 1978.

(26) Alario GASULLA: Modelo de Prevision de las Necesidades Energeticas a Medio y Largo Plazo: Applicacion a Catalunya, Barcelona (Espana), Instituto de Tecnicas Energeticas, February 1980

(27) DARTOIS, HOURCADE, LAGANDRE: Stratégies énergétiques alternatives pour le Tiers-Monde et besoins d'investissements, Paris, CIRED, 1980.

(28) HOURCADE, LAPILLONNE: Prospective de la demande d'énergie dans le Tiers-Monde: problèmes de modélisation, 2ème Journées Internationales d'Etudes "Econométrie et Crise de l'Energie", Aix-en-Provence, 10-11 decembre 1981 (forthcoming)

(29) FINON, LAPILLONNE: Long Term Forecasting of Energy Demand in the Developing Countries; Operations Research, 1982, (forthcoming).

PRICING, FISCAL INCENTIVES AND LEGISLATION AS ENERGY MANAGEMENT AIDS

Ramesh Bhatia
Institute of EconomicGrowth, Delhi-110 007, India

Demand for commercial energy grew rapidly in developing countries during the 1960s and, although more slowly, during the 1970s. Estimates by the World Bank /1/ show that in oil importing developing countries (OIDC) commercial energy consumption would grow by 6.3% per annum during 1980-90. Net oil imports in these countries are expected to increase to around 7.6 million barrels a day (mbd) as compared to 4.5 mbd in 1980. The striking feature of the projections is that, although their oil production is expected to nearly double by 1990 and their energy supplies from other sources will increase, the import expenditures of the oil importing developing countries in real terms are also expected to double, i.e. from 49.3 billion US (1980) dollars in 1980 to 111 billion US (1980) dollars in 1990.
These rates of oil consumption and import growth could be significantly reduced if both energy conservation and oil production measures were carried out to their full potential. Besides, maximum efforts could be made to encourage the use of new and renewable energy sources for oil in industry, transport and agriculture.

The World Bank report cited earlier has estimated that "developing countries could achieve substantial benefits by a broad program of demand management designed to increase the efficiency of energy use". The energy consumption in 1990, presently projected to 30.6 million barrels of oil equivalent a day (MBOED) could be reduced by as much as 4.6 MBOED, i.e. by about 15%. As may be seen from Table 1, more than

H. Neu and D. Bain (eds.), National Energy Planning and Management in Developing Countries, 299–316.

half of this saving is to be achieved from retrofitting and technical improvements in industry, electric power generation and transport. Interfuel substitution and scale economies, especially in households and industry, are expected to provide about 22% of the potential savings.

Various measures considered for demand management may be classified into three categories: Conservation, Interfactor Substitution and Interfuel Substitution. Conservation refers to a reduction in consumption of energy (or a particular fuel) which is the result of a reduction in the economic activity using energy. Interfactor substitution covers all those efforts where inputs (capital, skilled labor, etc.) other than energy are used as substitutes for energy. The third category of measures involve interfuel substitution where the emphasis is on substituting one form of energy (e.g. oil) by another form of energy such as coal, natural gas, biogas, electricity, etc. It is important to make this distinction among various types of measures because each of them has different implications in terms of requirements of scarce inputs such as capital, foreign exchange, skilled labour and management. Besides, relative prices of inputs other than energy become crucial determinants of the financial viability and acceptance of different measures.

A given strategy of demand management involving a mix of conservation interfactor substitution and interfuel substitution may use pricing mechanism or non-price measures including legislation as instruments. The new and renewable energy sources can play an important role if fiscal incentives are provided to the producers and consumers. Some of these issues are discussed in detail in the following paragraphs.

ENERGY PRICES

In a free-market economy, energy prices act as an allocative mechanism by providing appropriate signals to the consumers and producers. In a developing country, the role of prices may depend on a number of stated policy objectives and the special characteristics of energy demand and supply. Energy prices may have to be adjusted to meet the objectives of control or inflation, meeting basic needs of the poorer sections, regional development, environmental aspects, security of supply and balance of payments considerations. The issues of energy pricing become more complex in view of the role of the government as an owner, producer and consumer of energy resources/products as well as the authority to determine and administer prices.

Objectives of Energy Pricing

The objectives of energy pricing are related to the goals of macro-economic policy and goals of energy planning. The objectives of economic development would require that pricing policy should promote economically efficient allocation of resources during the planning period. It would require that the relative fuel prices should influence the pattern of consumption in the direction of least-cost mix of energy resources to meet future demands. However, there may be other socio-economic objectives which may suggest deviations from economically efficient prices. Some of the important objectives, as listed in Table 2, may be control of inflation, meeting the basic needs of a section of population, encouragement to domestic resource-use, security of supply, environmental considerations and profitability of public sector units. The relative importance of these objectives may vary from one country to the other and also over time. On account of the pervasive influence of changes in energy prices in economic activities, the governments seem to be concerned about large fluctuations in consumers prices. As discussed later, most of the important objectives of energy pricing can be achieved by adjusting the final product prices (through appropriate taxes/subsidies and may not require adjustments in resource prices or transfer prices.

Characteristics of Energy Demand and Supply

The special characteristics of energy demand and supply in developing countries can be briefly summarized as follows:

a) Traditional sources of energy (firewood, charcoal, crop residues, animal residues, human and animal power) meet a major share of the energy requirements, especially in rural areas. Although these resources are known to be important, the information base on the level and pattern of use of these resources is rather inadequate.+)

+) Useful further references in this important area would include studies reported in Hughart (1979), Knowland (1979), Briscoe (1979), Bajracharya (1979, 1980), Revelle (1976), Bhatia (1977, 1980), Manibog (1979), deLucia and Lesser (1980).

b) The level and pattern of energy consumption is determined more by supply constraints than by demand considerations. Consumers (households, industries) have to adjust their demand pattern according to availabilities of different fuels in different regions/periods even if their economic and technical preference may suggest fuel-mix.

c) User choices among alternative energy source may be determined by factors other than the relative prices of fuels adjusted for efficiency difference. +) Equally or more important considerations may be auxiliary costs of handling and storage, tax/subsidies on investment security on supply, uniformity of quality and the effect of fuel-mix on the quality of processed materials. Although significance of these factors would vary from one use to the other, it is important to consider that inter-fuel substitution may be affected by factors other than prices.

d) In many developing countries, the existing prices of capital, labour and foreign exchange may not necessarily reflect their "true opportunity costs" due to market imperfections and/or administrative controls (minimum wages, differential rates or ceilings on interest rates, import quotas, etc.). The governments may be concerned about maintaining a balance in relative prices of energy and non-energy inputs to avoid distortions in the rates of substitution of scarce non-energy inputs for energy inputs, especially in the industrial and transport sectors. Given the distortions in prices of non-energy inputs (whether justified or not), fixing energy prices which reflect their "true opportunity costs" (or replacement costs) may, in fact, provide wrong signals to the consumers and result in substitution of these scarce inputs for energy which need not be necessarily desirable.

e) Governments exercise direct influence through the ownership of energy resources, processing facilities and/or price controls. In practically all developing countries, electricity generation and distribution is government owned. In oil and gas production, refining and distribution as well as coal mining, both public and private organizations operate, often side by side. Countries may differ on account of share of public sector units in sub-sectors, the role of foreign oil/ mining companies in total operations and the issues of

---

+) For details see Munashinghe and Schramm (1980), Desai(1980).

federal finance. All governments exercise some form of price control during production, refining and transport/transmission of energy. In addition to direct involvement, indirect influence of the government occurs through taxes, import duties, subsidies, market quotas, taxes on energy-using equipment and government regulated investments. This pervasive role of the governments along with their socio-economic objectives make issues of energy pricing rather complex.

Components of an Integrated Framework for Energy Pricing

One of the important aspects of energy pricing in developing countries is that prices of different fuels at different stages should be analysed in an integrated framework. Figure 1 outlines the scope of such an integrated framework where links among energy sub-sectors and different stages of pricing have been shown. The first stage relates to pricing of energy resources such as coal, crude oil/natural gas, hydro and renewables for evaluating investment options and financial planning. The important question here is whether the prices used for valuation of output of different energy projects should be based on the opportunity cost of importing/exporting that output (or its substitutes). The second stage known as transfer pricing is concerned with fixation of prices which are paid by the organizations who purchase these energy resources for further processing, e.g. price of crude oil paid by refineries or price of coal paid by electricity generation firms. The main question to be decided at this stage is as to how different organizations share the "economic rent" available from the use of exhaustible resources. The issues are further complicated by the fact that some of these outputs (e.g. natural gas, coal) are also sold to consumers other than energy units (fertilizer plants, households, industries, etc.). The third stage relates to producer's prices, i.e. prices which are received by refineries and electric utilities for their output. The important question here is the role of differential prices of petroleum products received by a refinery in determining its output-mix. The issues relating to relative profitability of different units in the energy sector may also be discussed at this stage along with transfer pricing. The final stage relates to prices paid by the consumers of final outputs (e.g. kerosene, diesel, electricity, soft coke, charcoal, etc.). The important aspects of taxes and subsidies for different energy products are discussed at this stage.

It may be emphasized that the above discussion of various stages is for the purpose of analytical convenience and may not be necessary in countries where all these sub-sectors/stages are not important. Besides, the nature of organizations involved in various stages may differ for each fuel sub-sector: a few refineries (perhaps, government owned) for oil products against a large number of private units manufacturing charcoal/soft coke.

However, it is important to analyse the pricing issues in an integrated framework+) as outlined above, because:

a) The substitution possibilities differ from one fuel to the other and the elasticities of response to changes in prices of each fuel also vary.
b) The price of a given fuel to the consumer can be varied by changing the prices at one or all of the above stages mentioned above. The impact of higher import prices of crude oil can be passed on to the consumers either by increasing transfer prices to refneries or by changing the excise duties on oil products or both. Since different options have different implications on government revenues and profitability of energy units, it may be necessary to analyse the detailed impacts.
c) A given objective of energy pricing policy may not require changes at all stages. For example, keeping prices of a few or all oil products at levels lower than import costs could be done by fixing low or differential excise duties and does not require fixing a low transfer price of crude oil.
d) It is not necessary that a higher price of crude oil will result in higher product prices if there is a cushion in terms of excise duties which could be modified to keep the final product prices at desirable levels. It is important to realize that most of the objectives of energy pricing can be achieved by adjusting the final product prices through appropriate taxes and subsidies. As shown in Table 1 adjustment of consumer prices is sufficient to attain the objectives of meeting basic needs of some sections of the population, control of inflation, environmental considerations. This shows that distortions in pricing of energy resources, transfer pricing and producers' prices can be avoided if it is kept in mind that these prices do not have to be used for the macroeconomic objectives. Thus, for example, optimum invest-

---

+) For a detailed discussion, see Munasinghe (1980). Also see Siddayo (1980), Government of India (1969, 1979).

ment pattern in fuel producing sectors and optimum-mix of refineries can be ensured by setting output prices which reflect their opportunity costs or replacement costs without considerations of their impact on consumers. In fact, a great deal of confusion in the arguments on energy pricing can be avoided if a clear distinction is made between crude oil/gas prices used in evaluation of benefits from an investment project and prices (transfer) to be charged to the user industries/ units. In the same way, relative product prices faced by the refineries can be adjusted to provide incentives for producing desired product-mix (e.g. maximizing kerosene and diesels at the cost of furnace oil) rather than be constrained by the level of consumer prices. +)

Energy Pricing and Demand Management

The success of price policy in achieving the desired objectives of conservation and interfuel substitution depends on the special characteristics of energy demand in developing countries. The available experience shows that an increase in oil prices (or other energy prices) does not bring about the desired conservation or fuel-shifts for one or more of the following reasons:

i) In developing countries, the level and pattern of energy consumption is determined more by supply constraints than by demand considerations. Consumers (households, industries) have to adjust their demand pattern according to availabilities of different fuels in different regions/ periods even if their economic and technical preference may suggest a different fuel-mix. Non-availability of relatively cheaper fuels may be on account of lack of production capacity or transport bottlenecks or both. To the extent such shortages can be foreseen and output/investment plans are modified accordingly, such factors may not act as constraints to interfuel substitution in the long-run.

ii) Under choices among alternative energy sources may be determined by factors other than the relative prices of fuels adjusted for efficiency differences. Equally or more

---

+) For example, in India, it would be useful to fix a high price for kerosene/diesel and a very low price (about zero) for fuel oils to encourage refineries to set up secondary processing units (such as hydrocracking units) to convert fuel oils into kerosene/diesel. For details, see Bhatia (1976 and 1974).

important considerations may be auxiliary costs of handling and storage, tax/subsidies on investment, security of supply, uniformity of quality and the effect of fuel-mix on the quality of processed materials.[+)] Although significance of these factors would vary from one use to the other, it is important to consider that inter-fuel substitution may be affected by factors other than prices.

iii) Fuel costs form a relatively small proportion of the total costs of production in many industries. For example, data from India reported in the Working Group on Energy Policy, Government of India (1979, p. 98) show that average incidence of energy costs as a percentage of value of production in industry was only 6.8% in 1975-76. There were a number of industries, accounting for as much as 45% of the total industrial production, where energy costs as a percentage of value of production is less than 3%. Industries where incidence of energy costs was higher than 15% accounted for only 9% of the total industrial output. With energy costs forming a relatively small proportion, the scarce management inputs and capital are likely to be used for improvements other than fuel conservation and substitution. Similarly, fuel charges form a small proportion of the total costs in rail and road transport costs.

[+)] Schramm and Munasinghe (1980) have analysed the interrelations of technical, economic, social and organizational aspects of user choices among alternative energy sources with a case study on tobacco curing in Thailand. The alternatives examined are: woodfuel, lignite, lignite briquettes and diesel - both at current and projected price levels, using both market and shadow prices. The results show that the use of lignite briquettes in place of diesel oil would be economic both in terms of social and financial considerations if 1979 petroleum-product world prices are assumed to prevail in the future. The study emphasizes that user choices among alternative energy sources will usually be determined only partially by the financial costs of each alternative fuel compared on a net heat content basis, but more significantly by the "auxiliary costs". They also point out that work habits and the availability of experienced operators may significantly affect use if the choice is between customarily-used energy materials and new alternatives of unknown characteristics.

iv) Apart from fuel costs being a small proportion of total costs, it is possible for the industrial units and transport operators to pass on the increases in fuel costs to the consumers. This happens due to protected markets for domestic producers, lack of internal competition, general shortages of goods and transport capacity. Since consumers have little choice in switching to other goods/modes, the entrepreneurs/transporters have no incentive to spend scarce capital and management resources in reducing energy consumption.

v) Supply bottlenecks and inefficiencies in management of public transport do not permit commuters to shift from one mode of transport (e.g. cars, scooters) to another, energy-efficient, mode of transport (diesel buses, local trains, etc.). Besides, higher gasoline prices do not result in significant reduction in gasoline consumption because private cars are used mainly by people who are not sensitive to higher operating costs either because of their high incomes or because they are in a position to pass on higher costs to the government or other consumers.[+)]

vi) Many consumers may not be aware of the technical possibilities of conservation and/or interfuel substitution. Even where the knowledge may exist, there may be difficulties in finding funds for investment, technical expertise to carry out the modifications and in obtaining required equipment and spare parts. Thus, lack of an organizational set up along with financial bottlenecks may result in non-implementation of some good schemes of energy conservation/substitution.

[+)] As it is gasoline consumption forms a relatively small proportion of the demand for oil products in a developing country. Most of this is consumed by the government in defence or civil or construction activities or by public sector undertakings where there is no incentive to reduce costs. The private sector enterprises or businessmen are also in a position to pass on higher costs to the government in the form of lower taxes and, thus, have no incentive to economise.

LEGISLATION AND NON-PRICE INSTRUMENTS

Non-price instruments include:

a) rationing of fuels which are in short supply or load-shedding of electricity at the time of shortage;

b) direct regulations such as maximum speeds for automobiles/trucks, reduction in working days, control on use of air-conditioners or other equipment, regulation of temperatures, etc.;

c) legislative regulations such as the establishment of efficiency standards for energy-using equipment, the imposition of an obligation to indicate energy efficiency and to circulate and reutilize waste heat;

d) withdrawal of restrictions on captive power generation when total energy concept (using both steam and power) is adopted by an industrial unit.

The efficacy of these instruments would vary from one country to the other depending on the use of private cars, machinery for implementation of government measures/legislation and the extent to which government can control the activities of large numbers of producers. Since in many developing countries consumption of gasoline is mainly for essential purposes such as defence, government work, public transport (taxis, auto-rickshaws) and other essential activities, there is little scope for reduction in these uses. Private car is still a luxury in many countries and is usually owned by high income groups who may not respond to either price increases or conservation measures. Rationing is generally advocated to give a fixed quantity of gasoline per vehicle so that consumption of private car owners can be restricted. However, due to practical difficulties of administering a rationing system where leakages would be inevitable, this measure may have to be used in emergency situations of extreme shortages. In fact, one may advocate a fixed quota of gasoline at lower prices for taxi (and autorickshaw) operators so that private car owners find it economic to use taxis and, thus, conserve gasoline. The fixation of maximum speed limits is not likely to conserve much fuel because most of the vehicles do not operate at high speeds due to conditions of roads, traffic congestions and weather conditions. The experience of car pools shows that it works quite well immediately after a substantial increase in gasoline prices. However, when the enthusiasm wears out, the difficulties of running car pools by senior officers and executives, who have varying work-hours, come to the surface and the system breaks down. On account of these factors, there is not much scope for conservation of gasoline/diesel on account of non-price considerations unless the extreme step of rationing is taken.

One measure which is likely to be very effective in oil conservation is strict adherence to fuel-efficiency criteria in motor vehicles at the time of giving permission to ply vehicles such as cars, taxis, trucks, buses, etc. It has been found that relatively simple measures connected with the maintenance of automobiles can save 10-15% of oil consumption. However, it would be necessary to provide efficient vehicle maintenance services equipped with required instruments in big cities, towns and highways so that operators are not left at the mercy of unscrupulous officials.

Legislative regulations such as the establishment of efficiency standards and display of efficiency rating of energy-using equipment would be very useful wherever it can be implemented. In countries where these are imported or are manufactured in large-scale units, it would be easy to implement these measures. However, the measure is not likely to be very effective in some countries where energy-using equipments such as electric motors, diesel engines, fans, irons, stoves, heaters, etc. are manufactured by a large number of small industrial units spread over different small towns/cities. Nevertheless, this measure is expected to be very effective in the long run and, hence, efforts should be made in this direction as soon as possible.

FISCAL INCENTIVES

As was discussed earlier, the extent of energy conservation achieved depends on the actions of a large number of consumers: households, entrepreneurs, managers, transport operators and government officials. These consumers respond differently to different signals and react in a variety of ways, usually unknown. The consumer is to be made aware of the existence and financial viability of a variety of measures/technologies for fuel conservation. He has to be convinced that these measures are, in fact, effective and these technologies would work. Even when he is convinced of the technical and financial viability of the measure/technology involved, he has to be provided with an incentive to adopt that measure. Besides this incentive in terms of some form of subsidy or tax concession, it may become necessary to make provisions for financial assistance. All these services would require a well-coordinated organizational set-up which could provide the necessary technical and financial inputs to the consumers.

One of the most important conditions for a successful implementation of the program is knowledge and availability of various fuel-efficiency measures and new technologies. This knowledge and availability has to be very specific, e.g. a specific process for better utilization of steam or process heat or availability of a solar system for industrial process heat or a biogas plant. It is not sufficient to know that these technologies exist, it is important to convey that such technologies/units can be installed if so desired by an entrepreneur/farmer. The information on each process/technology should be complete, giving data on technical, economic and financial aspects; listing maintenance and special input requirements as well as giving details of agencies who have the experience to install and maintain the related equipment.
In addition to the listing of available technologies/processes, it is important to carry out detailed economic evaluation of available technical alternatives from the point of view of the society. It is necessary to rank each available alternative using the framework of Social Benefit Cost Analysis (SBCA). +) This would require that for each scheme/process/technology, a detailed quantification and valuation of direct as well as indirect benefit and costs should be done. Care should be taken to quantify the savings in oil consumption as well as the requirements of capital, foreign exchange, scarce raw materials/inputs, skilled labour and management. The technique of SBCA can also be used to reflect objectives of development such as employment creation, income distribution, regional development and meeting basic human needs. Environmental impacts, including health benefits and costs, of each technology/process can also be reflected in the SBCA. A complete analysis would require the use of shadow prices for valuation of inputs and outputs and put explicit weights on other objectives.

Even if a technology/process is economically desirable from the society's point of view, it is necessary that it would also be financially viable proposition. Hence, a cost benefit analysis

---

+) The technique of Social Benefit Cost Analysis (SBCA) is well-known and has been used in evaluation of renewable energy technologies. For theoretical aspects, reference may be made to UNIDO (1972, 1978), Little and Mirrlees (1974), Squire and Van der Tak (1975). Applications of SBCA can be found in French (1979), Bhatia (1977), Bhatia and Niamir (1979), Barnett et al. (1978), de Lucia and Lesser (1978), Ghate (1980), Parikh (1980), Turner (1979).

would also be required from the point of view of the user. This result may not coincide with the result of SBCA because market prices may deviate from shadow prices due to the presence of price controls, taxes, subsidies, etc. In the case of a divergence in the recommendations about a technology/process, it may be necessary to suggest use of tax/subsidy mechanisms to bring the financial viability in line with economic desirability. For example, if biogas used for lighting and irrigation is considered socially profitable at the current and projected shadow prices of diesel and kerosene, then this fact should be reflected in the private profitability of biogas units. For socio-political reasons it may be considered desirable to keep kerosene prices low through subsidies and also keep diesel prices low to avoid diversion of kerosene to diesel by consumers. Under these circumstances, a consumer who faces low kerosene and diesel prices is going to calculate his savings of kerosene/diesel at market prices and, hence, would find their substitution by bio. gas relatively less profitable than is the case when shadow prices are used. Thus, it may become necessary to provide an outright subsidy on capital cost and/or provide subsidized capital (as soft loans) so that private profitability moves in the direction of social profitability. Unless this is done, there is no motivation for the farmer to install a biogas plant. Similar examples can be given in the application of solar energy for process heat in industrial units or shift of goods from road to railways to save diesel or improvement in the efficiency of kerosene stoves. Given the subsidized price of an oil product, its substitution by another fuel or by capital (or skilled labour) is not likely to be financially profitable unless similar subsidies are also provided for the substitutes through lower prices, lower interest rates, tax concessions, etc. Thus, the process of determination of relative prices for various energy products also involves a corresponding set of decisions on tax/subsidies mechanisms so that desirable forms of substitutions are encouraged. It is necessary to recognize that consumers' decisions would be motivated by their private profitability calculations and their view of returns from alternative uses of their scarce funds of manpower resources. Given the inherent uncertainties in the use of new processes or renewable energy technologies, an entrepreneur is understandably reluctant to commit scarce resources without substantial tax incentives/subsidies. However, availability of these tax incentives and/or other subsidies would be only a necessary condition for the motivation of the entrepreneur/consumer and not a sufficient condition to induce investments. His decision would be crucially dependent on his views regarding the availability of technology/process, uncertainties in operation, problems of maintenance and spares and his confidence in the

ENERGY RESOURCE PRICES	TRANSFER PRICES	PRODUCERS' PRICES		CONSUMERS' PRICES (TAXES/SUBSIDIES)
CRUDE OIL/GAS	OIL REFINERIES - Crude price	OIL REFINERIES - Product price	Taxes + Subsidies (T+S)	Oil Products prices
COAL-MINING	Electricity generation - Coal/oil price	Elect. utilities - Electricity prices	T+S	Electricity tariffs
HYDRO POWER	Coal gasification or liquefaction - coal	Gas prices	T+S	Prices for coal gas
RENEWABLES (land, labour)	Fuelwood Cowdung Crop residues	Prices for biogas, charcoal & p.gas	T+S	Prices for biogas, charcoal, pyrolysis gas

Fig. 1 - Integrated framework for energy pricing

TABLE 1 - Developing countries: potential savings in commercial energy consumption, 1990 (million barrels a day of oil equivalent)

	Projected consumption*	Savings from				
		Pricing policies	Taxes and regulations	Retrofitting & technical improvements	Interfuel substitution & scale economies	Total reduction
Electric power**	6.5	0.1	(.)	0.5	0.1	0.7
Agriculture	1.5	(.)	(.)	0.1	(.)	0.1
Households	5.9	0.3	0.1	0.1	0.4	0.9
Transport	7.5	0.1	0.1	0.6	0.2	1.0
Industry	8.7	0.2	0.2	1.1	0.3	1.8
Other	0.5	0.1	(.)	(.)	(.)	0.1
Total	30.6	0.8	0.4	2.4	1.0	4.6

* Based on "High Case" projections of GNP and 3% per year increases in oil prices as in World Development Report, 1980, a ratio of energy consumption to GNP growth of 1 : 1 : 2 and a price elasticity of energy consumption of –0.3%.

** Includes energy consumed in generation, station use, losses in transmission and distribution.

(.) Less than .05 the unit shown.

Source: Zorld Bank, Energy in Developing Countries, 1980.

TABLE 2 - Stages and objectives of energy pricing

Objectives	Stages			
	Pricing of energy resource	Transfer pricing	Producers' price	Consumers' price (including taxes/ subsidies)
Control of inflation				X
Basic needs				X
Profitability & efficient management of public sector units		X		
Environmental considerations				X
Optimum investments in fuel producing sectors	X			
Security of supply	X			
Encouragement to domestic resources	X			
Optimum product mix of refineries			X	
Adequate government revenue and Federal finance		X		X

organizations who are sponsoring the technology/process. Thus, a program of energy conservation and interfactor and interfuel substitution in developing countries would require careful studies of changes in demand in response to changes in relative prices of fuels and other factor inputs. There is need to determine energy prices in an integrated framework so that optimum pattern of interfuel substitution is encouraged. A suitable system of fiscal incentives and tax/subsidy mechanisms would be required to encourage substitution of oil by new and renewable energy sources. It is considered that an efficient organizational set-up which coordinates the work of financial, executive and production agencies is a necessary condition for the successful implementation of the programs of oil conservation and adoption of renewable energy sources.

REFERENCES

A. Barnett, L. Pyle and S. K. Subramaniam, Biogas Technology in the Third World: A Multidisciplinary Review. Ottawa, Int. Dev. Res. Centre, 1978.

Ramesh Bhatia, The Oil Crisis: An Economic Analysis and Policy Imperatives, Economic and Political Weekly, July 27, 1974.

Ramesh Bhatia, Technological and Locational Choices in Petroleum and Fertilizer Industries; A Multiprocess Interregional Programming Model for India, Indian Economic Review, April 1976.

Ramesh Bhatia, Energy Survey Methodologies: A Framework for Measuring Non-Conventional Energy Sources in Developing Countries, Paper presented at the ESCAP/IEA Workshop on Energy Statistics, Karachi, Oct. 1980.

Ramesh Bhatia, Energy and Rural Development in India: Some Issues in W. Lockerets (Ed.) Agriculture and Energy, Academic Press, 1977.

Ramesh Bhatia, Economic Appraisal of Biogas Units in India: Framework for Social Benefit-Cost Analysis, Economic and Political Weekly, August 1977a.

Ramesh Bhatia and M. Niamir, "Renewable Energy Sources: The Community Biogas Plant", Mimeo (draft), (revised 1980), Cambridge, MA; Harvard Univ., 1979.

Ramesh Bhatia, Energy Alternatives for Irrigation Pumping: Some Results for Small Farms in Bihar, in Int. Energy Studies, Ed. by R.K. Pachauri, Wiley Eastern, 1980a.

Deepak Bajracharya, Rural Energy Utilization Patterns: Some Observations from Pangama, Nepal, Mimeo, IDS, Sussex, Jan. 1980.

John Briscoe, The Political Economy of Energy Use in Rural Bangladesh, Harvard Univ. Environm. System Program, August 1979.

R.J. De Lucia and M. Lesser, Energy from Organic Residues in Developing Countries: Present Use, Potential and Observation on Feasibility, paper presented at the World Bioenergy Conf., Atlanta, 1980.

A.V. Desai, Interfuel Substitution in the Indian Economy National Council of Applied Economic Research, New Delhi, India (Mimeo), 1980.

P. B. Ghate, "Irrigation for Very Small Farmers: Appropriate Technology or Appropriate Organization?", Economic and Political Weekly, pp. A-161-171, 1980.

Government of India, Report of the Working Group on Energy Policy, New Delhi, India, 1979.

Government of India, Report of the Working Group on Oil Prices, New Delhi, India, 1969.

David Hughart, Prospects for Traditional and Non-Conventional Energy Sources in Developing Countries, World Bank Staff paper no. 346, August 1979.

Bill Knowland and Carol Ulinski, Traditional Fuels: Present Data, Past Experience and Possible Strategies, USAID, Sept. 1979.

I. M. D. Little and J. Mirrlees, Project Evaluation in Developing Countries, 1974.

Fermando Manibog, An Economic Evaluation of Renewable Energy Systems for Philippine Rural Area, draft, Sept. 1979.

M. Munasinghe and G. Schramm, Interrelationships in Energy Planning: The Case for Tobacco Curing Industry in Thailand, Mimeo, The World Bank, 1980.

M. Munasinghe, "An Integrated Framework for Energy Pricing in Developing Countries", Energy Journal, pp. 1-30, 1980a.

J. K. Parikh, Energy Systems and Development, Constraints, Demand and Supply of Energy for Developing Regions, 1980.

Roger Revelle, Energy Use in Rural India, Science, 192, 1976.

G. Schramm and M. Munasinghe, Interrelationships in Energy Planning: The Case of Tobacco Curing Industry in Thailand (Mimeo), The World Bank, USA, 1980.

C. M. Siddayao, "Petroleum and Coal Pricing Policies", Mimeo, Honolulu: prepared for the Asian Development Bank Re- tional Energy Survey by the East West Center, 1980.

L. Squire and Van der Tak, Economic Analysis of Projects, Johns Hopkins Press, 1975.

Wallace Turner, Energy Resources and Economic Development in India, 1979.

UNIDO, A Practical Guide to Project Appraisal in Developing Countries, 1978.

World Bank, Energy Options and Policy Issues in Developing Countries, Bank Staff Working Paper No. 350, August 1979.

World Bank, Energy in Developing Countries, 1980.

World Bank, Alcohol Production from Biomass in Developing Countries, Sept. 1980 (1980a).

COORDINATED USE OF PRICE AND NON-PRICE POLICY TOOLS FOR ENERGY DEMAND MANAGEMENT AND CONSERVATION

Mohan Munasinghe *

Senior Energy Advisor to the President of Sri Lanka
Ministry of Power and Energy, Colombo 2, Sri Lanka

ABSTRACT

The policy tools available for optimal supply-demand planning and management include pricing, physical controls, technical methods (including research and development), and education and propaganda. Since these tools are interrelated, their use should be well coordinated. To meet the many national objectives, energy pricing policy must be formulated in two stages. In the first stage, a set of prices which strictly meets the economic efficiency objective is determined, based on a consistent and rigorous framework. The second stage consists of adjusting these efficient prices (established in the first step), to meet all the other objectives. The adoption of a given energy conservation measure is economically justified if the resulting economic benefits exceed costs. An illustrative example is analyzed.

1. DEMAND MANAGEMENT AND ENERGY PLANNING

Modern societies require increasing amounts of energy for domestic, industrial, commercial, agricultural, and transport uses. Arrayed against these energy needs are the short-term, depletable fossil fuel supplies--petroleum, coal, and natural gas,

---

* The author is also Senior Economist-Engineer, World Bank, Washington, D.C., presently on a leave of absence. The views and opinions expressed in this paper are the author's and do not necessarily reflect those of the Government of Sri Lanka or the World Bank.

H. Neu and D. Bain (eds.), National Energy Planning and Management in Developing Countries, 317–340.

as well as the longer-run, renewable energy sources. The latter include, water, nuclear, solar, geothermal, wind, tidal, and biomass including traditional or noncommercial fuels such as wood and animal waste. (1) To ensure that the energy needs of the economy will be efficiently met, policymakers in many countries are beginning to realize that energy investment and pricing decisions should be carried out on an integrated basis, e.g., within the framework of a national energy master plan that addresses both supply and demand management issues.

Before discussing demand management and conservation it is necessary to briefly discuss what is meant by national energy planning. The broad underlying rationale is to make the best use of energy resources to promote socio-economic development and improve the welfare and quality of life of citizens. Therefore energy planning is an essential part of overall national economic planning, and should be carried out and implemented in close coordination with the latter. However, in energy planning, the principal emphasis is on the comprehensive and disaggregate analysis of the energy sector, with due regard for the main interactions with the rest of the economy. In strictly technical sense, the energy planners' role might be confined to seeking the least-cost method of meeting future energy requirements. However, energy planning also includes a variety of other objectives including reducing dependence on foreign sources, supplying basic energy needs of the poor, reducing the trade and foreign exchange deficit, priority development of special regions or sectors of the economy, raising sufficient revenues to finance energy sector development (at least partially), ensuring continuity of supply and price stability, preservation of the environment, and so on.

In general, energy planning requires analysis at the following three hierarchical levels in relation to fundamental national objectives: (a) links between the energy sector and the rest of the economy; (b) interactions between different sub-sectors within the energy sector; and (c) activities in each individual energy sub-sector. The steps involved in the planning procedure usually include energy supply and demand analysis and forecasting, energy balancing, policy formulation, and impact analysis, to meet short, medium and long-range goals. Initially, these activities may be carried out at a relatively simple level and later as data and local analytical capabilities improve more sophisticated techniques, including computer modelling, could be implemented. The institutional structure should also be rationalized by setting up a central energy authority (CEA), or ministry of energy, with its principal focus on energy planning and policy-making. The execution of policy, and day-to-day operations, would remain the responsibility of line agencies such as the electricity utilities or petroleum corporations that already exist in practically all countries.

Demand and supply management permit the energy policymaker to maintain the energy supply-demand balance, thus avoiding major economic disruptions and reductions in national welfare. Demand management includes all means of influencing the magnitudes and patterns of energy consumption. Supply management includes identification and optimal exploitation of all energy resources, investment planning, transformation, refining and distribution of energy, and so on.

Price is most effective as a long-run policy tool. From the viewpoint of economic efficiency, the price indicates to the suppliers, the consumers' willingness-to-pay and use-value of energy; while to the consumers, it signals the present and future opportunity costs of supply based on various energy sources. Physical controls are most effective in the short-run when there are unforeseen shortages of energy. All methods of physically limiting consumption such as load shedding and rotating power cuts in the electricity sub-sector, as well as reducing the supply of gasoline or banning the use of motor cars during some periods, are included in this category. Technical means include, on the supply side, the least cost or cheapest means of producing a given form of energy, the best mix of fuels, research and development of substitute fuels such as wood-alcohol for gasoline, and on the demand side, introducing higher efficiency energy conversion devices such as better stoves for woodfuel, etc. Education and propaganda include, on the supply side, efforts to make citizens aware of external diseconomies such as pollution, and supportive of re-afforestation schemes to preserve the environment, and on the demand side public education for energy conservation.

2. INTEGRATED PRICING FRAMEWORK

Pricing is a very important tool especially in the long run. As discussed below the pricing and investment decisions should be closely related. However, energy supply systems, e.g., electricity generation, transmission and distribution, oil and gas wells and pipelines, coal mines, forests, etc., usually require large capital investments with long lead times and lifetimes. Therefore once the investments decision is made, usually on the basis of the conventional least-cost method of meeting demand by sub-sector, with due regard for interfuel substitution possibilities, there is a lock-in effect with respect to supply, and thus prices should be related to the long-run horizon of planning. On demand side also, energy conversion devices, e.g., motor cars, gas stoves, electric appliances, machines and so on, are expensive relative to average income levels and have relatively long lifetimes, thus limiting the ability of consumers to respond to changes in relative fuel prices in the short run.

The objectives of energy pricing are closely related to the goals of energy planning, but are more specific. First, the economic growth objective requires that pricing policy should promote economically efficient allocation of resources both within the energy sector and between it and the rest of the economy. In general terms, this implies that future energy use would be at optimal levels, with the price (or willingness-to-pay of the consumer) for the marginal unit of energy used reflecting the incremental resource cost of supply to the national economy. Relative fuel prices should also influence the pattern of consumption in the direction of the optimal or least-cost mix of energy sources required to meet future demand. Distortions and constraints in the economy necessitate the use of shadow prices and economic second-best adjustments, as described in the next section.

Second, the social objective recognizes the basic right of all citizens to be supplied with certain minimum energy needs. Given the existence of significant numbers of poor consumers and also wide disparities of income, this implies subsidized prices, at least for low income consumers.

Third, the government would be concerned with financial objectives relating to the viability and autonomy of the energy sector. This would usually be reflected, by pricing policies which permitted institutions (typically, government-owned) in the different energy sub-sectors to earn a fair rate of return on assets and to self-finance an acceptable portion of the investments required to develop future energy resources.

Fourth, energy conservation is also an objective of pricing policy. While prevention of unnecessary waste is an important goal, there often are other reasons underlying the desire to conserve certain fuels. These include the desire for greater independence from foreign sources (e.g., oil imports), the need to reduce the consumption of woodfuel due to deforestation and erosion problems, etc.

Fifth, we recognize a number of additional objectives, such as the need for price stability to prevent shocks to consumers from large price fluctuations, the need for simplicity in energy pricing structures to avoid confusing the public and simplify metering and billing, and so on.

Finally, there are other specific objectives such as promoting regional development (e.g., rural electrification) or specific sectors (e.g., export-oriented industries), as well as other socio-political, legal and environmental constraints.

Because the objectives mentioned above are often not mutually consistent, a realistic integrated energy pricing structure must be flexible enough to permit tradeoffs among them. The latter procedure is more ad-hoc, with the extent of the adjustments being determined by the relative importance attached to the different objectives. In the next section, we discuss the importance of shadow pricing and develope the economic framework which permits the efficient pricing of energy. The second stage adjustments due to non-economic factors are discussed in the next section.

3. SHADOW PRICING AND ECONOMIC EFFICIENCY

Shadow pricing theory has been developed mainly for use in the cost-benefit analysis of projects. (2) However, since investment decisions in the energy sector are closely related to the pricing of energy outputs, for consistency the same shadow pricing framework should be used in both instances. Shadow prices are used instead of market prices (or private financial costs), to represent the true economic opportunity costs of resources.

In the idealized world of perfect competition the interaction of atomistic profit maximizing producers and atomistic utility maximizing consumers yield market prices which reflect the correct economic opportunity costs, and scarce resources including energy will be efficiently allocated. However, in the real world, distortions due to monopoly practices, external economies and diseconomies (which are not internalized in the private market), interventions in the market process through taxes, import duties and subsidies, etc., all result in market prices for goods and services, which may diverge substantially from their shadow prices or true economic opportunity costs. Therefore, shadow prices must be used in investment and output pricing decisions, to ensure economically efficient use of resources. Moreover, if there are large income disparities, we will see later that even these "efficient" shadow prices must be further adjusted, especially to achieve socially equitable energy pricing policies for serving poor households.

It is important to realize that lack of data, time and manpower resources, particularly in the LDC context, will generally preclude the analysis of a full economy-wide model when making energy-related decisions. (3) Instead a partial approach may be used, where key linkages and resource flows between the energy sector and the rest of the economy, as well as interactions among different energy sub-sectors are selectively identified and analyzed, using appropriate shadow prices such as the opportunity cost of capital, shadow wage rate, marginal opportunity cost for different fuels, and so on. In practice, surprisingly valuable results may be obtained from even relatively simple models and assumptions.

To clarify the basic concepts involved in optimal energy pricing we first analyze a relatively simple model. Next the effects of more complex features including short-run versus long-run dynamic considerations, capital indivisibilities, joint output cost allocation, quality of supply, and price feedback effects on demand are examined. The process of establishing the efficient economic price in a given energy sub-sector may be conveniently analyzed in two steps. First, the marginal opportunity cost (MOC) or shadow price of supply must be determined. (4) Second, this value has to be further adjusted to compensate for demand side effects arising from distortions in the prices of other goods, including other energy substitutes. From a practical viewpoint, an optimal pricing procedure which begins with MOC is easier to implement because supply costs are generally well defined (from technological-economic considerations), whereas data on the demand curve is relatively poor.

Suppose that the marginal opportunity cost of supply in a given energy sub-sector is the curve MOC(Q) shown in Figure 1. For a typical non-traded item like electricity, MOC which is generally upward sloping is calculated by first shadow pricing the inputs to the power sector and the estimating both the level and structure of marginal supply costs (MSC) based on a long-run system expansion program. (5) For tradable items like crude oil and for fuels which are substitutes for tradables at the margin, the international or border prices of the tradables (i.e., c.i.f. price of imports or f.o.b. price of exports, with adjustments for internal transport and handling costs) are appropriate indicators of MOC. (6) For most developing countries, such import or export MOC curves will generally be flat or perfectly elastic. Other fuels such as coal and natural gas could be treated either way depending on whether they are tradables or non-traded. (7) The MOC of non-renewable, non-traded energy sources will generally include a "user cost" or economic rent component, in addition to the marginal costs of production. The economic value of traditional fuels are the most difficult to determine because in many cases there is no estabished market. However, as discussed later, they may be valued indirectly on the basis of the savings on alternative fuels such as kerosene, the opportunity costs of labour for gathering firewood, and/or the external costs of deforestation and erosion.

Thus, for a non-traded form of energy, MOC is the opportunity cost of inputs used to produce it plus a user cost where relevant, while for a tradable fuel or a substitute, MOC represents the marginal foreign exchange cost of imports or the marginal export earnings foregone. In each case, MOC measures the shadow priced economic value of alternative output foregone, because of increased consumption of a given form of energy. After identifying the correct supply curve, we next examine demand-side effects, especially second best corrections which capture interactions between different energy sub-sectors. (8) This second step is just as important as the first one, and therefore it will be examined in some detail.

In Figure 1, the market priced demand curve for the form of energy under consideration is given by the curve PD(Q), which is the willingness-to-pay of consumers. Consider a small increment of consumption dQ at the market price level p. The traditional optimal pricing approach attempts to compare the incremental benefit of consumption due to dQ, i.e., the area between the demand curve and X-axis, with the corresponding supply cost, i.e., the area between the supply curve and X-axis. However, since MOC is shadow priced, PD must also be transformed into a shadow priced curve to make the comparison valid. This is done by taking the increment of expenditure p.dQ and asking the question: "What is the shadow priced marginal cost of resources used up elswhere in the economy if the amount p.dQ (in market prices) was devoted to alternative consumption (and/or investment)?

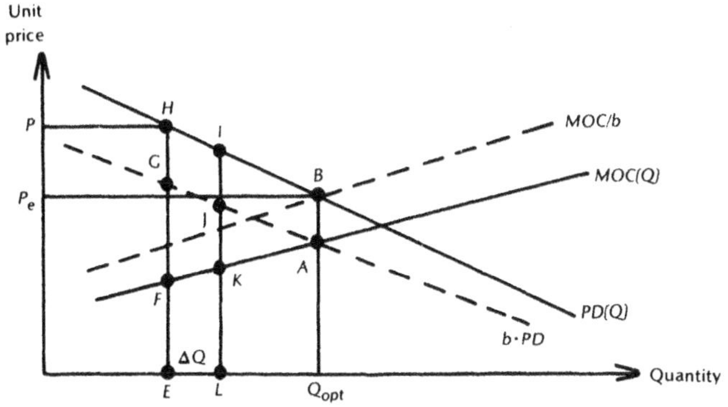

Figure 1. Efficient Energy Pricing using Shadow Prices

Suppose that the shadow cost of this alternative pattern of expenditure is b(p.dQ), where b is called a conversion factor. Then the transformed PD curve which represents the shadow costs of alternative consumption foregone is given by b.PD(Q); where in Figure 1, it is assumed that $b < 1$. Thus at the price p, incremental benefits EGJL exceed incremental costs EFKL. The optimal consumption level is Qopt, where the MOC and b.PD curves cross, or equivalently where a new pseudo-supply curve MOC/b, and the market demand curve PD intersect. The optimal or efficient selling price to be charged to consumers (because they react only along the market demand curve PD, rather than the shadow priced curve b.PD) will be: P_e = MOC/b, at the actual market clearing point B. At this level of consumption, the shadow costs and benefits of marginal consumption are equal, i.e., MOC = b.PD. Since b depends on user specific consumption patterns, different values of the efficient price p_e may be derived for various consumer categories, all based on the same value of MOC. We clarify the foregoing by considering several specific practical examples.

First, suppose that all the expenditure (p.dQ) is used to purchase a substitute fuel, i.e., complete substitution. Then the conversion factor b is the relative distortion or ratio of the shadow price to market price of this other fuel. Therefore p_e = MOC/b, represents a specific second-best adjustment to the MOC of the first fuel, to compensate for the distortion in the price of the substitute fuel. (9) Next, consider a less specific case in which the amount (p.dQ) is used to buy an anverage basket of goods. If the consumer is a residential one, b would be the ratio of the shadow price to the market price of the household's market basket (here, b is also called the consumption conversion factor). The most general case would be when the consumer was unspecified, or detailed information on consumer categories was unavailable, so that b would be the ratio of the official exchange rate (OER) to the shadow exchange rate (SER), also called the standard conversion factor (SCF). (10) This represents a global second-best correction for the divergence between market and shadow prices averaged throughout the economy. (11)

5. ADJUSTMENTS TO EFFICIENT ENERGY PRICES

Once efficient energy prices have been determined, the second stage of pricing to meet social, financial, political and other constraints, must be carried out.

We note that efficient energy prices deviate from the prices calculated on the basis of financial costs, because shadow prices are used instead of the market prices. This is done to correct for distortions in the economy. Therefore, the constraints which force further departures from efficient prices (in the second stage of the pricing procedure) may also be considered as distortions which impose their own shadow values on the calculation. See M. Munasinghe and J. J. Warford, op cit., for details.

Socio-political or equity arguments are often advanced in favor of subsidized prices of "lifeline" rates for energy, especially where the costs of energy consumption are high relative to incomes of poor households. Economic reasoning based on externality effects may also be used to support subsidies, e.g., cheap kerosene to reduce excessive firewood use and prevent deforestation, erosion, etc. To prevent leakages and abuse of such subsidies, energy suppliers must act as discriminating monopolists. Targeting specific consumer classes (for example, poor households), and limiting the cheap price only to a minimum block of consumption is easiest to achieve, in practice, for metered forms of energy like gas or electricity. Other means of discrimination may also be required such as rationing, licensing, etc. (12) All these complex and inter-related issues require detailed analysis.

The concept of a subsidized "social" block, or "lifeline" rate, for low income consumers has another important economic rationale, based on the income redistribution argument. We clarify this point with the aid of Figure 2 which shows the respective demand curves for energy AB and GH of low (I_1) and average (I_2) income domestic users, the social tariff p_s over the minimum consumption block 0 to Q_{min}, and the efficient price level p_e. All tariff levels are in domestic market prices. If the actual price $p = p_e$, then the average household will be consuming at the "optimal" level Q_2, but the poor household will not be able to afford the service.

If increased benefits accruing to the poor have a high social value, then, although in nominal domestic prices the point A lies below p_e, the consumer surplus portion ABF multiplied by an appropriate social weight w could be greater than the shadow price of supply. The adoption of the block tariff shown in Figure 2, consisting of the lifeline rates p_s, followed by the full tariff p_e, helps to capture the consumer surplus of the poor user, but does not affect the optimum consumption pattern of the average consumer. (13) In practice, the magnitude Q_{min} has to be carefully determined, to avoid subsidizing relatively well-off consumers; it should be based on acceptable criteria for identifying "low income" groups, and reasonable estimates of their minimum consumption levels (e.g., sufficient to supply basic energy requirements for the household). The level of p_s relative to the efficient price may be determined on the basis of the poor consumer's income level relative to some critical consumption level. (14) The financial requirements of the energy sector would also be considered in determining p_s and Q_{min}. This approach may be reinforced by an appropriate supply policy (e.g., subsidized house connections for electricity, special supply points for kerosene, etc.).

Financial Viability

The financial constraints most often encountered relate to meeting the revenue requirements of the sector, and are often embodied in criteria such as some target financial rate of return on assets, or an acceptable rate contribution towards the future investment program. In principle, for state-owned energy suppliers, the most efficient solution would be to set price at the efficient level and rely on government to subsidize losses or

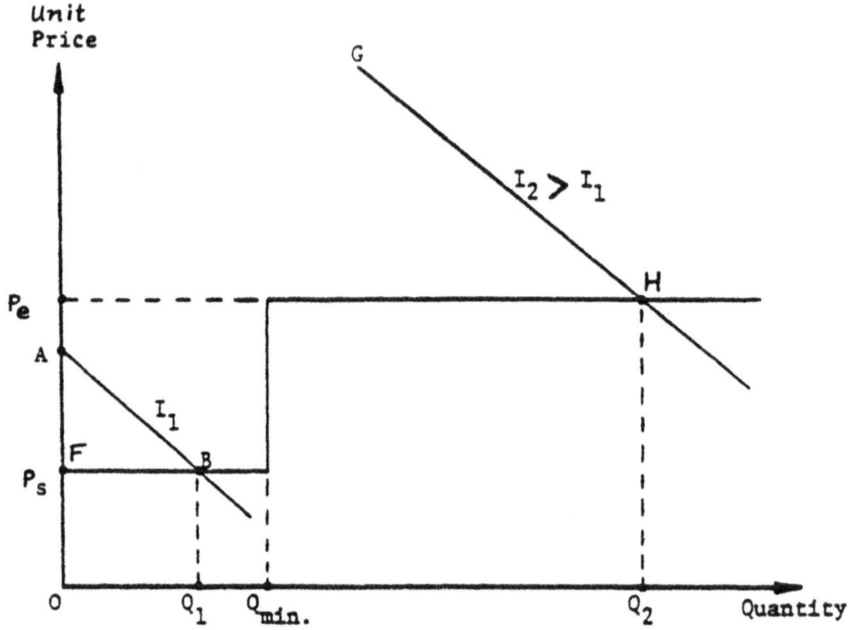

Figure 2. Economic Analysis of the Social or Lifeline Rate

tax surpluses exceeding sector financial needs. In practice, some measure of financial autonomy and self-sufficiency is an important goal for the sector. Because of the premium that is placed on public funds, a pricing policy which results in failure to achieve minimum financial targets for continued operation of the sector, would rarely be acceptable. The converse and more typical case, where efficient pricing would result in financial surpluses well in excess of traditional revenue targets, may politically unpopular, especially for an electric utility. Therefore in either case, changes in revenues have to be achieved by adjusting the efficient prices.

It is intuitively clear that discriminating between the various consumer categories so that the greatest divergence from the marginal opportunity cost based price occurs for the consumer group with the lowest price elasticity of demand, and vice versa, will result in the smallest deviations from the "optimal" levels of consumption consistent with a strict efficiency pricing regime. (15) Generally, in many countries the necessary data for the analysis of demand by consumer categories is rarely available, so rule-of-thumb methods of determining the appropriate tariff structure have to be adopted. However, if the energy sub-sector exhibits increasing costs (i.e., marginal costs greater than average costs), the fiscal implications should be exploited to the full. Thus for example electric power tariffs (especially in a developing country) constitute a practical means of raising public revenues in a manner which is generally consistent with the economic efficiency objective, at least for the bulk of the consumers who are not subsidized, while at the same time helping to supply basic energy needs to low income groups. Similar arguments may be made in the petroleum sub-sector where high prices for gasoline, based on efficiency externality and conservation arguments, may be used to cross-subsidize the "poor-man's" fuel-kerosene, or diesel used for transportation. (16)

Other Considerations

There are several additional economic, political and social considerations that may be adequate justification for departing from a strictly efficient pricing policy. The decision to provide commercial energy such as kerosene or electricity in a remote rural area, which often also entails subsidies because the beneficiaries are not able to pay the full price based on high unit costs, could be made on completely non-economic grounds, e.g., for general socio-political reasons such as maintaining a viable regional industrial or agricultural base, stemming rural to urban migration, or alleviating local political discontent. Similarly, uniforn nation-wide energy prices are a political necessity in many countries, although this policy may, for example, imply subsidization of consumers in remote rural areas where energy transport costs are high, by energy users in urban centers. However, the full economic benefits of such a course of action may be much greater than the apparent efficiency costs which arise from any divergence between actual and efficient price levels. Again this possibility is likely to be much more significant in a developing country than a developed one, not only because of the high cost of energy relative to incomes in the former, but also because the available administrative or fiscal machinery to redistribute incomes or achieve regional or industrial development objectives by other means is frequently ineffective.

The conservation objective (to reduce dependence on imported energy, improve the trade balance, and so on) usually runs counter to subsidy arguments. Therefore it may be necessary to restrict cheap energy to productive economic sectors which need to be strengthened, while in the case of basic energy needs of households, the energy price could be sharply increased for consumption beyond appropriate minimum levels. In other cases conservation and subsidized energy prices may be consistent. For example, cheap kerosene might be required, especially in rural areas, to reduce excessive woodfuel consumption, prevent deforestation, erosion, and so on.

It is particularly difficult to raise prices to any where near the efficient levels where low incomes and a tradition of subsidized energy have increased consumer resistance. In practice, price changes have to be gradual, in view of the costs which may be imposed on those who have already incurred expenditures on energy using equipment and made other decisions, while expecting little or no changes in traditional energy pricing policies. At the same time, a steady price rise will prepare consumers for high future energy prices. The efficiency costs of a gradual instead of a sharp price rise can be seen as an implicit shadow value placed upon the social benefits that result for this policy.

Finally, owing to the practical difficulties of metering, price discrimination, and billing, and the need to avoid confusing consumers, the pricing structure may have to be simplified. Thus, the number of customer categories, rating periods, consumption blocks, and so on, will have to be limited. Electricity and gas offer the greatest possibilities for structuring. The degree of sophistication of metering depends on the net benefits of metering, problems of installation and maintenance, and so on. In general, various forms of peak electricity pricing (i.e., using maximum demand or time-of-day metering) would be particularly applicable to large medium and high voltage industrial and commercial consumers. However, for very poor consumers receiving a subsidized rate for electricity, a simple current limiting device may suffice, because the cost of even simple kWh metering may exceed the net benefits (equal to savings in supply costs due to reduced consumption, less the decrease in consumption benefits). For electricity or gas, different charges for various consumption blocks may be effectively applied with conventional metering. However, for liquid fuels like kerosene, subsidized or discriminatory pricing would usually require schemes involving rationing and coupons, and could lead to leakage and abuses.

5. NON-PRICE TOOLS AND ENERGY CONSERVATION

The non-price tools of demand management including physical controls, technical methods, legislation, education and propaganda, etc. are often closely linked with energy conservation. Therefore, we discuss them together in this section. Conservation is an important element of demand management--in fact, we define economically desirable policies below as those that help establish energy demand at optimal levels.

In its most general sense energy conservation can be defined as the deliberate reduction in the use of energy below some level that would prevail otherwise. Usually, such reduction requires some tradeoffs in terms of comfort, convenience or the use of additional capital or labor.

Some conservation can be brought about simply by reducing or eliminating certain energy-using activities. Foregoing Sunday pleasure driving, using lower thermostat settings and shutting off appliances and lighting fixtures when not directly needed are typical examples. Other conservation measures may require the substitution by either capital or labor. Examples are the re-circulation of heat in industrial processing, the energy-saving reductions in the weight of vehicles by better engineering or lighter materials, the use of more sophisticated controls to minimize energy consumption per unit of output or work performed, or the use of improved insulation.

An important sub-set of effective energy conservation measures consists of the substitution of some form of costly, or scarce energy resource by some other that is more readily available. Examples are the use of coal instead of fuel oil in heat processes, the use of natural gas instead of petroleum products for power plants in countries in which gas is plentiful compared to oil, or the use of alcogas instead of gasoline for transportation. In a physical or technical sense, (as measured by btu's of Kcal.'s consumed) such substitution may not "save" energy, or even result in increased energy use per unit of output. In an economic sense, however, such substitution may make perfectly good sense, given the economic scarcity values of the alternative fuels.

The pursuit of energy conservation as a goal raises the issue of up to what point the reduction of energy consumption is socially beneficial or desirable. Common sense indicates that while "wasteful" energy use should be discouraged, there is a limit beyond which conservation measures may become too costly in terms of foregoing other resources or useful outputs, thereby causing more harm than good. In order to define desirable conservation levels more precisely, both the technical and economic feasibilities must be evaluated. As will be argued, technical feasibility is a necessary, but not a sufficient condition for determining the desirability of specific conservation measures. As argued elsewhere in this book, the principle objective of a given policy should be the maximization of the welfare of a society's citizens over time. If aggregate consumption or production (e.g. gross domestic product or GDP) is taken as a proxy for aggregate welfare, then welfare maximization implies the use of scarce resources such as energy, capital, labor and land in such a way that output is maximized. It also implies that the value of output, at the margin, must be equal to or greater than the costs of the resources utilized in its production. (17) Reductions in the use of energy will contribute to this goal, provided that the added costs of such conservation measures or process changes do not outweigh the value of the energy savings achieved.

Energy Conservation Economics

In a market economy free of distortions the basic rule of energy conservation should be that the total system's cost of any energy-using activity per unit output, not the energy costs only, should be minimized over time, while at the margin the value of the output produced should be just equal to someone's willingness to pay for it. Two conditions are of importance.

The first is that total system's costs must be minimized, not just energy costs. For example, a 40-mile per hour speed limit on freeways may minimize fuel consumption, but it may be economically inefficient because of the greater time requirements for moving freight and people, which would result in higher labor, transportation-equipment and freight-inventory costs. Hence, the appropriate amount of energy conservation at the margin is determined by the equality of the value of the energy savings on the one hand and the sum of the costs of the needed additional inputs plus the net reductions in the value of the output produced (if any) on the other, i.e.,

$$MB = MC_1 + MC_2 \qquad (6.5).$$

Where MB, MC_1 and MC_2 are the marginal energy savings, marginal additional input costs and marginal reductions in output values respectively.

The second condition is that this cost minimization should be achieved over the life-expectancy of the activity. (18) This means that resource input mixes (including the share of energy inputs) should be based on expected life-time costs, not just presently prevailing cost relationships. For example, if energy costs are expected to increase relative to other input costs or the value of output over time, greater substitution by non-energy inputs (i.e., higher levels of energy conservation) are called for. Thus the test of economic desirability of a conservation measure becomes:

$$\sum_{t=1}^{n} b_t/(1+r)^t > \sum_{t=1}^{n} (c_{1,t} + c_{2,t})/(1+r)^t$$

Where b_t, $c_{1,t}$ and $c_{2,t}$ are the respective annual energy savings, additional input costs and losses in output values in years t=1, 2-n, and r is the discount rate, all defined in terms of appropriate shadow prices. These optimum economic conservation principles can be illustrated by some simple examples.

Let us consider a particular end use for energy such as home lighting, and let us suppose there is a choice of two distinct types of light bulbs, incandescent and fluorescent. Both have the same economic cost, same lifetime, and provide light output of the same quality. If the fluorescent bulb uses less electrical energy than the incandescent one, then replacing the latter by the former is a conservation measure that results in an unambiguous improvement in economic as well as technical efficiency. In this case, using fluorescent bulbs instead of incandescent lamps reduces the economic resources expended to provide the desired output, i.e., lighting. Electrical energy has been conserved, with no change in other economic costs and benefits.

Next, let us suppose that the fluorescent bulb is more costly to install. There is a trade-off between the higher capital cost of the fluorescent lamp and the greater consumption of kilowatt-hours by the incandescent bulb. The relevant data to determine whether substitution of incandescent bulbs by fluorescent ones is economically justified, is summarized in Table 5.1. At this state we distinguish between the economic value (or opportunity cost or shadow price, as discussed earlier) of a good or service, and its market price. The former is relevant to decision-making from a national perspective and the latter is more appropriate from a private individual viewpoint. The national cost (based on economic values) of using the incandescent and fluorescent bulbs over their two year lifetimes are respectively: (19)

$$EC_I = 10.5 + 16 + 16/(1 + r) \text{ and} \qquad (6.7a)$$

$$EC_F = 32 + 4.4 + 4.4/(1+r) \qquad (6.7b)$$

Assuming an economic discount rate of:

$r = 0.1$, we find $EC_I = 41.0 > EC_F = 40.4$.

What we have done is to compare the energy cost saving of dineros (16-4.4) = 11.6 per year for two years against the increase in capital costs, dineros (32-10.15) = 21.5. What we find is that (16-4.4) + (16-4.4)/(1=r) > (32-10.5). Therefore the use of fluorescent lightbulbs and the associated reduction in energy consumption will improve economic efficiency as well as technical efficiency.

However, if $r = 0.2$, $EC_I = 39.8 < EC_E$ 40.1. In this case the conservation measure is no longer beneficial. This reduction in the relative value of conservation measures will always occur with increases in the discount rate, because increases in initial investment costs are traded off against the future cost-savings realized by conservation. This finding has important policy implications. Energy users who confront high opportunity costs of capital (as for example, most developing countries or low-income consumers) will find capital-cost-intensive energy conservation measure relatively less attractive than users who have access to low-cost sources of capital. This means that economically "optimal" conservation measures may differ significantly among different countries or user classes.

The analysis of optimal economic rates of conservation becomes more complex under conditions of projected relative price changes for energy inputs, and technically determined, fixed energy-capital-labor inputs over the life expectancy of given energy-using capital assests (such as a truck, or boiler or smelting oven). While it is usually possible to vary the physical mix of inputs (at least within certain ranges) in the design stage of new equipment or production processes, this is usually not possible (or at least quite difficult and costly) once the equipment has been installed. Hence, equipment conservation decisions should be made beforehand.

It can be shown that under these conditions the "optimum" economic rate of conservation (or energy utilization) will lie between the most efficient (lower) rate prevailing in the early life stage of the asset and the most efficient (higher) rate that would be most appropriate towards the end of its life. Furthermore, the optimum "average" rage will be the lower, the higher the applicable discount rate is.

To accommodate such relative price changes in the evaluation of the benefits from energy conservation, the earlier equation has to be changed to:

$$\sum_{t=1}^{n} b_t (1+g)^t/(1+r)^t > \sum_{t=1}^{n} (c_{1,t} + c_{2,t})/(1+r)^t$$

where g is the expected average annual rate of relative price increase of the energy input that is saved.

The Effects of Market Imperfections

So far the analysis has been based on the assumption of perfectly operating markets, with prices for all inputs and outputs including those for energy reflecting the long-run marginal production or opportunity costs. This, however, is rarely the case. Market imperfections, particularly in the pricing and availability of energy, abound throughout most countries of the world. Legal barriers, capital rationing or lack of credit facilities, inappropriate foreign exchange rates, price controls, externalities and many other factors interfere in the normal functioning of the energy market. Other complications arise from a lack of knowledge of available alternatives and lack of foresight with respect to future costs and prices and the lock-in effects of long-life facilities and equipment. All of these factors tend to distort rational choice patterns, with the result that energy use and conservation patterns diverge substantially from those found to be optimal in terms of overall economic efficiency criteria.

To illustrate the effects of these divergencies, let us return to the simple light-bulb example of Table 5.1 in the preceeding section. As we have seen, at a discount rate of $r = 0.1$ the use of fluorescent light bulbs proved to be superior in economic terms, while at a discount rate of $r = 0.2$ the incandescent light bulb was less costly.

The private costs (based on market prices) of using incandescent or fluorescent lighting respectively are as follows:

$$PC_I = 18 + 12 + 12/(1+r)$$

$$PC_F = 36 + 3.3 + 3.3/(1+r)$$

At a discount rate of $r = 0.1$ (the lowest rate used in our preceeding economic analysis) $PC_I = 40.9 < PC_F = 42.3$. This means that a rational consumer would prefer to use incandescent light bulbs, because this is the cheaper option. At any higher discount rate the advantage of the incandescent system over the fluorescent one increased further. Thus, as a consequence of the divergence between market prices and actual underlying economic costs economically inefficient energy use and conservation decisions are made by users.

Inappropriate pricing of energy resources is not the only reason for inefficient energy conservation decisions. In many developing countries the lack of foreign exchange resources forces governments to maintain strict import controls. As a consequence, it is often impossible for large energy users to import new, more energy-efficient equipment in order to replace existing ones, even though they are usually able to secure their share of high-cost imported fuel supplies in order to keep their existing fuel-inefficient equipment operating. In countries in which fuel prices are subsidized at the same time, there is little incentive for such equipment owners to press for appropriate changes in import policies.

A special problem exists with regard to energy consumption patterns of households that depend on the collection and utilization of traditional fuel resources such as firewood, charcoal and dung. As many studies have shown, the vast majority of these households depend on rather primitive cooking appliances which are usually little more than open fires. These are highly inefficient in terms of energy utilization and usually use only about 5% of the inherent heat energy of the fuel. Heavy population pressures, dwindling firewood resources, resulting in sharply increased costs of fuelwood gathering as well as increased soil erosion, reduced availability of crop residues from new short-stem, high yield crop varieties, all combine to make this one of the foremost and serious energy problems in the majority of developing countries.

The use of simple cooking stoves constructed of locally available materials at out-of-pocket costs barely exceeding $5 to $10 could improve energy efficiency by a factor of 4 to 5 in laboratory tests and perhaps by a factor of 2 or better in actual day-to-day household use.

6. PRICE AND NON-PRICE POLICY INTERACTIONS AND SUMMARY

In addition to appropriate pricing there are a wide variety of direct and indirect policy measures that can be taken to bring about desirable levels of energy conservation. Among them are direct regulations of energy uses, regulations of the use of energy consuming equipment and appliances, mandatory standards, mandatory information requirements about energy consumption rates, taxes and subsidies, appropriate infrastructure investments for energy saving facilities (e.g. better roads, railroads, marine shipping facilities) education and propaganda, and others.

To analyze some of the effects of such conservation-oriented policies let us first return to our light bulb example of the proceeding sections. As we have found, existing market prices have made it more attractive for users to opt for the incandescent light bulb system. To resolve this difference between optimal economic and private market choices, the first option policy makers might consider could be to raise the market price of electricity from 0.3 dineros per kWh to its economic value of 0.4 dineros per kWh. We now have: $PC_I = 48.5 > PC_F = 44.4$ and rational electricity consumers will make the correct decision in favor of fluorescent lighting. In addition, setting the electricity price equal to its marginal opportunity cost will also establish electricity consumption for non-lighting purposes at optimal levels.

Suppose that public resistance or other social pressures make it impossible to raise electricity prices. Let the economic value of an incandescent bulb be its cost of production or producer price, while the imposition of a government tax of dineros 7.5 determines the market price. Similarly, assume that an import duty of 4.0 dineros represents the difference in the c.i.f. import cost (dineros 32) and the market price of fluorescent bulbs. Instead of raising electricity prices, an alternative policy option might be to raise the tax on incandescent lightbulbs to dineros 9.5, making the market price dineros 20. In this case, $PC_I = 42.9 > PC_F = 42.3$, which encourages the desirable consumer decision. Reducing the duty on fluorescent bulbs to dineros 2 and lowering the retail price to dineros 34 would also yield a favorable result, since now: $PC_I = 40.9 > PC_F = 40.3$. Some combination of the tax increase and lowering of duty could also be used. From a general economic viewpoint, and ignoring effects outside the lightbulb market, reducing the import duty would be preferable to raising the producer tax because the former action reduces the divergence between market price and economic opportunity cost of fluorescent bulbs whereas the latter has the opposite effect and increases the market distortion in the price of incandescent lightbulbs.

Next, assume that the tax on incandescent lightbulbs cannot be increased because the legislation affects a much larger class of related products. Similarly, suppose that the import duty on fluorescent bulbs cannot be reduced because it would undercut the price of a high-cost local producer and drive him out of business. In this instance, some final options left to the energy policymaker might be to legislate that all incandescent lightbulbs be replaced by fluorescent ones, or to give a direct cash subsidy to consumers who adopt the measure, or to mount a major public education and propaganda campaign to bring about the required change. (20)

We may summarize the steps that an energy policymaker should take before adopting and implementing any given conservation measure. First, using economic opportunity costs consistent with the national viewpoint, he should establish whether the benefits of such an action exceed the costs. If this is the case, then the same test should be repeated, using market prices relevant to the appropriate consumer group, in order to establish whether a rational consumner would adopt the conservation measure. If this is not the case changes in energy prices, taxes or import duties on equipment, subsidies to consumers, legislation, and other policy options may have to be used to implement the conservation technique. In general, price changes that reduce the divergence between market prices and opportunity costs should be preferred. However, care should be exercised to ensure that these policy actions do not have adverse repercussions in other energy as well as non-energy markets. We discuss next, some complications that could arise in the assessment and application of conservation measures.

If the useful lifetimes of technological alternatives are different, then economic comparisons become somewhat more complicated. This would be the case in our earlier example if the lifetime of incandescent bulbs were to be only one year while that of fluorescent lamps might be 3 years. Two alternative approaches could be used to overcome this difficulty. In the first, the investment costs of each alternative would have to be annuitized over its lifetime at the appropriate discount rate and the associated energy consumption and other recurrent costs for one year would be added on. Then the total costs for each option would be compared. The second method would compare the full costs of each alternative over a much longer period, say 20 years, including the costs of periodic replacement of worn-out equipment.

reducing the import duty would be preferable to raising the producer tax because the former action reduces the divergence between market price and economic opportunity cost of fluorescent bulbs whereas the latter has the opposite effect and increases the market distortion in the price of incandescent lightbulbs.

Next, assume that the tax on incandescent lightbulbs cannot be increased because the legislation affects a much larger class of related products. Similarly, suppose that the import duty on fluorescent bulbs cannot be reduced because it would undercut the price of a high-cost local producer and drive him out of business. In this instance, some final options left to the energy policymaker might be to legislate that all incandescent lightbulbs be replaced by fluorescent ones, or to give a direct cash subsidy to consumers who adopt the measure, or to mount a major public education and propaganda campaign to bring about the required change. (20)

We may summarize the steps that an energy policymaker should take before adopting and implementing any given conservation measure. First, using economic opportunity costs consistent with the national viewpoint, he should establish whether the benefits of such an action exceed the costs. If this is the case, then the same test should be repeated, using market prices relevant to the appropriate consumer group, in order to establish whether a rational consumner would adopt the conservation measure. If this is not the case changes in energy prices, taxes or import duties on equipment, subsidies to consumers, legislation, and other policy options may have to be used to implement the conservation technique. In general, price changes that reduce the divergence between market prices and opportunity costs should be preferred. However, care should be exercised to ensure that these policy actions do not have adverse repercussions in other energy as well as non-energy markets. We discuss next, some complications that could arise in the assessment and application of conservation measures.

If the useful lifetimes of technological alternatives are different, then economic comparisons become somewhat more complicated. This would be the case in our earlier example if the lifetime of incandescent bulbs were to be only one year while that of fluorescent lamps might be 3 years. Two alternative approaches could be used to overcome this difficulty. In the first, the investment costs of each alternative would have to be annuitized over its lifetime at the appropriate discount rate and the associated energy consumption and other recurrent costs for one year would be added on. Then the total costs for each option would be compared. The second method would compare the full costs of each alternative over a much longer period, say 20 years, including the costs of periodic replacement of worn-out equipment. (21)

The two methods should give consistent results, assuming the same values are used for parameters such as the discount rate.

Another type of difficulty associated with changes in the benefits of consumption arises if either the quality or quantity of the end-product of energy use is different for the two alternatives. As an example of the first effect, consumer a comparison of electric versus kerosene lamps for lighting. In addition to the differences in equipment and fuel costs, the cost-benefit assessment of the two optins should also include a term to recognize that electricity is likely to provide lighting of a superior quality. (22) While the quantification, in monetary terms, of this qualitative superiority will be difficult, one measure might be the willingness-to-pay of the consumers for the different forms of lighting, usually represented by the area under the relevant demand curve.

Specific conservation measures such as rationing have a quality effect that must be taken into account. Let us consider the physical rationing of gasoline. In this case, the cost, or welfare loss to the consumer due to the reduction in the miles he can travel in his car must be added to the cost of implementing the rationing scheme and then compared with the benefits of reduced gasoline supply. Once again, the willingness-to-pay of gasoline users would be the appropriate measure of the foregone consumption benefit. However, in the long-run gasoline consumption could also be reduced by the introduction of a more fuel efficient automobile engine without (perhaps) requiring a reduction in the miles travelled. This shows that a reduction in energy consumption does not always imply a reduction in consumption benefits. A major focus of the appropriateness of conservation policies should be the service derived from the energy use.

Finally, the costs and benefits associated with externalities should be included in the economic cost-benefit comparison of alternatives. For example, improvements in tecnical efficiency or fuel substitution measures may give rise to pollution, as in the case of conversions from oil burning to coal-fired electric power plants. These additional "external" costs should be explicitly evaluated in the analysis.

FOOTNOTES

(1) Traditional or noncommercial fuels such as firewood are particularly important in developing countries, amounting to about 40 percent of total energy consumption in 1975. See for example: Mohan Munasinghe and Colin J. Warren, "Rural Electrification, Energy Economics and National Policy in Developing Countries," in Future Energy Concepts, Institution of Electrical Engineers, (London, January-February 1979).

(2) For general use of shadow prices in developing countries, see: L. Squire and H. Van der Tak, Economic Analysis of Projects, Johns Hopkins, Baltimore, 1975; and I.M.D. Little and J.A. Mirrlees, Project Appraisal and Planning for Developing Countries, Basic Books, New York, 1974. For more specific application to the energy sector, see: Mohan Munasinghe, The Economics of Power System Reliability and Planning, Johns Hopkins, Baltimore, 1979 b, Chap. 9.

(3) This holistic approach or general equilibrium analysis is conceptually important. For example, the efficient shadow price of a given resource may be represented by the change in value of aggregate national consumption or output, due to a small change in the availability of that resource. A more detailed discussion of general versus partial equilibrium in relation to energy sector analysis is given in Mohan Munasinghe, The Economics of Power System Reliability, Johns Hopkins Press, Baltimore, 1979.

(4) The same model will be modified in the next section to establish socially equitable subsidized prices for low income consumers.

(5) For a detailed discussion of the procedures used in the electric power sub-sector, see: Mohn Munasinghe, "Electric Power Pricing Policy", Staff Working Paper No. 340, World Bank, Washington, D.C., June 1979 a. In this sub-sector MSC is also called the long-run marginal cost (LRMC).

(6) We note that the use of border prices does not require the assumption of free trade, but implies that the numeraire or unit of value for shadow pricing is essentially uncomitted foreign exhange (but converted into local currency at the official exchange rate). For details, see L. Squire and H. Van der Tak, op.cit.

(7) A non-traded item is generallyi characterized by a domestic supply price that lies above the f.o.b. price of exports, but below the c.i.f. price of imports.

(8) The general theorem of the second-best shows, for example, that if the price of a given fuel is not set at it's MOC, then the efficient price of a close substitute also must diverage from it's own MOC. For a detailed discussion of the theory of the second-best in economics, see: D.M. Winch, Analytical Welfare Economics, Penguin Books, Harmondsworth, U.K., 1971.

(9) For example, MOC_{EL} could represent the long run marginal cost of rural electricity for (lighting), and the substitute fuel could be imported kerosene. Suppose that the (subsidized) domestic market price of kerosene is set at one half it's import (border) price, for socio-political reasons. Then b = 2, and the efficient selling price of electricity $p_e = MOC_{EL}/2$ (ignoring differences in quality of the two fuels, capital costs of conversion equipment such as light bulbs, kerosene lamps, partial substitution effects etc.: a more refined analysis of substitution possibilities would have to incorporate these additional considerations). It would be misleading however, to then attempt to jusify the subsidized kerosene price on the basis of comparison with the newly calculated low price of electricity. Such circular reasoning is far more likely to occur when pricing policies in different energy subsectors are uncoordinated, rather than in an integrated energy pricing framework. We note that all these energy sector subsidies must be carefully target to avoid leakages and abuses.

(10) Note that with the foreign exchange numeraire, conversion of domestic price values into shadow price equivalents by application of the SCF to the former, is conceptually the invese of the traditional practise of multiplying foreign currency costs by the SER (instead of the OER) to convert to the domestic price equivalent.

(11) For example, suppose the border price of imported diesel is Pesos 4 per litre (i.e., US¢20 per litre, converted at the OER, of Pesos 20 per US$). Let the appropriate SER which reflects the average level of import duties and export subsidies be Pesos 25 per US$. Therefore SCF = OER/SER = 0.8, and the appropriate strictly efficient selling price of diesel: $P^e = 4/0.8$ = Pesos 5 per liter.

(12) For example, a minimum ration of chep kerosene for households, or a special licence for trucks using subsidized diesel oil and ban on diesel-driven passenger cars.

(13) Ignoring the income effect due to reduced expenditure of the average consumer for the first block of consumption, i.e., up to Q_{min}.

(14) See: M. Munasinghe and G Schramm, Energy Economics, Demand Management and Conservation Policy, Van Nostrand Publ. Co., New York, 1983.

(15) See W. J. Baumol and D. F. Bradfor, "Optimal Departures from Marginal Cost Pricing," Am. Econ. Rev., June 1970, pp. 265-283.

(16) However, a number of undesirable side effects may follow, such as the practice of mixing gasoline with kerosene, substitution of diesel for gasoline, and so on. The income distribution efect may also be perverse, with the relatively wealthy diverting cheap kerosene or diesel for use in vehicles or in industry.

(17) For a detailed exposition of the market conditions and relationships that give rise to the efficient production and consumption of economic goods and services, see, for example: R. Henderson and D Quandt, Micro-economic Theory.

(18) More accurately, over the life expectancy of the fixed input components of the activity.

(19) As discussed earlier, the term economic value is used synonymously with opportunity or efficiency shadow prices, and it my differ from actual market prices.

(20) The evidence concerning the effectiveness of education and propaganda as a energy concervation policy tool is mixed. See for example: James M. Walker, "Voluntary Responses to Energy Conservation Appeals", Journal of Consumer Research, vol. 7, June 1980, pp. 88-92; and A.E. Peck and O.C. Doering, "Voluntarism and Price Response: Consumer Response to the Energy Shortage", The Bell Journal of Economics, vol. 7, Spring 1976, pp. 287-292.

(21) Strictly speaking the discounted scrap value of equipment left at the end of the 20 year period should also be deducted from the costs stream associated with the corresponding alternative.

(22) The ease and convenience of using a fuel, the danger from its use, its social acceptability, and so on, are all factors that may affect the consumers choice.

Table 5.1. Physical and Economic Data to Assess the Economic Efficiency of Energy Conservation for Lighting

		INCANDESCENT BULB	FLOURESCENT BULB
Installation Cost (Dineros)	Economic Value (Opp. Cost)	10.5	32
	Market Price	18	36
Physical Energy Consumption (kWh/year during 2 year lifetime)		40	11
Value of Energy Consumption (Dineros/year during 2 year lifetime)	Economic Value (Opp. Cost) a/	16	4.4
	Market Price b/	12	3.3

a/ Dineros 0.4 per kWh   b/ Dineros 0.3 per kWh

PROJECT ANALYSIS AND ENERGY POLICY

Mohan Munasinghe *

Senior Energy Advisor to the President of Sri Lanka
Ministry of Power and Energy, Colombo 2, Sri Lanka

ABSTRACT

The importance of systematic integrated national energy planning is emphasized. By identifying the key interaction between the energy sector and the rest of the economy, as well as within the sector itself, and using appropriate shadow prices, policymakers can make rational economic decisions from the national viewpoint. The stages of a typical development project cycle include: identification, preparation, appraisal, negotiations and financing, implementation and supervision, and evaluation. The basic economic criterion for acceptance of a project is that the net present value of benefits should be positive. Two case studies of investment decisionmaking involving solar PV pumping and electric power projects illustrate the close relationship between the investment and pricing decisions, and the interplay between shadow and market priced calculations in implementing the project decision.

1. INTEGRATED NATIONAL ENERGY PLANNING (INEP)

In recent years, decision makers in an increasing number of countries have realized that energy sector planning should be carried out on an integrated basis within the framework of a national energy master plan which determines energy policy, long-

* The author is also Senior Economist-Engineer, World Bank, Washington, D.C., presently on a leave of absence. The views and opinions expressed in this paper are the author's and do not necessarily reflect those of the government of Sri Lanka or the World Bank.

H. Neu and D. Bain (eds.), National Energy Planning and Management in Developing Countries, 341–368.

term planning. However, in practice investment planning and ranging from short-term supply and demand management decisions to pricing are still carried out on an ad-hoc and at best partial or sub-sector basis. Thus typically, electricity and oil sub-sector planning have traditionally been carried out independently of each other as well as of other energy sub-sectors. As long as energy was cheap such partial approaches were acceptable, but lately, with the international oil price acting as the messenger of rising energy costs, significant fluctuations in relative fuel prices, and sudden energy shortages, the advantages of an integrated energy policy have become evident.

We begin by discussing what is meant by integrated national energy planning (INEP) in the developing country context. The broad rationale underlying INEP is the need for an energy master plan (EMP) that will make the best use of energy resources, promote socio-economic development and improve the welfare and quality of life of citizens. Therefore energy planning is an essential part of overall national economic planning, and should be carried out and implemented in close coordination with the latter. However, in energy planning, the principal emphasis is on the detailed and disaggregate analysis of the energy sector, with due regard for the main interactions within the sector itself, as well as with the rest of the economy.

The scope of integrated national energy planning may be clarified by examining the hierarchical framework depicted in Figure 1. At the highest and most aggregate level, it should be clearly recognized that the energy sector is a part of the whole economy. Therefore, energy planning requires analysis of the links between the energy sector and the rest of the economy. Such links include the input requirements of the energy sector like capital, labour and raw materials as well as energy outputs such as electricity, petroleum products, woodfuel and so on. Other links are, the impact on the economy of policies concerning availability of supply, energy prices, taxes and so on, in relation to national objectives. Since energy permeates the whole economy, the energy sector is much more extensive than other sectors such as agriculture or industry. In this respect, the energy sector is like the "financial sector," with energy acting as the physical counterpart of money.

The second level of INEP treats the energy sector as a separate entity composed of sub-sectors such as electricity, petroleum products and so on. This permits detailed analysis of the sector with special emphasis on interactions among the different energy sub-sectors, substitution possibilities, and the resolution of any resulting policy conflicts, such as competition between kerosene and electricity for lighting, or woodfuel and kerosene for cooking. The third and most disaggregate level pertains to planning within each of the energy sub-sectors. Thus, for example, the electricity sub-sector must determine its own demand forecast and long-term investment programme, the woodfuel sub-sector must develop consumption projections and detailed plans for re-afforestation, harvesting of timber, and so on.

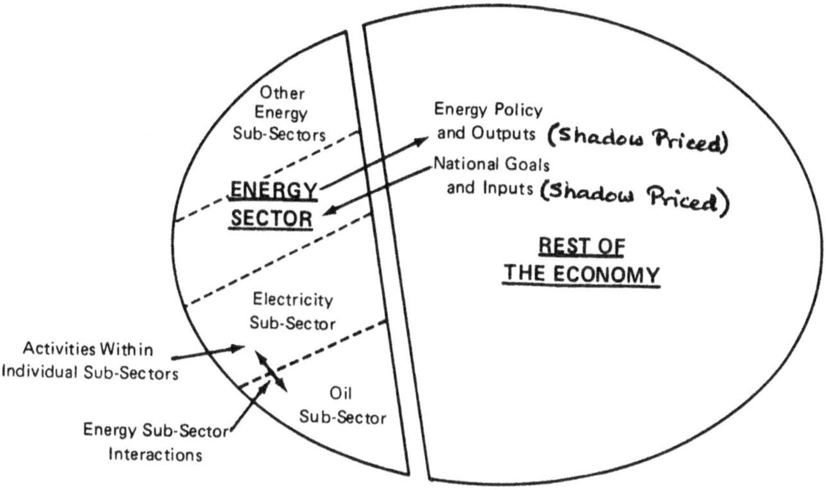

Figure 1. INEP: Hierarchy of Interactions

Integrated energy planning permits the development of a coherent set of policies which meet the needs of many interrelated and often conflicting national objectives. These objectives include: (1) determining the detailed energy needs of the economy to achieve growth and development targets; (2) choosing the mix of energy sources to meet future energy requirements in the cheapest way possible; (3) conserving energy resources and eliminating wasteful consumption; (4) diversifying and reducing dependence on foreign sources; (5) supplying basic energy needs of the poor; (6) saving scarce foreign exchange and reducing the trade deficit; (7) priority development of special regions or sectors of the economy; (8) raising sufficient revenues to finance energy sector development; (9) ensuring continuity of supply and price stablity; (10) preservation of the environment, and so on.

One of the most important aspects of INEP is investment planning. Therefore, in this paper, we will focus attention on how investment decisions are made in the energy sector. Other aspects such as demand analysis and energy pricing policy will be discussed, to the extent that they are related to and affect investment policy. In the next section the principal steps that have to be followed to prepare, accept and implement energy supply projects are outlined. Section III contains a summary of a shadow pricing framework suitable for use in developing countries. Practical case studies of commercial (electric power) and renewable (solar) energy projects are outlined in Sections IV and V. The paper ends with a section setting out the summary and conclusions.

2. PROJECT ANALYSIS

The successful processing of a development project usually involves several well defined steps. In order to place energy project decisions and economic cost-benefit analysis in its proper perspective, we summarize below the systematic approach used by the World Bank in a typical project cycle that includes: identification, preparation, appraisal, negotiations and financing, implementation and supervision, and evaluation. (1)

The Project Cycle

Project identification involves preliminary selection (by the borrowing country and the Bank) of potential projects that appear to be feasible and conform to national and sectoral development goals. In the preparation phase which may last one year or more, the borrower (often assisted by consultants) studies the engineering-technical, institutional, economic and financial aspects of a proposed project. At this stage, the Bank may provide staff guidance and financial assistance or help borrowers obtain other assistance to carry out the studies. Project appraisal consists of a comprehensive and systematic review of all aspects of the project, by Bank staff, including typically a four week field visit. They prepare an appraisal report that discusses comprehensively, the national and sectoral strategies, as well as the engineering-technical, institutional, economic and financial issues.

The appraisal report is used as the basis for negotiations at which the borrower and financier (i.e., the Bank) discuss the measures required to ensure the success of the project, and the conditions for funding. The resulting agreements are included in loan agreements which together with the appraisal report are considered and accepted by the Bank's Board of Executive Directors and the borrowing government. The borrower is responsible for implementing the project according to conditions mutually agreed on with the Bank, especially with respect to procurement of goods and services for the project in conformity with Bank guidelines that seek to ensure efficiency and economy in these procedures.

Supervision of the implementation process is carried out by the Bank through periodic field visits and progress reports from the borrower. The Bank has its own annual reviews of supervision reports to continually update and improve implementation. Evaluation is the final stage of the project cycle, following disbursement of the loan. Project performance audits are carried out by an independent Bank department involving review of previous project documents and field visits, where appropriate. This analysis yields valuable experience that helps improve the work at all stages of the project cycle for future projects.

Economic Cost-Benefit Criteria

Economic cost-benefit analysis of projects which is the topic of this paper is typically one of the important analyses carried out in the critical appraisal stage - the point at which a firm decision is made regarding the acceptance or rejection of the project. As mentioned above, the technical, institutional and financial aspects also need to be considered in project appraisal. We summarize next, some criteria commonly used in the economic or cost-benefit test of a project. (2)

The most basic rule for accepting a project is that the net present value (NPV) of benefits is positive:

$$NPV = \sum_{t=0}^{T} (B_t - C_t) / (1 + r)^t$$

when B_t and C_t are the benefits and costs in year t, r is the discount rate, and T is the time horizon.

As described later, B_t, C_t and r are defined in economic terms and appropriately shadow priced. In particular the shadow price of r is the accounting rate of interest (ARI). If projects are to be compared or ranked, the one with the highest (and positive) NPV would be the preferred one, i.e., if $NPV^I > NPV^{II}$ (where NPV^i = net present value for project i), then project I is preferred to project II.

The internal rate of return (IRR) is also used as a project criterion, and may be defined by:

$$\sum_{t=0}^{T} (B_t - C_t) / (1+IRR)^t = 0$$

Thus, the IRR is the discount rate which reduces the NPV to zero. The project is acceptable if IRR > ARI, which in most normal cases implied NPV > 0 (i.e., ignoring cases in which multiple roots could occur).

The benefit-cost ratio (BCR) is sometimes used:

$$BCR = [\sum_{t=0}^{T} B_t/(1+r)^t] \; / \; [\sum_{t=0}^{T} C_t/(1+r)^t].$$

If BCR > 1, the n NPV > 0 and the project is acceptable.

Each of these criteria has its strengths and weaknesses, but NPV is probably the most useful. In the case of energy projects, the benefits of two alternative technologies are often equal (i.e., they both serve the same need or demand). Then the comparison of alternatives is simplified. Thus:

$$NPV^{I}-NPV^{II} = \sum_{t=0}^{T} (C_t^{II}-C_t^{I}) \; / \; (1+r)^t;$$ since the benefit

streams cancel out. In other words the project which has the lower present value of costs is preferred since it has the higher NPV. This is called the least-cost alternative (when benefits are equal).

3. SHADOW PRICING

This section summarizes the main concepts involved in shadow pricing that would be relevant to the energy sector. Since the perspective is that of the national economy, rather than of a private power utility company, the discussion is particularly relevant to developing countries. Usually, when market imperfections occur, shadow prices may be used instead of distorted market prices to represent the true economic resource costs of various goods and services in an economy. These shadow or social accounting prices stem from welfare economic theory and cost-benefit analysis.

If perfect competition existed, a Pareto-optimal situation would result as atomistic profit maximizing producers and atomistic utility maximizing consumers interacted. In this ideal-world state, prices reflect the true marginal social costs, scarce resources are efficiently allocated, and for a given income distribution no individual can be made better off without making someone else worse off. (3)

Real world conditions are different, however. Distortions caused by monopoly practices, external economies and diseconomies, interventions in the market process through taxes, duties, and subsidies, and so on, result in market prices for goods and services that may be substantially different from their shadow prices or true economic values. If large income disparities exist, the passive acceptance of the existing skewed income distribution, may be socially and politically unacceptable. These considerations require appropriate shadow prices instead of market prices for economic analysis, especially in the developing countries where market distortions are more prevalent. As described below, a set of "efficiency (shadow) prices" may be developed to ensure that scarce investment resources are efficiently allocated, irrespective of the income distribution.

Consider a general equilibrium model of the economy in which the national goal is embodied in an acceptable objective function such as aggregate consumption. This consumption is to be maximized subject to constraints that might include limits on resource availabilities, distortions in the economy, and so on. Then, the shadow price of a given scarce economic resource represents the change in the value of the objective function, caused by a marginal change in the availability of that resource. In the more specific context of a mathematical programming macro-economic model, the optimal values of the dual variables that correspond to the binding resource availability constraints in the primal problem have dimensions of price, and could be interpreted as shadow prices. (4)

Two basic types of shadow prices exist. These involve whether or not society is indifferent to income distributional considerations. To illustrate this point, consider the simple national goal of maximizing the present value of aggregate consumption over a given time horizon. If the consumption of different individuals is added directly without distinction regarding income level, then the shadow prices derived from such a model are termed efficiency prices because they reflect the pure efficiency of resource allocation. When the consumption of the lower income groups becomes the objective function, this consideration is given a greater weight in evaluating aggregate consumption. Then, the resultant shadow prices are called social prices. (5)

There are two practical difficulties. First, some analysts object to using social pricing in project work on the grounds that the weighting of consumption benefits by income category is necessarily arbitrary and that income-redistribution goals are better realized through fiscal and other policy instruments rather than by biasing investment decisions. But acceptance of efficiency prices implies that equal weights are arbitrarily attached to benefits accruing to all individuals. Also, in many countries the government may have few practical policy tools available to influence the existing income distribution. (6) In practice the use of social versus efficiency prices depends on factors such as data availability and the commitment of the government to income distributional goals. Thus in the case studies, which are presented later, emphasis is placed on efficiency pricing because the weights to be used in social pricing are difficult to determine.

The second problem is that estimating shadow prices with the full-fledged general equilibrium model referred to in the theory is difficult, because many countries do not have the requisite data. Reanalyzing such a large model whenever new investment decisions were required would be wasteful and time consuming. Instead, partial equilibrium techniqes may be used that evaluate the impact of the change in the availability of a given resource in its impact on a few key areas, rather than throughout the economy (see Figure 1).

The partial equilibrium method has been criticized on the basis of nonmarginality, for its lack of simultaneity, and because of the judgemental element involved in choosing which linkages to analyze and which to neglect when the impact of some disturbance is to be evaluated. The objections can be overcome by using an iterative process of successive (convergent) approximations and by considering a sufficient number of linkages so that the partial equilibrium shadow prices closely approximate the corresponding general equilibrium values.

Numeraire

To derive a consistent set of economic prices for goods and services a common yardstick or numeraire to measure value is necessary, as illustrated by a simple example.

If one wishes to compare avocados with coconuts, the equivalent units might be either one avocado for one coconut, or one kilo of avocados for one kilo of coconuts. In the first instance, the common yardstick is one fruit; in the second, it is the unit of weight. Clearly, if the weights of the two types of fruits are different, the result of the comparison will depend on the numeraire used.

With a numeraire of economic value the situation is more complicated, because the same nominal unit of currency may have a different value depending on the economic circumstances in which it is used. For example, a rupee's worth of a certain good purchased in a duty free shop is likely to be more than the physical quantity of the same good obtained for one rupee from a retail store, after import duties and taxes have been levied. Therefore, it is possible to distinguish intuitively between the border-priced rupee, which is used in international markets free of import tariffs, and a domestic-priced rupee, which is used in the domestic market subject to various distortions. A more sophisticated example of the value differences of a currency unit in various uses arises in countries where the aggregate investment for future economic growth is considered inadequate. In these instances, a rupee of savings that could be invested to increase the level of future consumption may be considered more valuable than a rupee devoted to current consumption.

The choice of the numeraire, like the choice of a currency unit, should not influence the economic criteria for decision-making (except in relation to magnitude), provided the same consistent framework and assumptions are used in the analysis. For example, only one difference exists between a study using cents as units and one using rupees (one rupee = 100 cents). All magnitudes in the former system are one hundred times the values in the latter. Therefore, a numeraire may be selected purely on the basis of convenience of application.

A most appropriate numeraire in many instances is a unit of uncommitted public income at border prices. Essentially, this unit is the same as freely disposable foreign exchange available to the government, but expressed in terms of units of local currency converted at the official exchange rate. (7) The discussion in the next section is developed in relation to this particular yardstick of value. The border-priced numeraire is particularly relevant for the foreign exchange-scarce developing countries. It represents the set of opportunities available to a country to purchase goods and services on the international market.

Applying Shadow Prices

The estimation and use of shadow prices is facilitated by dividing economic resources into tradable and nontradable items. The values of directly imported or exported goods and services are already known in border prices, that is, their foreign exchange costs converted at the official exchange rate. Locally purchased items whose values are known only in terms of domestic market prices, however, must be converted to border prices, by multiplying the former prices by appropriate conversion factors (CF). Therefore, tradables and nontradables are treated differently.

For those tradables with infinite elasticities--of world supply for imports and of world demand for exports--the cost, insurance, and freight (c.i.f.) border price for imports and the free-on-board (f.o.b.) border price for exports may be used with a suitable adjustment for the marketing margin. If the relevant elasticities are finite, then the change in import costs or export revenues, as well as any shifts in other domestic consumption or production levels or in income transfers, should be considered. The free trade assumption is not required to justify the use of border prices since domestic price distortions are, in effect, adjusted by netting out all taxes, duties, and subsidies.

To clarify this point, imagine a household in which a child is given an allowance of twenty pesos a month as pocket money. The youngster may purchase a bag of lollipops from the grocery store at a price of two pesos. Since the parents want to discourage consumption of sweets, however, they impose a fine of one peso on each bag of lollipops. The fine is exactly like an import duty, and the child must surrender three pesos for every bag of candy (valued at its domestic price, inside the household). From the family's perspective, however, the total external payment for the item is only two pesos, because the one peso fine is a net transfer within the household. Therefore, the true economic cost, or shadow price, of the bag of lollipops (valued at its border price) is two pesos, when the impact of the fine on the distribution of income between parent and child is ignored.

Two types of nontradable economic resources are discussed. A nontradable is coventionally defined as a commodity whose domestic supply price lies between the f.o.b. and c.i.f. prices for export and import, respectively. Items that are not traded at the margin because of prohibitive trade barriers, such as bans or rigid quotas, are also included within the category. If the increased demand for a given nontradable good or service is met by the expansion of domestic supply or imports, the associated border-priced marginal social cost (MSC) of this increased supply is the relevant resource cost. If decreased consumption of other domestic or foreign users results, the border-priced marginal social benefit (MSB) of this foregone domestic consumption or of reduced export earnings would be a more appropriate measure of social costs.

The social optimal level of total consumption for the given input (Q_{opt}) would lie at the point where the curves of MSC and MSB intersect. The differences between MSB and MSC arise from price and other distortions that lead to nonoptimal levels of consumption $Q \neq Q_{opt}$. More generally, if both effects are present, a weighted average of MSC and MSB should be used. The MSB would tend to dominate in a short-run, supply constrained situation; the MSC would be more important in the longer run, when expansion of output is possible. (8)

Clearly the MSC of nontradable goods and services from many sectors can be determined through appropriate decomposition. For example, suppose one rupee's worth, in domestic prices, of the output of the domestic construction sector may be broken down successively into n components. This would include capital, labor, and materials, which are valued at rupees $a_1, a_2, \ldots, a_n$ --in border prices. Since the conversion factor of any good is defined as the ratio of the border price to the domestic price, the construction conversion factor equals:

$$\sum_{i = i}^{n} a_i.$$

The standard conversion factor (SCF) may be used with nontradables that are not important enough to merit individual

attention or lack sufficient data. The SCF is equal to the official exchange rate (OER) divided by the more familiar shadow exchange rate (SER), appropriately defined. Converting domestic priced values into border price equivalents by applying the SCF to the value, is conceptually the inverse of the traditional practice of multiplying foreign currency costs by the SER (instead of the OER), to convert to the domestic price equivalent. The standard conversion factor may be approximated by the ratio of the official exchange rate to the free trade exchange rate (FTER), when the country is moving toward a freer trade regime: (9)

$$SCF = \frac{OER}{FTER} = \frac{eX + nM}{eX(1-tx) + nM(1+tm)}$$

where X = f.o.b. value of exports; M = c.i.f value of imports; e = elasticity of export supply; n = elasticity of import demand; t_x = average tax rate on exports (negative for subsidy); and t_m = average tax rate on imports.

The most important tradable inputs used in the energy sector are capital goods and petroleum-based fuels. Some countries may have other fuels available, such as natural gas or coal deposits. If no clear-cut export market exists for these indigenous energy resources, then they cannot be treated like tradables. In addition, if there is no alternative use for the fuels, an appropriate economic value is the MSC of production, that is, of extracting gas or coal plus a markup for the discounted value of future consumption foregone, or "user cost". (10) If another high value use exists for this fuel, the opportunity cost of not using the resource in the alternative use should be considered the economic value of the fuel. The most important nontradable primary factor inputs are labor and land, the next subjects for discussion.

Consider a typical case of unskilled labor in a labor surplus country--for example, rural workers employed for dam construction. The foregone output of workers used in the electric power sector is the dominant component of the shadow wage rate (SWR). Complications arise because the original rural income earned may not reflect the marginal product of agricultural labor and, furthermore, for every new job created, more than one rural worker may give up former employment. (11) Allowance must also be made

$$ESWR = a.m. + c.u.$$

where m and u are the foregone marginal output and overhead costs of labor in domestic prices, and a and c are corresponding conversion factors to convert these values into border prices.

The appropriate shadow value placed on land depends on its location. Usually, the market price of urban land is a good indicator of its economic value in domestic prices, and the application of an appropriate conversion factor, such as the SCF, to this domestic price will yield the border-priced cost of urban land inputs. Rural land that can be used in agriculture may be valued at its opportunity cost, the net benefit of foregone agricultural output. The marginal social cost of other rural land is usually assumed to be negligible, unless there is a specific reason to the contrary. Examples might be the flooding of virgin jungle because of a hydroelectric dam that would involve the loss of valuable timber, or spoilage of a recreational area that has commercial potential.

The shadow price of capital is usually reflected in the discount rate or accounting rate of interest (ARI), which is defined as the rate of decline in the value of the numeraire over time. Although there has been much discussion concerning the choice of an appropriate discount rate, (12) in practice the opportunity cost of capital (OCC) may be used as a proxy for the ARI, in the pure efficiency price regime. The OCC is defined as the expected value of the annual stream of consumption, in border prices net of replacement, which is yielded by the investment of one unit of public income at the margin.

Usually the estimation of shadow prices on a rigorous basis is a long and complex task. Therefore, the energy sector analyst is best advised to use whatever shadow prices have already been calculated.

Alternatively, the analyst would estimate a few important items such as the standard conversion factor, opportunity cost of capital, and shadow wage rate. When the data are not precise enough, sensitivity studies may be made over a range of values of such key national parameters.

4. CASE STUDY: ELECTRIC POWER

The economic evaluation of an electric power project consists of three basic steps: (a) demand forecast; (b) least cost investment program; and (c) cost-benefit analysis. The investment decision also interacts with the output pricing policy as summarized in Figure 2.

Investment-Pricing Interaction

The load-demand forecast is made by systematically taking into account all factors that might have quantitative relationships with future electricity consumption such as price, income and economic growth, population growth, energy substitution, etc., as well as judgemental factors including effects of energy

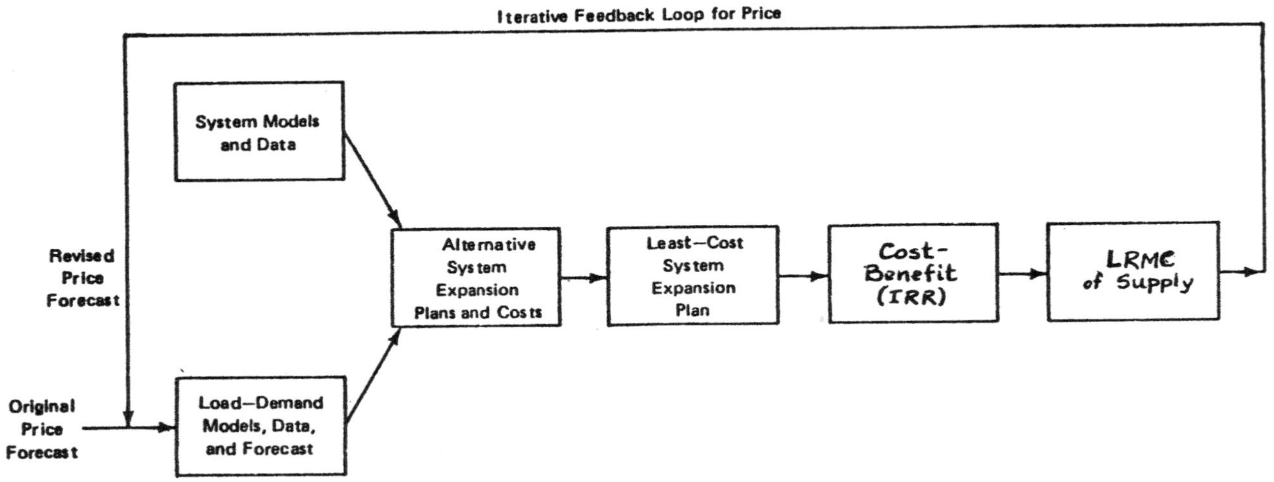

Figure 2. Interaction of Investment Decision, Price and Demand Forecast

conservation programs. The assumptions made regarding the future evolution of electricity price is especially critical. Several alternative long-run investment programs (or system expansion plans) are then identified to meet the demand growth, subject to various constraints including minimum reliability of supply standards, environmental and safety criteria, and so on. (13) The least cost or cheapest alternative (in present discounted value terms and using shadow prices) is selected as the optimal program.

However, economic theory indicates that the benefits (to society) of electricity consumption will be maximized if the output prices are set equal to the marginal costs of supply based on the least cost plan. (14) In practice, prices will be adjusted to reflect not only long-run marginal costs (LRMC) of supply, but also other financial, social and political criteria. Whatever future prices result, they must be compared to the prices assumed in making the original demand forecast. If there is inconsistency then the demand forecast may have to be adjusted and the investment plan reviewed again.

The above considerations indicate that the investment and pricing decisions are closely interrelated. (15) The simplified case study presented below illustrates this point, and also indicates how shadow prices are used in practice. The project is a hydroelectric plant with a 950 million m^3 of storage capacity providing about 7,500 GWh of energy in an average year, and about 5,350 GWh in a dry year. The cost of the associated transmission spur line and substation to link up with the interconnected national power gride, are included in the project.

Demand Forecast

The demand forecast for the customers to be served by the power utility over the period 1981-1990 was prepared on the basis of historical trends and recorded indices for each consumer category for each of the years from 1973 to 1980 (see Table 4.1). Trend curves were derived to project monthly and annual requirements which were subsequently adjusted to reflect population growth trends, sector economic conditions and market saturation effects. Residential consumption grew at an average annual rate of 11.5% in the period 1974-1980 and it is expected to grow at an average annual rate of 12% in the period 1981-1990. Also residential consumption augmented its share of the market from 36% in 1973 to 42% in 1980 as a result of increased urbanization and access to service, as well as expansion of services to semi-rural areas. This trend is expected to continue over the next few yers and by 1990 residential consumption would represent about 47% of total sales.

Commercial consumption grew at an average annual rate of 9% in the period 1974-1980 and it is expected to grow at an average

	Residential	%	Commercial	%	Industrial	%	Government & Pub.Lighting	%	Total Sales	%	Losses and Station Use	1/ %	Total Requirements
1973	782		367		742		292		2183		433		2616
1974	923		411		809		305		2448		426		2874
1975	1042	39	450	17	855	32	316	12	2663	100	695	21	3358
1976	1175		513		945		347		2980		625		3605
1977	1237		589		995		353		3174		836		4010
1978	1392		628		1104		351		3475		787		4262
1979	1549		622		1168		374		3713		968		4681
1980	1672	42	673	17	1263	32	401	10	4009	100	1139	22	5226
Average Growth rate (1974-1980) %	11.5		9.0		7.9		4.6		9.1		14.8		10.4
1981	1808		741		1420		441		4410		1341		5751
1982	2029		813		1556		477		4875		1489		6364
1983	2278		893		1706		516		5393		1628		7021
1984	2558		981		1869		559		5967		1741		7708
1985	2875	44	1077	16	2049	31	604	9	6605	100	1897	22	8502
1986	3233		1182		2246		654		7315		2107		9422
1987	3637		1298		2461		708		8104		2171		10275
1988	4094		1425		2697		766		8982		2285		11267
1989	4610		1565		2956		829		9960		2466		12426
1990	5193	47	1719	16	3240	29	896	8	11048	100	2670	19	13718
Average Growth rate (1981-1990) %	12.0		9.8		9.9		8.4		10.7		8.9		10.1

1/ As a percentage of total requirements.

Table 4.1 Actual and Forecast Sales and Gross Generation for System (GWH) annual rate of 9.8% in the period 1981-1990. Commercial consumption which represented 17% of sales in the period 1973-1980 is expected to decrease slightly to 16% in the period 1981-1990. Industrial consumption grew at an average annual rate of 7.9% in the period 1974-1980 and it is expected to grow at an annual average rate of 9.9% in the period 1981-1990. The industrial consumption share of the market which decreased from 34% in 1973 to 32% in 1980, is expected to decrease slightly to 29% by 1990. Government and Public Lighting grew at an average growth rate of 4.6% in the period 1974-1980, and it is expected to grow at an average growth rate of 8.4% in the period 1981-1990, as a result of future distribution system expansion. Government and Public Lighting consumption has shown a slightly decreasing share of total consumption and is expected to represent about 8% of total sales by 1990.

Other consumption, which includes station use (mainly pumping for storage) and losses, grew at an average annual growth rate of 14.8% in the period 1974-1980 and is expected to grow at about 8.9% in the period 1981-1990. Actual losses represented about 22% of total gross generation requirements in the period 1974-1980. Historical losses are relatively large and may contain a substantial amount of theft. This problem is presently being studied for the entire sector and remedial action would be implemented as a result of the study. The utility expects to reduce such losses gradually to about 15% by 1987, as a result of the measures to be implemented.

Least Cost Solution

Because of systems effects, the hydro project cannot be analyzed by itself. Instead, it must be clearly demonstrated using shadow prices that this hydro scheme is the next project which appears in the optimal or least-cost generation expansion program. We outline next how this least-cost program was selected from among the various other alternative programs.

The utility has prepared a first report to establish the least cost expansion for the sector for the period 1984-1988, on the basis of the economic comparison of incremental costs for several alternative expansion programs, including termal plants (both fuel-oil and coal-fired steam units), gas turbines and hydroelectric projects. The alternative expansion programs were designed to meet the sector energy and peak demand requirements through 1990.

The studies are carried out using a model with stochastic variables for both hydrology and demand, and with the restriction that under adverse hydrological conditions occurring less than once in 20 years, energy requirements would not have to be met.

The operation of the system is then simulated through computer models to estimate the operational costs of each alternative program and to compute the probability for the program of not meeting the system requirements. Adjustments are introduced to ensure that capacity deficits would have a probability of occurrence of less than 5% in any one year of the period. The alternative program representing the least-cost alternative is then selected as the System Expansion Program, representing the least net present value of all the costs associated to it, discounted at 13% (which represents the estimated opportunity cost of capital).

Sensitivity analysis of the program costs to the discount rate (10% and 15%) and to fuel prices (plus minus 20%) do not show any substantial changes in the merit order of the best and second-best alternatives. The analysis also shows that the results are insensitive to a reasonable range in variation of the standard conversion factor (corresponding to the shadow-priced exchange rate). Also, the results are insensitive to the use of other shadow prices since neither the timing of individual plants within a proposed expansion program, nor the ranking of the several alternative programs are affected whether conversion factors are or are not used. The utility annually reviews and revises if necessary the least-cost expansion program for the sector.

Cost-Benefit Analysis

In the cost-benefit calculation, it is once again impossible to isolate the effects of the hydro project from the rest of the system. Therefore costs and benefits associated with the utilities total investment program from 1981-91 are compared (see Table 4.2).

The investment costs are based on the least-cost generation program and the associated transmission and distribution expansion programs required to deliver power to consumers. Annual recurrent costs based on operation and maintenance (O&M) expenditures and fuel purchases are also included in the table. All market priced costs are converted to border (shadow) prices by applying conversion factors as indicated.

Ideally, benefits should be measured on the basis of the total willingness-to-pay of consumers for electricity supplied. In economic theory this is represented by the area under the demand curve OBCD in Figure 3, where the price is p_0 and consumption is Q_0.

Unfortunately the position of the demand curve is not easily determined. Therefore, we use the revenues represented by area OACD (or $p_0 \times Q_0$) as a minimum (proxy) measure of benefits, since this is readily available from billing information. With this approach the consumer surplus benefits ABC are neglected. (16) The

Year	Year No.	Investment Costs	O&M and Fuel Costs	Total Costs	Total Benefits from Incremental Sales Revenues 1/		
1981	1	17.6	-	17.6			
1982	2	90.9	-	90.9			
1983	3	73.2	-	73.2			
1984	4	117.5	-	117.5			
1985	5	133.6	-	133.6			
1986	6	214.0	-	214.0			
1987	7	173.2	12.7	185.9	89.7		
1988	8	175.8	80.7	206.5	161.0		
1989	9	109.5	57.7	167.2	231.1		
1990	10	145.3	98.9	244.2	309.4		
1991	11	196.5	.	295.4	.		
1992	12	0.0	.	98.9	.		
.	.	.	.	.	.		
.	.	.	.	.	.		
.	.	.	.	.	.		
2036	56	0.0	98.9		309.4		
		Present Values (Market Prices)					
Discount Rate						Net Present Value	
	8	862	659	1521	2137	616	
	10	768	464	1232	1516	284	
	12	687	339	1026	1114	88	
	14	617	255	872	842	-30	Rate of
	16	558	194	752	650	-102	return
	18	505	152	657	510	-147	13%
	20	460	120	580	407	-173	
		Present Values (Border Prices)					
Conversion Factor =		0.86	0.91		0.92		
Discount Rate							
	8	741	600	1341	1966	625	
	10	660	422	1082	1395	313	
	12	591	308	899	1025	126	
	14	531	232	763	775	12	Rate of
	16	480	177	657	598	-59	return
	18	434	138	572	469	-103	14%
	20	396	109	505	374	-131	

1/ Assumes 1981 average tariff level will continue unchanged in real terms.

2/ All costs and benefits in millions of Pesos (P 1 = US$1 at official exchange rate).

Table 4.2. Costs and Benefits for Expansion Program 2/

total benefits shown in Table 4.2 are actually the incremental sales revenues derived from the increased production of electricity made possible by the 1981-90 investment program. The market priced benefits are converted to border prices using the standard conversion factor (SCF = 0.92).

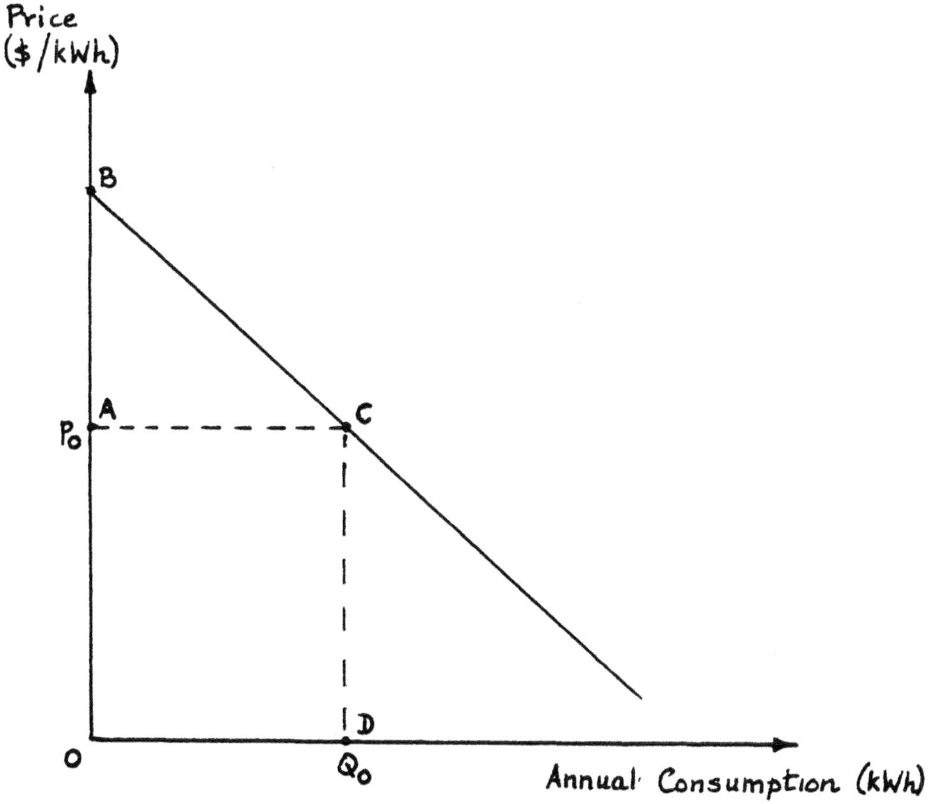

Figure 3. Benefit Measurement

On this basis, the internal rate of return (IRR) using shadow prices is about 14% (see Table 4.2), which compares favorably with the opportunity cost of capital (OCC) of 13%. Even with this minimum measure of benefits IRR > OCC, and therefore NPV > 0. When the neglected consumer surplus benefits are included, NPV will be even larger and the project and investment program are acceptable. The IRR in market prices is also calculated in the table, for comparison--it is quite similar.

A sensitivity analysis was carried out to estimate the impact on the IRR of possible changes in cost and revenues. These are summarized below:

Costs	Benefits =	85%	100%	115%
100%		11	15	17
115%		10	12	15

Figure 4: Internal Rate of Return on Investment Program (%)

If the cost of the program increases by 15%, the rate of return would be about 12%. If the program costs do not increase but benefits are 15% lower than estimated, the return would be about 11%. If the program costs increased by 15% and the benefits decreased by 15% the return on the investment program would be about 10%.

Next, we discuss another important conclusion we can draw regarding pricing policy, using the above analysis. An approximate measure of the average LRMC is provided by the following ratio:

$$LRMC \approx PVC(r)/PVE(r)$$

where $PVC = \sum_{t=o}^{T} C_t/(1+r)^t$ = present value of total costs (as shown in the table)

$PVE = \sum_{t=o}^{T} kWh_t/(1+r)^t$ = present value of stream of incremental output associated with above costs.

Assume that both PVC and PVE are evaluated using the opportunity cost of capital (r = OCC) as the discount rate (based on efficiency prices).

Suppose that the average electricity price level p used for estimating revenues is equal to LRMC, i.e., p = LRMC.

Therefore, $p = PVC(r) / PVE(r)$

or $p \cdot PVE(r) = PVB(r) = PVC(r)$

where PVB is the present value of benefits using incremental revenues as a proxy.

Clearly, from the above r = OCC = IRR by definition. Thus, since we use incremental revenues as benefits, and the IRR is close to the OCC, then we may conclude that, in the case study analyzed, the average price also approximates the LRMC (which is economically optimal). It is easy to see that if IRR $<$ OCC, then p $<$ LRMC and vice versa. Therefore, the value of IRR in this calculation serves a very useful purpose (although it understates the full benefits). The test IRR $\gtreqless$ OCC is not so much an accept/reject criterion for the investment program and project; it indicates whether the average tariff level is above or below the optimal or average LRMC level.

In summary, correct interpretation of the IRR value allows us to make a first round judgement regarding the adequacy of average price relative to LRMC. (17)

5. CASE STUDY 2: SOLAR (PHOTOVOLTAIC) ENERGY

The economic evaluation of a new renewable technology such as a photovoltaic (PV) irrigation pumping system introduces several complicating features into the analysis.

We assume that the agricultural area under consideration is so remote that electrification via the main power grid would be prohibitively expensive. The principal competitor of the PV system in this case is the diesel pump. Since the comparison is between two discrete alternatives, we need to only look at a single representative farm. But as discussed below, both economic and financial calculations need to be carried out, but carefully separated.

Consider a typical 1 hectare farm that requires about $20m^3$ per day of water with a head of 5m. Since the benefits are identical in both cases, we first seek to find the least cost alternative by comparing diesel and PV system costs. The basic data for a simplified hypothetical example are summarized in Table 5.1. (18)

Investment Decision

Let us compare the present value of costs (PVC) of the two alternatives over a 10 year period in (economic) shadow prices (Border rupees - BRs).

<u>Solar</u>: $$PVC_{SE} = 60{,}000 + \sum_{t=0}^{9} 1000/(1.1)^t = \text{BRs } 66{,}760$$

<u>Diesel</u> (19): $$PVC_{DE} = 36{,}000 + 36{,}000/(1.1)^5 + \sum_{t=0}^{9} [1{,}050/(1.1)^t + 1{,}200 \times (1.03/1.10)^t] = \text{BRs } 74{,}540$$

	Solar PV		Diesel	
	Market Prices (Dom. Rs.)	Shadow Prices (Bord. Rs.) 1/	Market Prices (Dom. Rs.)	Shadow Prices (Bord. RS.) 1/
Initial Cost	66,000	60,000	28,800	36,000 3/
Annual Maintenance Costs	1,100	1,000 2/	1,500	1,050 4/
Annual Fuel Costs	-	-	840	1,200 5/
Lifetime	10 years	-	5 years	-
Discount Rate (%)	15 6/	10 7/	15 6/	10 7/
Annual Inflation Rate (%)	10	-	10	-

1/ All foreign costs converted into border rupees (Rs.) at official exchange rate (OER) of US$1 = Rs. 20.
2/ All foreign exchange costs with 10% import duty.
3/ Import subsidy of 20% on diesel pumps provided to farmers.
4/ Conversion factor = 0.7 based on spare parts and labor.
5/ Subsidy of 33% provided on diesel fuel imported at US$0.3 per liter. International fuel price assumed to rise at 3% per annum in real terms.
6/ Bank borrowing rate for farm loans.
7/ Opportunity cost of capital used as proxy for ARI.

Table 5.1 Basic Data for Solar PV and Diesel Pumping Systems

On the basis of the above results, the solar PV system is the least cost alternative for irrigation pumping in this area.

Next, assume that due to the increased irrigation, the shadow priced value of the farmer's annual output of grain increases from BRs 10,000 to BRs 20,500 per hectare, based on the export price of grain. (20) The present value of benefits for the 10 year period is given by:

$$PVB_E = \sum_{t=0}^{9} (20,500 - 10,000)/(1.1)^t = \text{BRs } 70,970$$

Since NPV = $PVB_E - PVC_E - PVC_{SE}$ = BRs 4,210, there is a positive net benefit to the country by installing solar PV pumps in farms. These results are summarized in Table 5.2. Therefore, from a national viewpoint, the government is justified in taking a policy decision to encourage the use of solar PV systems by farmers, i.e., to stimulate demand and encourage this form of energy consumption.

Implementing the Investment Policy

However, farmers will make their decisions on irrigation pumping, on the basis of private financial costs and benefits. Therefore, let us analyse the present value of costs of the solar and diesel alternatives in financial terms, using market prices (Domestic rupees - DRs):

<u>Solar</u> (21): $$PVC_{SF} = 66,000 + \sum_{t=0}^{9} 1100x(1.10/1.15)^t = \text{DRs } 75,080$$

<u>Diesel</u> (22): $$PVC_{DF} = 28,800 + 28,800x(1.10/1.15)^5 + \sum_{t=0}^{9} [1500x(1.10/1.15)^t + 840 \times (1.03x1.10/1.15)^t] = \text{DRs } 72,110$$

Clearly, since $PVC_{SF} > PVC_{DF}$, if the average farmer wished to acquire a pumping system, he would prefer to purchase a diesel pump rather than a solar PV system. The market price signals are distorted so that the private individual's decision does not coincide with the national policy decision based on shadow prices (as summarized in Table 5.2). Therefore, the government must adjust the market prices or adopt other policies that will make the solar PV option attractive to farmers.

Item	A National Viewpoint: Shadow Prices (Border Rs.)	B Private Viewpoint (Initial): Market Prices Before Policy Changes (Domestic Rs.)	C Private Viewpoint (Interim): Market Prices After First Policy Change (Domestic Rs.)	D Private Viewpoint (Final): Market Prices After Second Policy Change (Domestic Rs.)
Solar PV Costs	PVC_{SE} = 66,760	PVC_{SF} = 75,080	PVC_{SF} = 75,080	PVC_{SF} = 75,080
Diesel Costs	PVC_{DE} = 74,540	PVC_{DF} = 72,110	PVC_{DF} = 85,070	PVC_{DF} = 85,070
Irrigation Benefits	PVB_E = 70,970	PVB_F = 73,670	PVB_F = 73,670	PVB_F = 78,000

CONCLUSIONS

Condition		Consequence
A. Shadow Priced Values:		
$PVC_{SE} < PVC_{DE}$	-	Solar PV is the economically preferred least-cost technology.
$PVB_E > PVC_{SE}$	-	Investment in solar PV is economically justified.
B. Market Priced Values Before Policy Change :		
$PVC_{SF} > PVC_{DF}$	-	Farmers will prefer diesel to solar PV pumps, financially.
C. Market Priced Values After First Policy Change :		
$PVC_{SF} < PVC_{DF}$	-	Farmers will prefer solar PV to diesel pumps, financially.
$PVB_F < PVC_{SF}$	-	Farmers will find solar PV pumps financially unprofitable to install.
D. Market Priced Values After Second Policy Change:		
$PVC_{SF} < PVC_{DF}$	-	Farmers will prefer solar PV to diesel pumps, financially.
$PVB_F > PVC_{SF}$	-	Farmers will find solar PV pumps financially profitable to install.

Table 5.2 Economic and Financial Tests for Investment Decision (all amounts are in present value terms)

First, the authorities could remove the 10 percent import duty on solar PV systems and spare parts. The PVC_{SF} = DRs 68,250 and the solar PV pump is the least cost alternative in both financial and economic terms. But suppose that this measure would also require the reduction of import duties on many similar photovoltaic and electric components, which the government is unwilling to consider.

A second policy alternative is to raise the price of diesel fuel until PVC_{DF} exceeds PVC_{SF}. But let us assume that there is a strong anti-inflation lobby that can block increases in the diesel fuel price on the grounds that this would raise transport costs too much.

A third approach might be for the government to legislate that farmers could no longer buy diesel pumps. This non-price policy option is cumbersome and has its own host of implementation difficulties. A fourth possibility is for the govenment to provide low interest agricultural loans or credits to buy solar pumps.

Finally the government could remove the import subsidy on diesel motors. This increases PVC_{DF} to DRs 85,070, and since this is greater than PVC_{SF}, the farmers will voluntarily prefer solar PV pumps to diesel pumps. Let us assume that this policy alternative is selected for implementation (see Table 5.2).

But we still have to consider the financial benefits aspect. The irrigation program increases the market value of farm output from DRs 8,500 to DRs 17,425 based on the guaranteed government purchase price for grain (which is 85 percent of the world market or export price). The present value of financial benefits for 10 years of output is:

$$PVB_F = \sum_{t=0}^{9} (17{,}425-8{,}500) \times (1.10/1.05)^t = \text{DRs } 73{,}670$$

Since PVB_F is less than the value PVC_{SF} = DRs 75,080, the average farmer will be financially worse off if he adopts the (least cost) solar PV technology.

The government must therefore make a further change in policy. For example, it could raise the domestic grain price up to 90 percent of the world market price. (23) Then the new PVB_F = 78,000 > PVC_{SF} and farmers will voluntarily begin using the solar PV systems (see Table 5.2).

Thus, this case study shows the demand management by implementation of a given investment decision (based on shadow prices) may require a related series of wide-ranging policy decisions by the government.

FOOTNOTES

(1) For a more detailed description, see: Warren C. Baum, "The World Bank Project Cycle", Finance and Development, December 1978, pp. 2-9.

(2) For details, see: Ezra J. Mishan, Cost-Benefit Analysis, Praeger Publishers, New York, 1976, ch. 25-30.

(3) See: Francis J. Bator, "General Equilibrium, Welfare and Allocation", American Economic Review, (March 1957), pp. 22-59, and Ezra J. Mishan, Cost Benefit Analysis, (New York: Praeger, 1976), part vii.

(4) See: David G. Luenberger, Introduction to Linear and Non-Linear Programming, Reading, Mass.: Addison-Wesley, 1973), chap. 4; and Peter G. Sassone, "Shadow Pricing in CBA: Mathematical Programming and Economic Theory", The Engineering Economist, vol. 22, (Spring 1977), pp. 219-33.

(5) See: Lyn Squire and Herman G. van der Tak,Economic Analysis of Projects, (Baltimroe, Md.: Johns Hopkins University Press, 1975), chap. 6.

(6) See: Partha Dasgutpa, Amartya Sen and Stephen Marglin, Guidelines for Project Evaluation (UNIDO), (New York: United Nations, 1972), chap.7; and Ian M. D. Little and James A. Mirrless, Project Appraisal and Planning for Developing Countries, (New York: Basic Books, 1974), chap. 4.

(7) Ian M. D. Little and James A. Mirrlees, Project Appraisal and Planning for Developing Countries (New York: Basic Books, 1974), chap. 9, and Lyn Squire and Herman van der Tak, Economic Analysis of Projects, Baltimore, Md: World Bank-Johns Hopkins University Press, 1975), chap. 3. A numeraire based on private consumption is advocated in: Partha Dasgupta, Amartya Sen and Stephen Marglin, Guidelines for Project Evaluation (UNIDO), (New York: United Nations, 1972). For a discussion of the different assumptions used in various approaches to shadow pricing, see Social and Economic Dimensions of Project Evaluation, Harry Schwartz and Richard Berney (eds), Washington, D. C.: Inter-American Development Bank, 1977); and Deepak Lal, Methods of Project Analysis, World Bank Staff Occasional Papers No. 16, (Washington, D. C.: World Bank, 1974).

(8) Mohan Munasinghe, The Economics of Power System Reliability and Planning (Baltimore: Johns Hopkins University Press, 1979), Chap. 9 and appendix B.

(9) For a more detailed discussion of alternative interpretations of the SCF and SER, see Lynn Squire and Herman G. van der Tak, Economic Analysis of Projects, and Bela Balassa, "Estimating the Shadow Price of Foreign Exchange in Project Appraisal", Oxford Economic Papers, Vol. 26 (July 1974), pp. 147-68.

(10) For a review of practical energy pricing rules, especially with respect to nonrenewable energy sources, see: Mohan

Munasinghe, "An Integrated Framework for Energy Pricing in Developing Countries," The Energy Journal, vol. 1, (July 1980), pp. 1-30.

(11) See John R. Harris and Michael P. Todaro, "Migration, Unemployment and Development", American Economic Review, vol. 60, (March 1970), pp. 126-42.

(12) See for example, Stephen Marglin, "The Social Rate of Discount and the Optimal Rate of Investment", Quarterly Journal of Economics, (1963); Ezra J. Mishan, Cost-Benefit Analysis, chap. 31-4; and Arnold C. Harberger, Project Evaluation, (London: Macmillan and Co., 1972).

(13) Ideally, the reliability level also should be optimized. For details, see: Mohan Munasinghe, "A New Approach to Power System Planning", IEEE Trans, PAS, vol. PAS-99, May-June 1980, pp. 1198-1209, also available as Reprint No. 147, World Bank, Washington, D. C.: and Mohan Munasinghe and Mark Gellerson, "Economic Criteria for Optimizing Power System Reliability Levels", The Bell J. of Econ., vol. 10, Spring 1979, pp. 353-65, also available as reprint No. 112, World Bank, Washington, D. C.

(14) For details, see: Mohan Munasinghe, "Principles of Modern Electricity Pricing", Proc. IEEE, vol. 69, March 1981, pp. 332-48.

(15) For details, see: Mohan Munasinghe, "Optimal Electricity Supply: Reliability, Pricing and System Planning", Energy Economics, vol. 3, July 1981, pp. 140-52.

(16) Consumer surplus benefits may be estimated in cases where grid supplied electricity is unavailable, in terms of the greater amounts consumers are willing-to-pay for alternative sources of energy, e.g., private autogeneration by industrial consumers or kerosene-use for lighting by rural households.

(17) We note that a more elaborate tariff study must be carried out to further check whether the existing tariffs conform to the structure of LRMC; for details see: Mohan Munasinghe, "Principles of Modern Electricity Pricing", op.cit., 1981.

(18) This example serves only to illustrate the principles involved. In actual practise, a range of parameters (e.g., different farm sizes, insolation rates, cropping patterns, system costs, etc.) would have to be considered.

(19) The compounding effect of the real increase in fuel prices of 3 percent per annum will partially offset the effect of discounting at the opportunity cost of capital of 10 percent.

(20) This increase in output value is net of any changes in costs of other imports such as fertilizer and labor.

(21) In the financial calculation, the compounding effects of the 10 percent annual inflation rate will partially offset the effect of discounting at the market interest rate of 15 percent.

(22) The compounding effects of the 3 percent per annum increase in real fuel prices is included.

(23) They may, however, have to also provide an equivalent subsidy to urban grain consumers to keep down their cost of living.

INSTITUTIONAL FRAMEWORKS FOR ENERGY PLANNING IN DEVELOPING COUNTRIES

Sergio F. Garribba

IEFE, Istituto di Economia delle Fonti di Energia
Via R. Sarfatti 25, Milano, Italy 20136

ABSTRACT

Energy issues are also governmental and institutional issues which involve the history, values and environment of the different developing countries. All energy development indeed is the result of organized effort channelled through institutions set up for specific purposes (government bodies, central and local administration, utilities, state enterprises, user associations, etc.). Questions concern both the roles and organizations which are the most efficient and adequate to deal with the diversity of energy problems. Formal institutions can be grouped according to several criteria. Three major groupings cater for energy policy and planning, management of energy production and demand, and for promotion and transfer of appropriate energy technology. It is apparent that, to a varying extent, institutions should be linked and their activities involved in all these groups. More precisely, energy policy and planning should be concerned with policy formulation, implementation of planned actions, allocation of tasks in participating institutions, monitoring of results and identification of correctives. On the other hand, in the chain of traditional and non-traditional energy sources, specific roles which must be covered involve energy supply, transportation and conversion, energy utilization and pricing policy. Finally, it is important to increase energy self-reliance by encouraging the development technology and a flow of technical information. In this perspective, a co-ordinated national programme directed through a specialized institution can be envisaged to ease the transition from oil and woodfuels.

H. Neu and D. Bain (eds.), National Energy Planning and Management in Developing Countries, 369–415.

1. INTRODUCTION

Energy planning as part of the economic policy of Less Developed Countries (LDCs) must seek to balance those measures which improve national longterm security against those which promote immediate socio-political objectives of equity and welfare, while promoting the utilization of domestic resources and preserving the essential character of the national environment. It has been made clear that during the next twenty years and probably also beyond that time span, the energy problems in LDCs entail a transition from oil and woodfuels. This double transition requires prompt policy actions which must develop in two directions, i. e. on the supply side, through the development of alternative and supplementary energy technologies, and on the demand side, through rationalization and improved efficiency. An appropriate infrastructure for energy transportation and distribution should also be enforced to cope with the changing environment (1).

Specifically, in the definition of its energy policies the developing World faces a tough dilemma: if the non-oil-producing Third World countries, in particular, increase their dependence upon oil, they will store up trouble for themselves for the next decade or the one after that, when oil supplies will begin to decline. Yet, if they do not use more oil, their development will suffer immediately. It is clear that the Third World will choose to continue to increase its oil consumption. For the modern industrial and commercial sector, a different course cannot be advised at this time, given the lack of proven and competing alternatives. For the rural sector and for parts of the residential sector, however, there may be room for a more cautious approach. Oil-based technology may not be always the most economical and convenient. Until this is clarified, it may be prudent to increase experiments with non-oil alternatives including those based on renewable and decentralized energy. Meanwhile any decisions in the direction of a major switch to oil in rural areas should be made very cautiously. The ability to decouple the incremental expansion of the national economy from (directly proportional) increases in energy consumption would assist by postponing the inevitable date at which additional energy resources must be searched for (2).

Energy policy implementation and programmes require varying strategies and efforts. Not only will these hinge upon the availability of primary resources and funds of LDCs, but also on their resources of human talent, technological capabilities and organization. It is the adequacy of the overall institutional framework in all its aspects that largely determines the quality and pace of development. As the energy system modernizes, the number of specific formal institutions multiplies, since it is only through new organizations and decision processes that complex modern technologies can be managed and integrated. The importance of the problem of adequate and suitable institutions becomes obvious if one considers the range of disciplines which is involved when deciding matters related to energy, as well as the trade-offs between the short and the medium or long term, or among different energy sources in their final uses.

The appropriateness of institutions to take these aspects into account in considering energy related problems is a question that is not easily answered. It is necessary to analyse the type of expertise, information inputs, and decision-making processes that may contribute to the solution of energy problems in the future. Against this, one can evaluate the existing institutional arrangements. Such analysis may reveal the existence of serious drawbacks and barriers that prevent the right decisions from being taken at the right time. There may be institutional weaknesses which give cause for conflicts within national institutions, whereby compromises are sought to satisfy special interest groups which do not necessarily result in the benefit of a given country.

Moreover, natural resources, funds and technical expertise can be imported, whereas the organization and institutions needed to find appropriate solutions for energy problems cannot. As a consequence, it must be stated that the real energy challenge for each developing country is to find ways of organizing itself to deal effectively with the problem of finding enough energy to maintain economic growth and accomplish the transitions that must be sought for.

2. INSTITUTIONAL FRAMEWORKS

The interplay of competing and often contradictory political, economic, social and regional forces within the constitutional systems of LDCs has produced a multitude of institutional types. This diversity is seldom determined by strict application of established organizational principles. Choices are influenced by factors reflecting divergent views about the proper sphere of government activity, tradition and innovation, programme importance and status, political and administrative autonomy and, most important, who should exercise control. The theoretical arguments frequently have little relevance to the real issues.

The assortment of institutional types adds to and compounds with the variety of economic situations. It has been remarked that the developing-country group of the South is much more heterogeneous than the North, with an enormous diversification of resource endowments and development conditions. At one end of the spectrum are countries with little economic activity beyond subsistence, agriculture and livestock tending. Conversely, there are countries such as Algeria, India, Mexico and the Philippines, with large and dynamic manufacturing and commercial sectors that are close to entering into the industrialized category (1). On the other hand, as far as energy policy is concerned, it must be noted that the Third World has for a long time neglected comprehensive energy planning, although the trend is changing rapidly. This lack of attention in developing countries to broad questions of energy policy is not surprising. In the past ten or twenty years the leaders of many developing countries had to turn their attention to the roots of backwardness and immediate pressing problems like food production, health care, education, establishment of basic infrastructures, employment, and the development of raw materials and export industries. To meet these needs, the developing countries during this same period increased their consumption of commercial energy - mainly oil - more than three times as quickly as the industrialized countries (although from a smaller base).

Even today, when there is a growing awareness of the broad importance of energy questions for economic development, the immediate and overwhelming concern of newly appointed energy ministries is not the question of long-term needs and supplies, but the question of how to pay for this month's petroleum imports or how to finance the construction of a new hydroelectric project. Notwithstanding their importance, these problems are only part of the central energy issue facing the people of the developing World.

In principle the taxonomy of institutions relevant to energy policy making and planning can be constructed according to several criteria. For instance, it can be thought of as a classification mode in terms of functions or roles, supply and demand sectors, jurisdiction. In this view, a number of divisions which do not pretend to be exhaustive can be traced, as in the following.

(i) Functions or roles

a) pathfinding and planning
b) policy enforcement and monitoring
c) governance over energy source supply, conversion, distribution and usage
d) price fixing
e) energy projects appraisal and management
f) energy technology research, development and demonstration
g) standardization and equipment testing
h) education, training and extension service.

These functions may be located in different centres or units, or in different offices within the same centre or unit. Often, some functions are clustered in a sectoral entity (cases are offered by national electric utilities, national oil companies, etc.) or others in a functional unit dealing with national administration. The assignment of political and financial powers will be particularly important.

(ii) Supply sectors

a) traditional energy resources
b) modern (or commercial) energy sources
c) renewable non-traditional energy sources.

(iii) Demand sectors

a) agriculture and rural systems
b) residential usages and urban systems
c) transportation
d) industry and services.

These and other sectors or issues may be ranked in importance variously for different countries, or even by different individuals. The ranking affects in part how these issues are examined institutionally and how the tradeoffs between sectoral goals are decided. Energy is related to each sector and decisions in one sector may compete with goals in another. Some issues (such as rural development or export support) may cut across several sectors and divisions. Moreover a hypothetical national enterprise (such as a national oil company) will have a say on supply, prices, investment decisions, allocation of products.

(iv) Jurisdiction

a) international
b) national

c) regional
d) local.

This division delineates the realm of operation and also specifies the area of influence. Each decision will be examined in relation to a specific range of impact. For instance, a decision to introduce diesel generators by a local electricity producer will have different consequences than a similar decision taken by a national agency.

It is patent that many, if not all institutions will be directly or indirectly involved in all the four groups. Although grouping should permit a rational approach to evaluating the adequacy and types of linkages, of decision and information flows, it is believed that the method of portraying such linkages and flows in abstraction leaves much to be desired (3). More rewarding results are obtained from the consideration of the roles, organization and linkages of institutions with reference to some main issues of energy policy and planning. In this view problems will be dealt with which emerge in relation with national energy policy governance (Chapter 3), supply and demand of commercial, traditional and renewable energy sources (Chapter 4 through 6), energy technology research and development (Chapter 7).

3. INSTITUTIONS FOR NATIONAL ENERGY GOVERNANCE

3.1 Need for a centralized planning and energy policy coordination

A key role of planning is to ascertain the choices which are available and evaluate their implications for national development. To do this energy planning must take shape in close coordination with the forces which define national priorities. Owing to the intersectoral and pervasive links of any energy policy, energy planning may become misleading when it ignores the social and economic decisions that are made continuously throughout the Country. For energy planning to be relevant at all, it should understand its proper place in the decision-making process. Energy planning as well as energy analysis can identify some of the possible long-term effects of any particular action; for instance, it could tell us that an emphasis on food self-sufficiency may require more oil, and this in turn will increase the amount of foreign exchange flowing out of the Country. Policies affecting rural to urban migration in turn can be shown to imply a certain demand for energy for the transport sector over the coming decades. An overview of the whole energy sector, in fact, may allow decision-makers to evaluate solutions to their various goals that would normally be missed when dealing with a specific subject.

National planning cannot happen too quickly. But it must be undertaken critically. Time must be taken to formulate hypotheses, to think, and to learn. Planning must be formulated initially around policy targets to which a variety of alternative actions and consequences can be related. At the same time, the context of discussion must be broad enough to allow connections to other related issues, such as welfare, environmental quality, and employment. In this context national energy planning can be understood as a process for assessing the economic conditions and prospects, setting national goals and priorities, and then letting institutions and market forces work. Furthermore, energy policy and planning show major social, cultural and political components which are not often discussed with respect to energy problems. Rather, energy problems are often represented as problems of resources, technology, or the infrastructure of supply. These are all perceived as technical or scientific approaches not directly involving social and political values, or institutions.

In the light of all these remarks, it becomes rather obvious to advocate a type of institutional arrangement relying upon some entity characterized by strong centralized drive and coordination capability. Such an entity should be established within the Executive branch of the Government. There is an ever-present danger that innovative institutions which challenge accepted norms, demand new skills and approaches, and create conflicts with agency constituencies, will be poorly assimilated into the "system" and their purposes muffled or distorted. One way to kill a programme is to house it in a hostile or unsympathetic environment.

Thus, smooth and covered solutions could be conceived which would simply consist of a formal or informal gathering of various ministers or operational staff to discuss specific energy policy issues as well as the assignment of duties that may shift from one ministry to another. However, this denial of centralization seems to apply to the complexity of energy planning as much as it applies to the political fragility of the Government. Creation of a new organization or agency, assuming the form of an interministerial body, or Commission, or new Ministry, is likely to present fewer problems than reform of an old one and enables the legislative power to finesse competing jurisdictional claims. Compromise arrangements are indeed always possible, and programmes or institutions can take root and grow within established departments under some circumstances if protected during the developmental period by a self-contained, relatively autonomous status.

3.2 The environment for national energy policy-making

This Section will focus on some of the major differences which are found in two, so to speak, extreme institutional frameworks. A first framework relies upon some Energy Planning Commission (or Interministerial Advisory Board) which responds to the President (or to the Chairman of the Council of Ministers), whereas a second framework hinges upon the establishment of a Ministry of Energy. Clearly, the Commission itself can be viewed as a transition stage towards the design of a more stable or permanent organization such as a dedicated Ministry. At least in principle, nevertheless, each of the two solutions may have its own merits. The following table summarizes the major differences between the two solutions.

Ministry	Commission
Larger	Smaller
Shorter term of office	Longer term of office
Less flexible rules	More flexible rules
Narrower constituency	Broader, more varied, constituency
Policy specialists	Policy generalists
More press and media coverage	Less press and media coverage
Power less evenly distributed	Power more evenly distributed
More prestige	Less prestige
More expeditious in debate and analysis	Less expeditious in debate and analysis
More reliance on staff	Less reliance on staff

Organizational arrangements are not neutral to energy policy. Organization is one way of expressing national commitment, influencing programme direction, and ordering priorities. Institutional location and environment, administrative arrangements and type of organization, can raise significant questions concerning the distribution and balance of power. Certainly, the factor of size explains much about why the two solutions differ. Because it is larger, the Ministry is a more structured body than the Commission (Fig. 1). The Commission, as a result, seems a more personal and individualistic institution than the Ministry (Fig. 2). For instance, the Commission can be

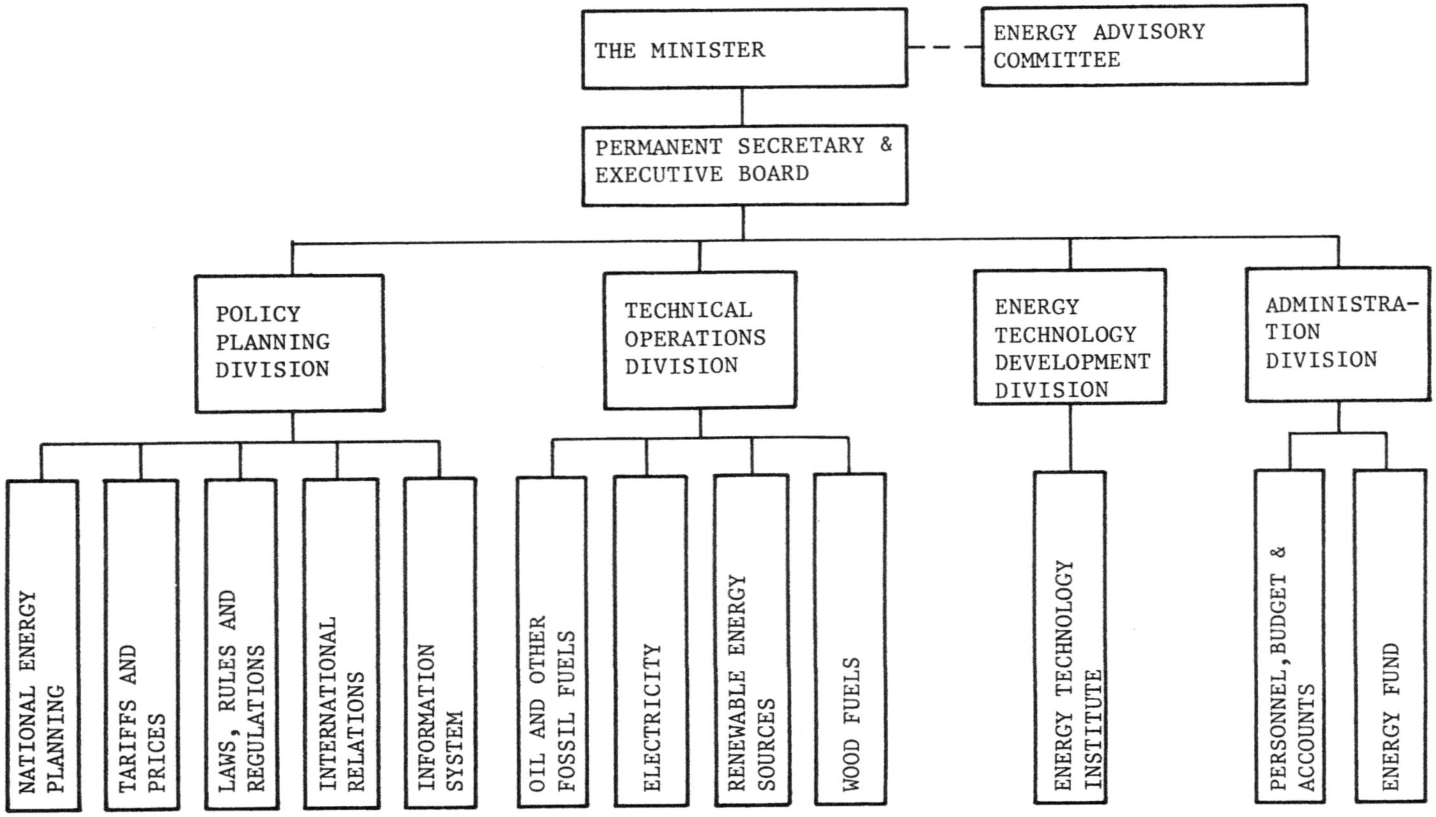

Figure 1 - Postulated organizational chart for the Ministry of Energy, its Divisions and Branches.

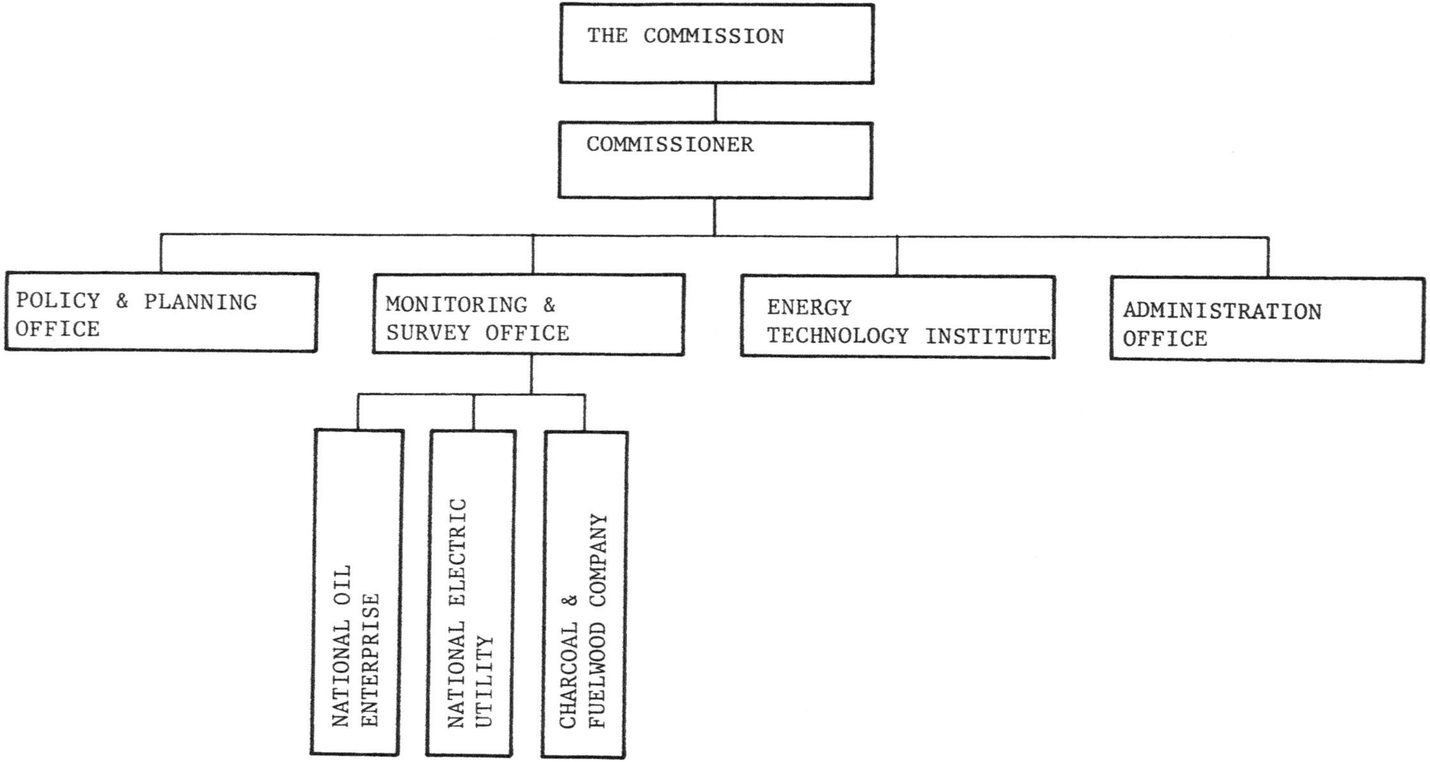

Figure 2 - Possible organizational chart for an Energy Planning Commission and its Offices.

thought of as an adequate solution for countries where managerial capability is scarce or the size of problems to be afforded does not deserve the importance of a full structured ministerial organization. A Commission would function to a large extent by unanimous consent, in effect adjusting or disregarding its rules as it goes along. Another difference between the solutions is that Ministers tend to be known as subject matter "specialists", while Commissioners tend to be "generalists". If the Commission is the Nation's great forum, then a Ministry would be its workshop. Power to influence policy could be more evenly distributed in the Commission than it is in the Ministry. There might be also many similarities between the Ministry and Commission. Both solutions could be essentially equal in power and share similar responsibilities in lawmaking, oversight and representation. They both may have heavy workloads, decentralized structures and somewhat parallel units.

An example of a possible institutional framework and organization determined by a Ministry of Energy will be considered more extensively in the following Sections and Chapters. It is clear that several remarks and suggestions would apply also to the case of the Energy Commission. In the choice of names for the different functions and institutional subjects we adopted a somewhat arbitrary nomenclature (for instance we considered Divisions and Branches within the Ministry, interministerial commissions and committees, public enterprises or corporations, national energy technology institute and so on). Therefore, if reference has to be made to a specific Country, it will be necessary to translate and find correspondences as required by national practices and traditions.

3.3 Possible roles and organization of a Ministry of Energy

Co-ordination, updating and setting up of policy guidelines are assignments of the Ministry of Energy (MOE), whose relations with other Ministries must be defined according to the experience and rules of behaviour of the Executive. This may not be sufficient, however. It would seem that in the interests of the Country's efficient energy demand and supply management, the Ministry of Energy should be invested with sufficient power and authority not only to provide guidlines, but also to co-ordinate, promote and control actions. In addition, the role of Local Administration in providing some services and implementing solutions should not be dismissed. In LDCs, many energy policy actions indeed involve rural areas requiring a substantial new public infrastructure. In this respect local development constraints should be viewed as a national concern.

These terms indicate that the Ministry of Energy is to be invested with power and authority not only to provide guidelines, but also to co-ordinate, to control and to promote actions on national energy matters. "Inter-alia", the roles of the Ministry could be as follows:

a) Supervision, control and direction of all energy-related affairs of the Country. Responsibility extends to the need to ensure that authorities and organizations involved in the implementation of national energy policy are properly managed and able to reach the goals which have been established.
b) General planning of financial resources in consultation with other Ministries, and administration of special funds.
c) Promotion of the capabilities of local industries and creation of adequate infrastructures for the most appropriate exploitation of domestic natural, labour and capital resources to the extent that they may contribute to the solution of the Country's energy problems.

Because of its very nature, energy policy requires a good deal of interministerial co-operation and integration. Now interministerial relations are usually conducted through discussions among the Ministries concerned, at official levels, on a purely informal basis. In consideration of the strategic nature of energy policy for the Country, it might be appropriate to set up a special Energy Advisory Committee. This Committee could be chaired by the Ministry of Energy and could include all Ministries closely concerned with energy matters (i.e. Ministry of Planning, Ministry of Finance, Ministry of Industry, Ministry of Agriculture, Ministry of Transportation, etc.), and perhaps external experts.

Although the function of general energy policy in consultation with other Ministers lies with the Minister of Energy, the Advisory Committee for Energy would be responsible for joint development planning and would be concerned with policy co-ordination at all levels.

Some aspects and characteristics of the operation of the Ministry of Energy deserve clarification and definitions. It is generally accepted that in its operational branches, an organizational structure must be flexible enough to adapt to the changing environment. Since the Ministry of Energy would be a newly created organ to cope with a highly dynamic situation, its roles and functions will need to be fulfilled in stages as the circumstances evolve. The full development of its powers would depend on the achievement of effective control on critical energy policy matters and will be limited by the need to endow the Ministry of Energy with personnel having adequate technical expertise and by the ability to act as a cohesive body.

Generic criteria towards a more precise definition of the roles and functions of the Ministry may be:

a) Transfer and delegation of powers in order to act with speed and flexibility with the general aim of meeting changing circumstances in the political and economic arena as effectively as possible.
b) Internal coordination. Powers are conferred or delegated for specific purposes, but there is always a general need for consultation and central co-ordination of activities. This should be expressly provided for as a general rule, in such a way that when powers are conferred for special purposes, these will nevertheless remain subordinate to more general policy objectives.
c) Possibility of sufficient control over the exercise of power, so as to ensure that the Ministry remains responsive to the needs of the Country and that rules, regulations and provisions are put into effect.

3.4 The organization of the Ministry

The organization of the Ministry of Energy would be under the responsibility of the Council of Ministers and the Minister. For instance, the Council of Ministers and the Minister would decide on the establishment of Divisions and responsibilities among Divisions. It would be responsibility of the Minister of Energy to follow up Council of Ministers decisions, to introduce and defend political issues and to ensure that action is taken and policies are implemented. The Minister plays a vital role in interministerial co-ordination. Much of this co-ordination is achieved through informal personal contacts and the exercise of influence rather than authority. The ability to this end may be limited by the fact that the Office of the Minister might have only a small staff.

The General Director or Permanent Secretary would have an essential role in the management of the Ministry. The authority granted to his status gives him a position of superiority which is important in his relations with other civil servants within the Ministry. He may and does take the initiative in raising specific matters concerning the working of particular aspects of energy policy. In addition, he might chair regular meetings of the heads of Divisions (Executive Board) where matters are discussed regarding the enactment and implementation of energy policy common to the various Divisions and where responsibilities are shared and assigned. The Office of the Permanent Secretary could also be responsible for relations with the peripheral Administration, to the extent that this Administration is the agent of the implementation of energy policy in the Country at large.

The organization of the Ministry of Energy shown in Fig. 1 is built around four principal Divisions:

- administration
- policy planning
- technical operations
- energy technology development.

The Divisions are organized in Branches, each one having authority over distinct areas or subjects. According to the criteria which have been outlined, action and responsibility of the Divisions and their Branches could develop as follows.

3.5 The policy planning division

The basic role of the Policy Planning Division is to advise the Executive and the Minister of Energy on the strategy for energy development. Within this framework the Division is responsible for setting global and sectoral targets, as well as identifying optimal growth paths. The Division should ensure that, in planning for energy, the Ministry works towards those development goals which have been established for the Country. At the same time it should ensure that the Ministry is informed of any incompatibilities with development goals that may arise, for example, from limitations in the supply of energy. It is the responsibility of the Division to verify that implementation of national energy objectives is undertaken on schedule. The Division should be on the alert to detect ahead of time any difficulties or deviations from planned developments, and promptly report for appropriate action.

The main functions of the Policy Planning Division can be identified as follows:

a) national energy planning and monitoring
b) tariffs and prices
c) laws, rules and regulations
d) international relations
e) information system.

These functions may be appropriately allocated within five corresponding Branches. Thus, besides national energy planning, four distinct policy making functions ought to be considered.

First, the definition of prices and tariffs. This function is centred around economic and financial analysis to assess the cost of energy produced from different sources and technologies. Needless to say, prices and tariffs are important corner stones of any strategy of energy conservation and fuel substitution. Attempts can be made to focus energy taxation on energy resources, especially primary energy supplies. Taxation on hydroelectric, and geothermal power, coal, wind and solar energy, firewood, and industrial or community wastes should be in general avoided. In practice, taxation should focus on oil products, charcoal and imported electricity. Two specific aims of taxation could be, for instance, to diminish the use of wood for charcoal production and to promote efficient public transport.

Conversely, possibilities of granting tax reliefs for the more productive combustion equipment required in industrial processes and domestic usages should be investigated, in order to accelerate self-financed conversion and replacement. In order to increase efficiency in the use of wood and other domestic fuels, possibilities could be studied to amend the tax laws in such a way that deductions for improved ways of using domestic fuels are broadened in scope. It is needless to say that, in addition to taxation, all other suitable means should also be investigated.

In the taxation of energy, the goal of preserving the competitiveness of national industry must deserve the highest consideration. The work of studying the effects of energy taxation, tariffs and price criteria in greater detail is a key to the success of energy policy. Therefore, it may be advised that a Special Price Committee linked with other Ministries is appointed for this purpose.

Energy prices and tariffs policy will be formulated in such a way that they promote the rational use of energy and increase supply of domestic energy. Furthermore, tariffs will be intended to maintain sufficient possibilities for self-financing on the part of energy companies. The intention should be to maintain a long-term basis in marginal energy pricing so that energy suppliers, consumers and the Government can take the long-term objectives of energy policy into account in making their plans and arrangements. The ultimate purpose is to price energy in such a way that changes in the prices of imported energy are in the proper ratio to changes in final product prices (1).

Second, the formulation of laws, rules and regulations. This Branch should be concerned with identification of appropriate rules for behaviour for statutory organizations, public corporations and private enterprises, to the extent that they are involved in energy matters. The establishment of rules and regulations and the enacting of specific laws will naturally require consultation with other Ministries concerned. A major task of the Branch would be that of preparing legal proposals and providing special expertise in the legislative process.

Third, international relations. National energy policy remains ill-defined in the absence of a coherent assessment of the international context. The success or failure of energy planning in a given Country is intimately connected with its economic and political relations with other Countries. International relations in the energy sector include two critical aspects: the supply of raw materials and technology, and the negotiations for aid and international technical cooperation. As regards the supply of raw energy materials and technology, attention should be directed to three main areas. First, the development of effective friendly relations with other Countries can provide circumstances of security against perturbations and unforeseen tensions in external markets; second, joint industrial, agricultural and commercial ventures can be initiated in order to rationalize and facilitate effective use of energy resources; third, diversification of supply partners and aid donors can be emphasized to reduce the hazard that external interests act to the detriment of national energy policy and economic betterment. Diversification can also be employed as a way of maintaining and practicing non-alignment. Finally, foreign relations can be organized and directed towards maximizing technical cooperation in the energy sector among LDCs and between them and other nations in the World.

As regards the negotiations of aids and international technical cooperation, it is important that two features, at least, be made clear. The first is the asymmetrical connotation of the act of giving and receiving aid, especially when the donor is an industrialized Country. The second is the need to arrange and build up substantial local competence in all critical areas where technical assistance from outside is concentrated. Thus external aid and technical cooperation should be negotiated for and accepted, provided that they are consistent with national capabilities and national interests.

Fourth, the information system. This Branch would be responsible for the collection, analysis and possible publication of statistical data that relate to energy supply and demand in the Country. The responsibility of the statistical function could be shared with other Ministries directly concerned. An additional responsibility relates to the adoption, updating and management of appropriate energy policy models constructed to test and evaluate energy strategies and the effect on the national energy system of variations in the values of external variables.

3.6 The Technical Division

As already mentioned, effective planning requires the strict co-ordination of many activities within the Ministry. To be precise, energy policy proposals depend on intimate knowledge of the sectors involved, and their relationships with the overall industrial, agricultural and economic development of the Country. It seems mandatory therefore that the closest possible co-operation be continuously maintained between the Policy Planning Division and the Technical Division and, in addition, that co-ordination also exist with other governmental bodies.

The main functions of the Technical Division might refer to four technical areas and corresponding Branches:
a) oil products and other fossil fuels
b) electricity
c) renewable (non-traditional) energy resources
d) traditional energy resources (woodfuels).

The reasons behind this aggregation of functions into Branches would be threefold. First of all, it seems necessary to avoid the idea that the establishment of many public enterprises (or other statutory bodies for energy matters) is a sufficient condition for development or the solution to administrative problems. There is indeed a real danger that the proliferation of corporate bodies may complicate rather than simplify the solution of administrative problems by the creation of a bigger, more complex and more extensive bureaucracy. The dangers of proliferation of corporate bodies are sometimes aggravated by the pressures and requirements of foreign aid donors, who often feel that their investments will be more secure if they are channelled into, and administered by, a body separate from the Government in which they can have some representation or influence. Doubts can be entertained whether aids which result in a multiplicity of administrative institutions are wholly fit or admissible.

In this context, there is a second reson for the preference of aggregation of functions. It is advisable that organizations dealing with specific energy projects reduce the number of interfaces to a minimum, so that the few specialists who are available are concentrated in strong multi-purpose units. Both efficiency and flexibility would be thereby increased. In conclusion, it seems necessary to keep the administrative system as simple as possible by relying upon existing resources which should be stimulated to operate according to viable economic, technical and organizational criteria. Finally, there is the fact that where different functions are amenable to integrated planning and management, it is desirable that they be collected together in one centre for appropriate co-ordination.

Given these premises, a few additional considerations can be made regarding the specific roles and functions of the four Branches envisaged.

a. Oil and other fossil fuels

In essence, the problems to be dealt with concern the definition, enactment and implementation of a national policy for fossil fuels. Clearly, oil has a central role for the years to come for it can be anticipated that coal will assume increasing importance only in the medium term perspective, while natural gas can rarely be considered an accessible option. The functions performed by the Branch can be grouped into classes:

(i) Co-ordination of activities carried out by private enterprises (i.e. multinational oil companies) and possible public enterprise.

(ii) Inspection and supervision to ensure that the laws of Government and the regulations are brought into effect, that general guidelines are followed and, moreover, that they correspond fully with the dynamic environment to which they are supposed to apply.

(iii) Promotion and approval of energy projects which are in the national interest.

b. Electricity

This Branch should exercise authority on generation, transmission and final delivery of electricity. Special problems are related to rural electrification and the criteria according to which priorities will be identified in extending the electric grid to remote areas and to settlements where the cost of connection to the grid, and the cost of supply electricity do not find an immediate economic justification. As in the case of the Oil and other Fossil Fuels Branch, there will by necessity be functions related to co-ordination, inspection and supervision, project appraisal and approval. It seems obvious that the opportunity for the Branch to endow itself with an increasing capacity to handle new commitments will be also defined by the extent to which the national utility becomes progressively integrated in the future.

A question of utmost importance is the selection of different projects taking into account the availability of domestic resources (primary energy resources, engineering and manufacturing capabilities, labour force) with the aim of achieving a sufficient degree of self-reliance and growth of national industries.

c. Renewable energy sources

Renewable energy embraces a large number of technologies. Marketing and deployment of technological options as they become commercially available can be performed through a variety of means. A few facts cannot be ignored:

(i) Renewable energy technologies require some kind of decentralized organization for their deployment and adaptation to very different territorial situations.

(ii) Many renewable energy technologies find their best use and exploitation when integrated with the development of rural areas. The solution of problems of agricultural growth are dependent on adequate supplies of energy. At the same time, agricultural development can be appropriately directed to fulfilling objectives of energy self-sufficiency at acceptable costs.

(iii) Because of strong affinities between different renewable energy technologies, resource requirements for their multiple deployment tend to be smaller than the sum of resources required for the independent deployment of single technologies.

These facts, taken together, suggest the importance of economies of scale in organization and financing. Such problems find a useful framework in the existing organizational infrastructure to provide a vehicle for the adoption and development of renewable energy technology as technical solutions become available and convenient.

Possible cases of conflict may arise between this Branch and others within the Technical Division as for example with the Electricity Branch, when the exploitation of renwable energy resources is through technologies based on power production. In this case, the conflict could be resolved by adopting a threshold (say 50 - 100 kW), with plants below the threshold falling under the responsibility of the Renewable Energy Sources Branch.

d. Wood fuels

A major issue for LDCs is connected with the rationalization and eventual substitution of woodfuels in final uses. Energy policy to this effect has potential affinities with other renewable energy technologies deployment. However, some basic differences should not be ignored. Apart from specific planning, organizational and training issues which will be discussed later, there are the subtle aspects of devising incentives to motivate and reshape patterns of behaviour in neglected sectors and regions.

In conclusion, like the area of renewable energy sources, traditional energy is and will in the future be likely characterized by a number of subjects at the local level and by a variety of companies dealing with different systems and types of equipment. Therefore the Branch will need to operate with basic responsibilities in co-ordination, inspection and supervision, promotion and approval of projects.

The primary function of the Technical Division is to provide the Country with the instruments and means to secure adequate supplies of energy. The necessity to co-ordinate with the activities and responsibilities pertaining to the Policy Planning Division are self-evident and have already been stated. In the broader framework of activities under the Ministry of Energy another element of co-ordination of fundamental importance relates to new and developing technological opportunities. The availability of appropriate technological solutions and designs are essential ingredients of any energy policy. As a consequence, a close and continuous relation should be established and assured between the Technical Division and the Energy Technology Development Division.

3.7 The Energy Technology Development Division

A distinctive aspect of energy policy is the need to seek and identify appropriate technologies. In general, technology must be adapted to the specific industrial, organizational and social context. In the interest of promoting broad diffusion of new technological opportunities, is the need to dispose of sufficient degree of expertise to generate the necessary skills which are required in design, construction, operation and maintenance activities of energy systems and equipment.

Ensuring that energy research and technology development is of adequate volume and versatility is an important prerequisite for implementing the targets set for the energy policy. The volume of the effort and the necessary funds should place emphasis on the transition from oil and traditional energy sources (woodfuels) and upon their rational use. Transition in energy production and efficient use of energy entail processes of substitution of limited and costly physical resources with new technology and organization, which in turn mean capital investments and creation of a special infrastructure. This infrastructure combines skilled manpower with social and econoöic systems which are accustomed to handle innovation. LDCs frequently lack the resources of skill, finance and organization to explore and absorb, over a wide front, the technologies that are appropriate to their own specific needs and conditions. The outcome could be reliance on imported technology in the form of capital goods, manufacturing processes and services originally designed for a developed social and economic structure, but without the social infrastructure and organization needed to adapt and modify such technology to the local context. This process of technology acquisition may create a one-sided dependence in which a given Country comes to rely almost exclusively on industrialized nations for advanced technology, as well as the know-how and the management of the technology. More important, at least in the short term, can be a loss of control over critical decisions affecting national energy policy resulting from too great a dependence on imported energy technology, particularly through a reliance on packaged turnkey contracts.

Successful technology development requires proper orientation and co-ordination of national efforts. To this purpose, the Energy Technology Development Division may offer viable means for industrialization and commercialization of new technological opportunities. The Division can most appropriately be endowed with the function of orientation and co-ordination of national research, technology development and demonstration of energy sources. More precisely, the functions of the Division would be:

(i) To determine priorities for scientific and technological activities in the Country, in relation to national energy policy and international commitments.

(ii) To ensure co-operation and co-ordination between the various agencies (or centres) involved in the national energy research and development effort.

(iii) To promote contacts with corresponding scientific organizations in other Countries.

(iv) To ensure the application of the results of energy technology development activities.

(v) To test and demonstrate technical solutions which are new for the Country and require considerable amount of work in order to render them appropriate for local implementation.

(vi) To review and advise on the programmes and budget for energy research and development, and related scientific activities proposed by national centres, universities and public corporations.

In this perspective, a co-ordinated national programme directed through a specialized organization, or institute, or centre under the Ministry of Energy seems needed to assist in initiating and expanding national technological capacity, as well as promoting technologies that would increase energy self-reliance. An important function of this specialized organization, henceforth referred to as (National) Energy Technology Institute, would be to strengthen planning and policy making capabilities by encouraging the flow of technological information and industrial innovation, particularly that associated with energy. The horizons of such an Institute could conceivably stretch beyond national boundaries, to include whole regions. Clearly, the Institute would not be intended to replace or duplicate any existing organization in the field, but it should draw on the work of existing capabilities, to carry out research and development and provide essential address and integration.

3.8 The Administration Division

The main functions of the Administration Division are:

a) budget and accounts
b) personnel and training
c) energy fund.

Given their nature, these functions could possibly simply be allocated in the Office of the General Director (or Permanent Secretary) without the need to create a special Division within the Ministry.

The limitations of domestic financing are a crucial bottleneck, and a realistic policy cannot be prepared in the absence of well established financial relations. This aspect calls for close co-ordination between the Administration Division and the Policy Planning Division. Planning is concerned mainly with overall goals and the real resources of the Country over a broad time horizon. Financial administration is concerned with monetary resources emphasizing the more immediate future. To the planner, the financial resources allocated in the energy sector by Government and budgeted by the Administration Division are essentially a means of moulding the Country's energy development. On the other hand, the system must be sufficiently responsive that the process of resource allocation is based at least in part on the planner's assessment of the situation.

An essential role in this countrywide financing scheme is assigned to the Energy (Development) Fund and to energy price and tariff policy. The Energy Fund would be used and administered by the Ministry of Energy to finance energy research, development and demonstration through the national Energy Technology Institute and other organizations or energy companies acting according to the national interests. The Fund would also serve to support energy projects and initiatives which cannot be entirely self-sustaining, or which await for different types of financial contributions. The Energy Fund would be made of national balance outlays, revenues resulting from some energy activities, and especially energy taxes and external aid.

4. INSTITUTIONS IN THE COMMERCIAL ENERGY SECTOR

4.1 Oil fuels

Many Third World economies are now moving into an energy-intensive phase of development (the production of goods, especially basic materials and heavy manufactured goods), while the industrialized nations are moving toward a less energy intensive phase (the production of services). If there are domestic oil resources, LDCs would be totally dependent on foreign sources for their oil needs. This type of unconditional dependence is disturbing. Further concern stems from the possibility of unforeseen price increases of crude and from the need to secure, under all likely circumstances, adequate quantities of oil products to the Country.

There is, however, no a priori reason why LDCs should not be able to respond through changes in the structure of the economy to absorb even high levels of oil dependence, given sufficient time for adaptation. As a recommendation of energy policy it is generally suggested that the dependence on imported oil be properly managed and kept under control, recognizing that oil is bound to remain an essential energy resource in the years to come.

Proposals have been recently circulated where the extensively debated idea of a direct intervention on the oil market takes the form of a proper national organization. In this view, an important role is emerging for state petroleum enterprises which might assume the form of a National Oil Company or Corporation. As such, the structure, the functions, management and accountability of this national oil organization will be determined by the general theory and practice followed for state or public enterprises in a given Country (4).

In the context of national energy policy it may be worthwhile remembering that the thrust towards a state petroleum enterprise derives from various factors. Some are as follows:

a) Government-to-government deals in the oil supply market tend to assume growing importance and government-backed companies provide the best instrument to handle oil contracts under privileged conditions.
b) Multinational oil companies may have not the willingness to intervene in some sectors of interest for the Country. Also they might be no longer prepared to invest to the same extent as in the past in the marginal areas. Therefore the risk has to be shared, or taken either directly or indirectly by consumer Country governments.
c) Some situations may require the institution of agencies of an operational nature, for instance in the acquisition of emergency stocks and/or supply arrangements in time of acute crisis.

Conversely, two aspects call for a prudent approach:

a) All sectors of the oil industry are capital-intensive, which means that fixed costs (in the extreme case of pipelines over 90%) account for a high proportion of total cost. Surplus capacity tends to raise average cost.
b) The strength of multinational companies lies in their vertical integration, in the economies of scale they have in all sectors of the oil industry, and in their easy access to crude oil supplies.

As a consequence, in establishing a national oil enterprise the Country will need to be prepared for a rather long period of coexistence of nationally-controlled operations with operations carried out by the major oil companies. The entry of Government into an oligopolistic industry such as oil may face barriers and take some time, especially in a country with limited petroleum consumption. Should this be the case a degree of co-ordination and integration with the international majors should be sought.

Economic and technical barriers which are innate to specific characteristics of petroleum operators would consist of effective integration, economies of scale, vast capital requirements, and high risk factors in the upstream operations. Exploration for hydrocarbons, the development and conservation of discovered reserves, and production itself would be left until last. Even the most experienced of the state petroleum enterprises did not venture thus far into exploration significantly, except with the cooperation of multinational oil companies. Conversely, supply contracts, refining, pipelines operation and bulk distribution set the least constraints against entry, since they could be entirely state controlled, or partly owned through joint ventures. Therefore, although the state petroleum enterprise would be bound to have a modest beginning, it may progressively grow into a large fully integrated organism.

In this context it should be noted that, due to the paramount importance attached to the partnership and service contracts to be concluded with foreign companies and the understandable concern that such agreements should harmonize with the national economic development objectives, it may be required that no such contracts can be deemed valid without the ultimate approval of the Executive. This could be a legitimate manifestation of government supervision and control on the activity of any prospective National Oil Corporation, due to its strategic character and role. However, the Ministry of Energy (or the Commission) should supervise, control, co-ordinate and trace guidelines for the activities of the state oil enterprise and other non-national oil enterprises operating within the national boundaries.

In order to fully achieve the basic policy objectives for which it is proposed and might be created, the national petroleum enterprise must be allowed the widest range possible of petroleum related activities to be carried out either directly, or through subsidiary companies. In this respect, measures of centralization would generally seem to be more conducive to efficiency and conservation of managerial, technical and financial resources than a plurality of parallel and unrelated entities. In fact, the petroleum enterprise should be granted a wide range of functions and not confined to one or a few specified aspects of the petroleum industry. The form of integration and co-ordination of the national oil enterprise and foreign companies would require clarification and analysis on a case by case basis.

4.2 Electricity

Careful planning of the electric sector is of great importance since the decisions to be taken involve the commitment of large resources, with potentially serious economic risks for the electrical utilities and the national economy as a whole. Therefore, leading objective in power system planning is to determine an optimum strategy for long-range expansion of the generation, transmission and distribution system adequate to supply the load forecast within a set of technical, economic and social constraints.

An important feature in the supply of electricity, which could result in significant economic benefits is the degree of co-ordination and integration in planning and operation of the electric system. In several LDCs there are at least two historical reasons which worked against co-ordination in the past. First, the relatively stable capital investments and operation expenses experienced until the early 70s inhibited an active change in the concept of individual planning and control toward co-ordination. Second, there have been numerous managerial, accounting, and economic issues that made close co-ordination in planning and operation a complex problem.

This has changed considerably and, particularly with the higher economies of scale of new hydro and coal-fired plants, co-ordination arrangements are likely to spread in the future, both in width and depth. The generating plants and the power lines that connect them and deliver power to urban centres and large users appear more and more to require some sort of single national management system to govern their expansion. Bulk power supply (i.e. electricity generation and transmission) is in general better handled by a single integrated national company or utility which could fully use the available technology and minimize the difficulties. Some financial and other organizational changes would also be needed to make such a system work. Needless to say that this co-ordination effort, if absent, should take place in steps and shall be effected under close and continuing co-ordination and guidelines provided by the Ministry of Energy (or the Energy Commission).

As a further policy recommendation, it is advisable that the National Electric Utility should have a responsibility for the design, construction and operation of all power stations which may be installed in the Country. Because of the somewhat conflicting objectives in water usage manifested by irrigation projects, it may be anticipated that co-ordination between the National Utility and local bodies in site selection would also demand a contribution from the competent offices of Ministry of Energy, Ministry of Agriculture, Ministry of Water Resources, etc. From the viewpoint of general national interest, the worst result of multiple, unco-ordinated licensing of plants is that there is no chance to select the best site in the Country and require that the plant be located there. There may also be no way to involve the public in such basic questions as whether a given plant is really needed, or whether fundamental alternatives to plant or system development have even been considered. As a result, power could be more expensive and the environment inadequately protected. Environmental problems raise serious concerns about how decisions to locate and build electric facilities should be made, and about who should have the final say.

A strengthened system of power plant licensing should rely primarily, although not exclusively, upon the Ministry of Energy because electric industry tends to have operations encompassing the jurisdiction of several Ministeries. For these reasons it can be recommended that plant siting and authorization issues are normally examined by an interministerial regulatory unit under the supervision (or authority) of the Ministry of Energy.

LDCs tend to put increasing resources into rural electrification and development of decentralized power generating units or non-interconnected grids. Most rural communities are often without electricity, and in those that do have it, the low-income groups cannot afford it. On the other hand, projects tend to be characterized by comparatively high costs. Revenues and financial returns are low at least in the early years. Information about consumer response is scarce,making forecasting difficult.

Project justification and rural electrification financing schemes must therefore start from a careful analysis and measurement of benefits, which would be once more performed under the responsibility of the National Utility. One particular problem of identifying and appraising projects is posed by the very large number of potential initiatives to be considered and evaluated. Another connected problem is to provide some criteria to measure the economic costs of investments when social aims are strong and economic returns deferred in time. The National Utility should make a decision whether to provide electricity from the grid or from local generators. This decision will depend on a number of factors: for example, the expected level and time of growth of demand; the utilization of the investment; the distance from the main network; and the difficulty of access, which can affect costs enormously (5).

It is believed that a National Electric Utility concentrating on building a fully integrated generation and transmission system could realize large opportunities for savings, help keep the price of electricity as low as possible, and be more responsive to environmental concerns. Furthermore, this vertically integrated utility would allow the creation of an engineering group, which may become competent in the design, construction and inspection of all types of power plants and power transmission and distribution facilities or equipment. This group, which may be referred to as National Power Engineering Unit, seems to respond to a strategic objective for the Country. The group could be organized as a multipurpose interdisciplinary task force. A first task is to endow the Country withe the capability of selecting power plant options effectively and of evaluating bids. This kind of capability would allow greater freedom in the choice of sources of technology, and thus lead to the encouragement of a more balanced development of local industry, quite apart from the competitive climate of several purchasing procedures. Such freedom implies, of course, improved local expertise in the preparation of specifications and in the negotiation of contracts, and this in itself would be an important step in improving the technological capabilities within the Country. In this spirit, given a power supply policy, the Power Engineering Unit can be required to determine how these policies could be implemented, ascertain their technical feasibility and practicability, evaluate the implications and possible impacts of each policy, and organize this information in a format that would assist the Ministry of Energy or the Government in taking the appropriate steps.

A second task is to include in the Power Engineering Unit contract managing, architectural and engineering design skills. It would then become possible to introduce larger elements of local content manufacturing in the plant construction. This policy definitely helps local industry both to provide work and to provide the established technologies and quality standard. However, this influence could not be exploited to the full, partly because local industry may not think it worthwhile to compete for the work unless it can see sufficient, continuing demand, and partly because of the restrictions that are frequently imposed by tying the aid to specific sources of equipment supply.

5. INSTITUTIONS FOR TRADITIONAL ENERGY SOURCES

Regardless of the World oil crisis, one of the basic problems of the LDCs appears to be the uncontrolled consumption of wood (and in some areas of agricultural residues and forest wastes) as a primary energy source. Wood indeed cannot be easily replaced by other traditional or non-traditional energy sources because of both the typology and the territorial distribution of users. With increasing woodfuel consumption by the rural population, further enhanced by the shift to charcoal (which stands long distance hauls better than fuelwood as this becomes locally too scarce to satisfy fuel demand), the potential supply of wood biomass is destined to decline in many LDCs, unless serious countermeasures are taken to avoid this (6). A factor that aggravates this decline is increasing population pressure on good agricultural land, and the consequent continued clearing of forest cover either through organized settlements or illegal squatting.

Thus any effort intended to restore an environmental equilibrium where wood production and use find a proper balance, should fit local situations and should be based on flexible approaches, which combine various elements of solutions in meeting local needs and possibilities. To this end programmes will be closely integrated in rural development efforts: they should involve local populations through an active and responsible participation. The urgency and magnitude of the task require that institutions, manpower, finance, research and development should be built up as an integral part of rural development and national energy policy. To this purpose co-ordination should be sought and maintained between the Ministry of Energy (or the Energy Commission) and other Ministries concerned with agriculture, forestry and national environment. Energy indeed in the form of firewood, charcoal, agricultural residues and animal power is both a by-product of the rural system and an input to it.

The Ministry of Energy should ensure that national forest policy priorities include the production of fuelwood as a major objective, equal in importance to water and soil conservation or timber production. National energy requirements should be part of an integrated forest management system. The Administration should review existing legislation with respect to land tenure, land use and forestry practice, and licensing regulations which affect the access to fuelwood resources, and revise it as necessary. A special task will be the training of the required foresting staff and technicians. It is also important that the Ministry of Energy with other governmental institutions initiates or intensifies popular education and extension programmes in order to raise population awareness and to stimulate the capability of rural communities to develop, manage and efficiently use their fuelwood resources.

The improvements in the efficiency of conversion and utilization technologies of woodfuels and agricultural residues are of great importance as they represent a substantial potential for energy and resources conservation. Such improvements range from the seasoning and pre-processing of fuelwood, to the more efficient production of charcoal and improvements in wood-burning stoves and kilns. It must be remarked, however, that as far as the cost per unit of useful energy is lowered, the improvements may have the undesired effect of increasing wood demand. In this last respect, it may be possible that the Ministry of Energy intervene by a proper taxation policy and control on the distribution system of fuelwood and charcoal.

In principle, to rationalize the national woodfuels production and utilization system it might be advised that the Government take steps to organize and monitor distribution and marketing in order to assure an adequate and stable access to supplies. A proposal which would deserve careful weighing is that central Government (or local Administration) establish some sort of Charcoal and Fuelwood Corporation or Company for supplying fuelwoods, making and marketing charcoal. Clearly, this organization would coexist with the innumerable private producers. It would offer, however, the possibility of proceeding towards a system where prices of fuelwood and, most important of all, of charcoal are regulated and charcoal production and commerce is properly taxed. Moreover, a more efficient distribution system would allow a much more equitable exploitation of wood resources in the Country. For instance, distant supplies may remain untouched because of the cost of transportation and the lack of infrastructure. Conversely, urban centres place heavy pressure on surrounding resources and suffer from the disappearance of trees. The constraints inherent to organizing the distribution more efficiently are the lack of storage facilities to ensure stable supplies, the equipment required for large-scale distribution, the inadequate transport and communication services and the disrupting impact of heavy concentrated demand on local production for rural communities. Relieving measures can consist of a better control of the distribution and pricing mechanisms, the establishment of storage facilities, the improvement of transportation, especially from more distant sources.

6. INSTITUTIONS FOR RENEWABLE NON-TRADITIONAL ENERGY SOURCES

Although the energy situation in LDCs demands a transition to exploitation of (non-conventional or) non-traditional technologies for conversion and use of renewable energy resources, it is not probable that a shift can be quickly engineered. The capital cost and organizational infrastructure required by the deployment of these technologies are often not easily available or affordable. In principle, renewable energy sources encompass a variety of technologies including direct uses of solar energy, wind energy, small-scale hydro resources, alcohol fuels production from biomass, generation of biogas from agricultural residues and other things. Common traits are territorial dispersion and the relatively low level of technology involved, which explain the emphasis given to these energy resources in the rural sector. The increase in productivity of agriculture and the improvement in the quality of life in rural areas tend to demand a corresponding increase in their energy input. Therefore, for the rural sector to become less dependent on progressively depleting resources in its energy supplies, LDCs should rely upon the contribution which may be given by renewable energy sources. It should also be noted that to a certain extent deployment and adoption of renewable energy technology could be compatible with dispersion of rural settlements and the low income level of rural population.

Territorial dispersion, however, calls for the introduction of appropriate technology, organization and institutions which can be effected also on a small scale, provided two prerequisites: (i) simple and reliable solutions are implemented; (ii) construction of equipment and maintenance may rely upon local forces and capability. In this view, the importance should not be attached to sectoral programmes and projects, rather there is a need to develop an integrated energy-generating system and a co-ordinated rural energy planning; such planning is needed to make the best use of the limited resources available and for the management of natural resources. In all cases care must be taken, to ensure properly that these systems do not become so complex that they cannot be managed by the local people, who should be fully involved in and trained for these projects. Demonstration projects should be initiated as early as possible so that problems of large-scale application and local optimization can be studied on the basis of field data and local experience (7).

A policy for the renewable energy sector needs definition of priority requirements, as well as the provision of suitable structures for the implementation of the resulting programme. This programme should be implemented with particular reference to the aspects of RD and D, and training for preparing experts in production activities, as well as commercial and technical assistance personnel. Training shall be inserted in a general framework of promotional activities. In fact, the deployment of renewable energy technologies must go step by step with the creation of a national Institute for Energy Technology and its operation.

Specifically, a number of constraints and barriers affect the development and implementation of rural energy projects, and there is a consequent need for special institutions and management innovations. Some typical constraints include:

a) Absence of sectoral co-ordination in national planning of renewable energy technology development;
b) Lack of an extension service to stimulate the innovation and promote the growth of design and manufacturing capabilities;
c) Inadequacy of local planning authorities and their inability to deal with the problems of supporting the demand;
d) Inefficiency of administrative mechanisms with regard to providing day-to-day maintenance for domestic energy supply systems.

Forms of vertical and horizontal co-ordination are therefore needed in the various governmental agencies which deal with rural energy problems in order to maximize benefits to communities through the implementation of projects on the basis of available human and financial resources. A programme for the supply of energy to low-income rural populations has a number of policy implications, one of which concerns the need to create an effective mechanism for the promotion of renewable energy sources and the rational use of commercial fuels. It appears that the real issue of energy supply is an institutional problem, not a technological one. Thus Governments should focus their attention on the need to create comprehensive national and local institutions to deal with rural energy problems.

High initial capital investment is perhaps the most crucial factor in the introduction of non-traditional renewable sources of energy. Greater efforts and long-term commitments at the national and local (i.e. regional) levels are required for the development, transfer, adaptation and dissemination of technologies for the exploitation of renewable energy sources. The traditional economic criteria used to determine the feasibility of large-scale, capital-intensive energy projects should be reviewed in order to take into account other objectives such as improving the quality of life of the rural population and increasing the productivity of low-income settlements.

This means that the Ministry of Energy and the above mentioned Energy Technology Institute should deserve attention in co-ordinating (i) the various research development and demonstration activities, (ii) the undertakings by other Ministries concerned with agricultural policy, industry and national education. As a means of enforcing a national policy of exploitation of domestic renewable energy sources the idea could be put forward of establishing some kind of national (renewable energy sources or rural energy) corporation or enterprise operating as an interface between available technology and energy users.

7. ENERGY TECHNOLOGY RESEARCH AND DEVELOPMENT

7.1 The proposal for establishing a national dedicated organization: a case study

Given the nature and importance of energy technology, it may be suggested that, whenever a sufficient amount of resources can be made available, some kind of (National) Energy Technology Institute (NETI) is established to assist Central and Local Administration in the process of acquisition, improvement and invention of technologies appropriate to face the

Country's energy problems. On the other hand, relevant advisory and monitory roles, together with technology co-ordination and promotion functions could be delegated to some Energy Technology Advisory (or Oversight) Unit (ETAU). Should it be deemed unsuitable to rely upon such a type of Unit, its roles might be simply transferred to the Energy Technology Development Division at the Ministry of Energy (or Energy Commission), to which the Energy Institute belongs.

In activities and initiatives carried on at the Energy Technology Institute attention should focus, at least in a first phase, on the development of equipment and other products aimed at yielding concrete and rapidly utilizable results. For instance, research on domestic energy sources, primarily the production, harvesting and use of wood and charcoal, would be made effective. Practicable solutions need to be identified which can be deployed with significant success throughout the different regions and types of settlements. Also the viability of renewable energy technology frequently depends on appropriate integration with the agriculture system and food processing industry (examples are biogas production, wind pumps, solar crop dryers). Research should then be directed with priority in this area. In all cases attention should be paid to standardization problems, information and training services. Information and training appear of fundamental importance in securing the dissemination of the results obtained to the existing or potential enterprises and organizations, which may become (or should become) interested in their exploitation. But at the same time it should be remembered that technological innovation must be seen as a means and not an end.

Given the strategic importance of the idea of the Energy Technology Institute, in the next few sections an approach will be briefly described which has been considered in connection with the Energy Consultancy for the Ministry of Energy of Kenya performed by Fidimi in co-operation with Cesen (Project No. 4505.016.32.29). During Spring 1981 we had indeed the opportunity to discuss several of the ideas that are herewith presented with Mr. Mwiraria, Permanent Secretary of the Ministry of Energy, and with Mr. Nyoike and Mr. Wairegi, Heads of the Planning Division and Technical Division.

Primary functions of the proposed Energy Technology Institute would be:

(i) research, development, demonstration and testing;
(ii) standardization and licensing;
(iii) information and consultancy services, technical training.

These functions should be performed on energy technologies that are in the interest and scope of the energy policy of the Country. Research, development, demonstration and testing activities refer essentially to technologies that are new to the Country's context, whether or not technologies may be already available and deployed in other countries and/or refer to traditional or non-traditional exploitation of energy resources, in which case the organization of a special technical effort is a prerequisite in obtaining applicable solutions transferred to or tuned with a changing local context.

Beyond innovation, transfer and adaptation of energy technology, there are the problems of providing technical advice to the Ministry of Energy, defining national standards and design criteria, licensing and inspection of energy related equipment and systems. Licensing and inspection should determine whether energy facilities are designed, constructed and operated in compliance with established criteria, provisions and regulations.

Furthermore, means should be procured to help the successful application of newly developed and tested technology on a broad scale. Consideration ought to be given to the process of dissemination of general information, design and operational data about energy facilities and solutions to be promoted. In this view, technical consultancy and training refer to assistance and education that may be offered to both technical and managerial personnel in the national enterprises.

To perform its tasks, the proposed Institute should be endowed with adequate structure and resources. Ideally, the Institute should cover all energy sectors, although as an immediate objective it may most profitably concentrate its effort on the more urgent needs that arise in relation to the development of renewable energy resources and particularly in relation to the rational use of woodfuels and their possible substitution in end uses. The medium-to-long term objectives of the Institute are to strengthen and enforce the national effort in energy research, development and demonstration activities; promote testing programs and pilot scale experiments to evaluate, select and standardize specific solutions on the basis of adequate substantiation; define priorities for technology transfer and development in the Country; ensure liaison with other international organizations and inform, train and qualify personnel to design, construct and operate energy equipment and systems.

Conversely, the early establishment of an Energy Technology Advisory Unit might fulfil aims which include review and assessment of RD and D activities of Government and non-Government organizations and enterprises that are operating in the field of energy technology; preparation of a work plan for the co-ordination of all national activities in the sector. Particularly, the creation of an Advisory Unit may help to ensure that energy technology policy is carried on in close adherence to guidelines and specifications emerging from the national energy plan. In fact, the Advisory Unit must be the element that represents the interests of the national community in the process of formulation of plans and evaluation of results. The functions attributed to the Advisory Unit will also reflect the necessary integration with the guidelines that can be proposed and enacted by some National Science and Technology Research Council. It is worthwhile remarking that among other functions, the Energy Technology Advisory Unit should:

- help the formulation of the Government's energy technology research, development, demonstration and testing policy;
- determine the general directives for the preparation of the programme of the Institute;
- provide for the co-ordination needed between the various governmental and non-governmental organizations;
- maintain liaisons with foreign and international bodies or organizations with similar interests;
- assist in the evaluation of international co-operation linked to energy technology transfer.

7.2 The energy technology problem in LDCs: search for co-ordination and planning

The role and functions of the Energy Technology Institute should be suitably embedded in the existing organizational and technical endowment of the Country. In consideration of the situation of most LDCs' resources, the Institute should be concerned with seeking the most rational management of the limited available scientific and technical expertise and infrastructures. This underscores the need for both realistic policy of technical resource management and the establishment of an appropriate institutional framework within which to implement programs.

At present in LDCs there exist few or unsufficient national mechanisms for realizing an effective co-ordination and planning in the area of energy RD and D, as may be granted and performed through the proposed Energy Technology Institute. The situation is unsatisfactory and likely to worsen unless the means are established for concerted action. Five issues deserve consideration:

a) Most international initiatives for technological development have so far failed to come to grips with the reality of LDCs' local situations.

b) There exists as yet no viable mechanism for incorporating technology, as it becomes available and practicable, into the national industrialization and commercialization process.

c) There is in most LDCs an obvious lack of assistance to the Central (and Local) Administration in identifying and promoting sound energy technology development strategies within the framework of national development planning and within the limits set forth by local circumstances and resources. Limitations indeed stem from lack of skilled labour, ad hoc infrastructures and financial resources. Moreover, in a long-term view, environmental safety and protection could also become a source of concern in selecting and licensing energy systems and equipment.

d) There is an urgent need to promote policies, actions and institutional arrangements for upgrading technical staff. The emphasis on training for energy technology comes from the need to provide competence in design, operation and maintenance of specific energy facilities through application in the field rather than through study in the laboratory. Training is absolutely necessary when diffusion of energy technology must rely upon existing organizations lacking the necessary technical expertise.

e) As LDCs can only rely upon limited quantities of internal primary energy resources, it is necessary that technological capabilities be developed to the utmost. Because of the general shortage of trained personnel and technical know-how, some countries may find insurmountable limits to their energy development. Thus co-operation with similar or neighbouring Countries is viewed as a strategic issue. Wide-scale regional co-ordination can be then envisaged where projects are started and worked out on a shared basis and the exploitation of the results is granted to each Country participating in the joint technological programme or organization.

The establishment of the Institute should aim at minimizing interministerial organisation, avoid duplicating existing institutions and organizations and be geared to the Country's productive context. To this purpose, any unnecessary proliferation of structures should be avoided. The ultimate success of the initiative of the Institute depends on the degree of co-ordination with policy and actions which may be undertaken by other Ministries with due account for the economic policy of the Government and the international commitments of the Country.

Interministerial co-ordination could be assured by the participation of a number of Ministries in the establishment of the Energy Technology Advisory Unit and the Energy Technology Institute. All the Ministries and national organizations or enterprises should concur, who have authority over subjects with direct or close relations to energy matters and technology, such as the Ministries of Agriculture, Industry, Transport, Housing and Public Works, the national electric utility, the national oil enterprise and so on.

7.3 Organization and management structure of the Institute

The management responsibility of the Institute could rest upon the Management Board and the Director. In consideration of the recognized need to ensure the best co-ordination between the Ministry of Energy and the Institute, it seems advisable that the Head of the Energy Technology Development Division at the Ministry of Energy be appointed as the Chairman of the Management Board. Furthermore, the Heads of the Planning Division and the Technical Division should be among the members of the Board.

The functions of the Management Board would be determined according to the system of national provisions and rules that hold for public institutes and centres. Nevertheless, it may be useful to sketch out some of the implicit assumptions that relate specifically to energy matters. In this context, the Board should:

a) periodically formulate and bring up to date the long and short term plans and submit them to the Energy Technology Advisory Unit and Ministry of Energy;
b) evaluate the achievements of the Institute's activity, enact and put in force the execution of programmes and projects;
c) participate in the formulation and evaluation of the budget;
d) present to the Minister of Energy periodical reports on the situation of energy technology in the Country and suggest measures suitable to reach the objectives and targets of the national energy policy;
e) standardize the formulation and evaluation of energy technology research, development and demonstration projects;
f) help in the formulation of national energy and technology policy;
g) participate in the definition of the regional and subregional integrated demonstration programme or pilot experiments;
h) promote and orient energy information dissemination and training by means of the structures of the Institute.

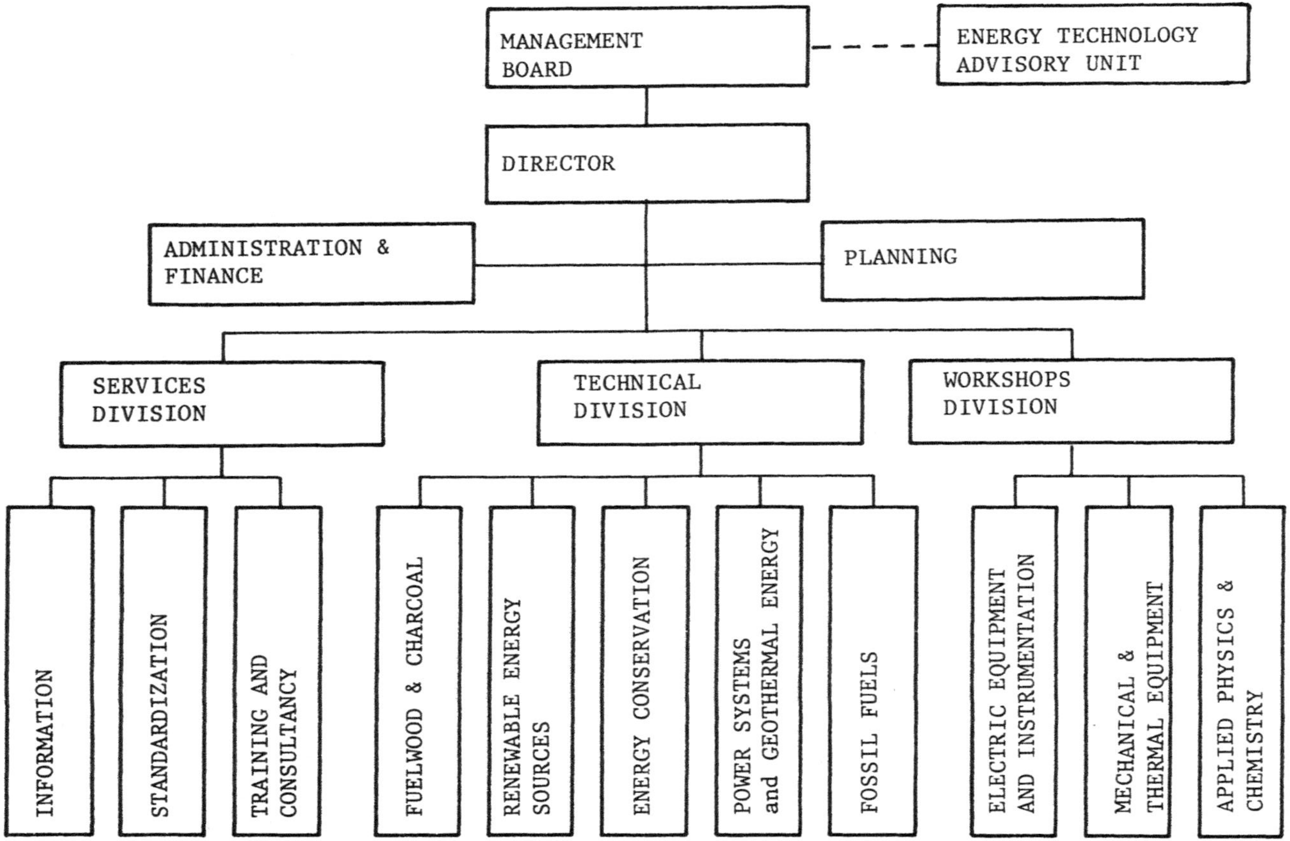

Figure 3 - Postulated organizational chart for the Energy Technology Institute and its Divisions.

The Director has the role of executive official: planning and administration will be two of his major concerns. To this end, a Technology Planning Official and an Administration and Finance Official can be situated in staff positions (Fig. 3).

The Technology Planning Official should provide assistance to the Director and perform basic work on:

a) identification of priorities and lines of action in the Institute by means of preliminary screening and selection among competitive energy technology development programmes;
b) evaluation and review of final results of demonstration experiments carried on a precommercial or pilot scale at the Institute (and possibly at other national organizations);
c) search for contacts with entrepreneurs (and public utilities or services) which may be interested in (or affected by) the exploitation and commercialization process of new or upgraded energy systems and equipment.

Clearly, its judgements and decisions the Technology Planning Official shall take into account the general objectives of the national energy policy and the guidelines given by the Ministry of Energy, and search for the most appropriate technology.

The purpose of the Administration and Finance Official is to collect and administer funds, make payments as required and update the book of accounts.

The Director would manage the following three Divisions (Fig. 3) directly:

- Services Division which may be constituted of the three Offices of Information, Standards, Technical Training;
- Workshops Division which is made up of the specialized Workshops of Applied Physics and Chemistry, Electrical and Instrumentation Equipment, Mechanical and Thermal Equipment. The scope of the Workshops Division is to make experiments and measurements, contribute to the manufacture of components and assembly of equipment for the different energy projects of RD and D programs carried out by the Technical Division. Workshops will be endowed with specialized capabilities which can apply to different energy sectors;
- Technical Division which divides itself into a number of Branches each one developing its responsibility over a specific type of energy source or problem area.

7.4 The Services Division

In the Services Division, the Information Office provides technological information, consultancy services and keeps records of technology transfer. The information service responds to the diversity of factors and actors involved in the complex mechanisms whereby technical know-how is generated, transformed and assimilated for various levels and kinds of decision problems. In this respect, the Information Office of the Institute should identify the types and sources of information, and collect the documents required in the process of introduction, adaptation and operation of new and/or upgraded energy technology. The information service may be classified against a scale of progressively-tightening restriction of access. The scale would progress through successive stages of supply of basic energy data about the Country (solar, wind maps), instructions regarding imported or national energy equipment and machinery, commercial market information, consulting advice and, most difficult of all, drawings, manufacturing instructions, patent and licence documentation.

Clearly, most of the energy technologies used in LDCs are transferred from abroad and are not entirely the results of domestic innovation. Transfer of technology is a difficult and controversial process which may occur in a variety of ways. In principle, energy technology can be brought into the Country through both international commercial transactions and non-commercial mechanisms. It may be emphasized that often these mechanisms become a valuable asset and have an economic impact if they are used in combination and interrelate with commercial transfers, such as licensing agreements, management contracts and technical services agreements, operation of purchased machinery and equipment which embodies technology imitation, and replicas of the product through inverse engineering. As a result, it may be advisable that the Information Office be able to arrange a continuing energy technology transfer survey, where the aim is maintaining control over or being actively involved in the entire energy technology transfer process.

On the other hand, the Technical Training and Consultancy Office has an essential task in achieving effective transfer and adaptation of new technological solutions in the productive context of the Country. Whether the Institute employs its own staff or relies upon consultants or contractors either to perform RD and D and testing activities, or to design, build and operate prototypes of energy systems and equipment, personnel with different qualifications will be needed to carry on the day-to-day activity of the Institute. To this end, there will be an internal plan for personnel recruiting and training. But what is of utmost importance, is that programs are devised to train national experts and technicians. The aim is that of providing the Country with crews who are able to design, construct and exploit new or upgraded energy systems and equipment. As it is necessary to modify energy facilities to match the indigenous resources and market needs, technical attitudes and the organization of industry may also have to be modified to suit the special characteristics of new products and situations. In practice co-operation on training between the Institute and local entrepreneurs may base itself upon formal and informal links and hopefully take advantage of the existing educational system. It is important to recognize, however, the benefit which could be gained through training at the plant or at the experimental facility. Indeed this type of training would effectively help to bridge the gap which often exists between technical and/or scholar apprenticeship and industrial levels of technology.

Next are functions of the Standards Office, which derive from and enact a technical role assigned to the Ministry of Energy. The functions played by the Office have a threefold aim. First, assurance must or should be given that energy facilities (i.e. energy systems and equipment) can be built and operated with minimal impact for the environment. Second, there is the need to develop and adopt regulations, guides, standardization procedures that may allow the best exploitation of natural resources of the Country and assure that performances of energy facilities are not in violation of product liability criteria and protection of workers and general public. It is needless to say that this task will be performed within the guidelines set forth by the national organization for standards wherever it exists. The Standards Office could also co-ordinate the Country's participation in general national and international standardization activities for their connections with the area of energy. Third, the Office would inspect facilities and equipment to identify and propose corrective actions for conditions which may adversely affect the physical and social environment.

7.5 The Technical Division: its sectoral and territorial structure

The activities of the Energy Technology Institute connecting RD and D to energy technology could be distributed among a few Branches within the Technical Division, each one covering a technical area or sector. The reason behind this type of sectoral organization is the necessity to ease the commercialization process of energy technologies through mechanisms of transfer that plunge directly into the national productive system. To this end, the Energy Technology Institute in its operation should look for ways to promote local capabilities through contracting of individuals or bodies, covering engineering design, pilot plant construction carried on with the co-operation of national enterprises and training for special energy production projects and energy utilization schemes.

Viewed together, Technical Branches of the Institute could be as in Fig. 3, namely:

- traditional energy sources (woodfuels, animal power, etc.);
- renewable (non-traditional) energy sources (solar energy, wind energy, biogas, minihydro resources, etc.);
- power systems (hydro, oil-fired, coal-fired power plants, transmission and distribution equipment, etc.);
- fossil fuels (oil, coal, etc.);
- energy conservation (by sector of use).

Obviously, this is only a list of suggestions where no attempt has been made to indicate possible weightings of sectoral areas. Nuclear energy, for example, has been omitted for it would be of interest for some countries only in a rather long-term view. As for the relations with the Ministry of Energy it must be said that Technical Division and Branches should maintain close and continuous links with the corresponding Branches of the Ministry of Energy, where subject matters covered refer to.

Furthermore, it may be appropriate to endow the Technical Division at the Ministry of Energy or the Energy Technology Advisory Unit (or both) with a monitoring group which can evaluate RD and D activity or projects carried on at Energy Technology Institute. Monitoring should provide an evaluation of whether or not projects are implemented and executed or operated in the way they were originally designed. The reasons for monitoring projects are obvious. First, proper administration of human and financial resources requires evidence that what was paid for and allocated was actually put into effect. Second, there may be concern with the social, economic and environmental impact or outcome of a specific project and the need to compare this impact or outcome with the ones of other projects planned or carried out by the various Branches and other organizations.

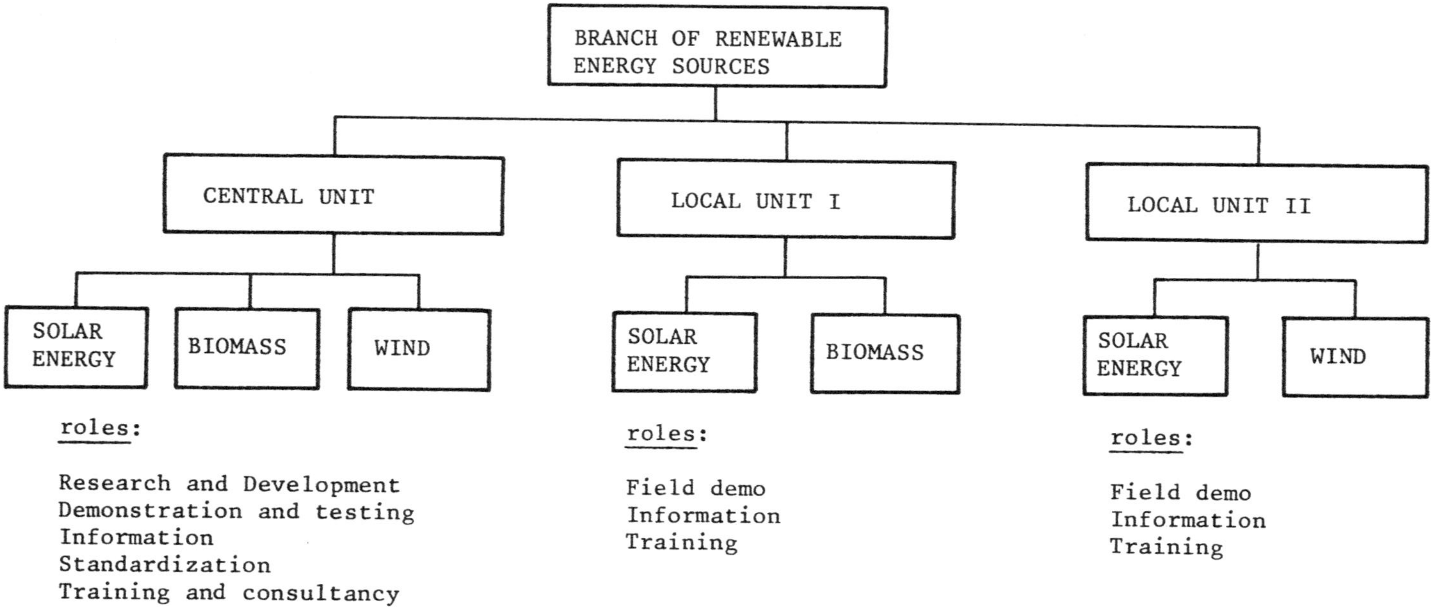

Figure 4 - Organizational chart of a Technical Branch in the Energy Technology Institute. The example displayed refers to the Branch of Renewable (Non-Traditional) Energy Sources, where a number of Local Units can be postulated.

Given the pervasive nature of energy technology, it seems appropriate that the Energy Technology Institute (and Energy Technology Advisory Unit) maintain close co-ordination links with all governmental bodies and organizations dealing with scientific research, development, demonstration and testing in the energy sector. On the other hand, liaison of the Institute with national corporations and enterprises could also take the form of participatory initiatives, where the marketing and diffusion of specific technical solutions are undertaken on a shared basis. Links with the manufacturing and industrial production sector may also find support. Initiatives will differ in extent and content, depending on the scale of the problems and resources available. For instance, the implementation of an industrial estate energy programme involves a substantial investment of public funds. In view of other Government commitments, it is necessary to consider whether technology marketing and diffusion objectives could be attained by other means. For example, innovation in energy technology may be fostered by proper legislation, remission of corporation tax and various duties, availability of cheap credit, etc.

In some LDCs, because of the varied territorial situation and local features involved in the deployment of specific energy technologies, it can be suggested that Branches of the Technical Division at the Energy Technology Institute are organized through both Central Units and Special or Local Units in the Regions. The Institute's regional decentralization has a double end:

a) promotion of direct local participation in the process of technological change;
b) adaptation and demonstration with reference to the specific local context.

The appropriateness of this type of structure can be recognized in the case of renewable energy technologies, and woodfuels conservation and substitution. With these technologies local determinants frequently assume a dominant place in any programme of research, development, demonstration and commercialization.

In principle, a structure of the Institute where Central and Local Units coexist and integrate (Fig. 4) offers the opportunity of optimizing project selection and implementation in a way that is coherent with effective local potential and requirements as well as local sensitivity. In this respect, Regional or Provincial Development Committees may be assigned the role of proposing energy projects and experiments for inclusion in the plans of the Energy Technology Institute. The proposed decentralised structure of the Institute is, furthermore, perfectly adapted to the functions of testing and evaluation of the results of local energy demonstration projects, and the function of training and transmitting technical information.

Clearly, a necessary premise of a decentralized structure is the existence of strong influential links with the centre and sound central technical units. The Central Unit should be capable of performing all such fundamental research, development and demonstration efforts as can be carried out without undue dispersion of productive opportunities and resources. The Central Units of the Institute would have the opportunity of taking advantage from obvious economies of scale in infrastructures and services. The degree of decentralization of RD and D will depend on the specific topic and, particularly, on the trade-off between the advantages of economies of scale on the one hand and the advantages of local specificity on the other.

REFERENCES

(1) J. Dunkerly, W. Ramsay, L. Gordon and E. Cecelski, Energy Strategies for Developing Nations, Johns Hopkins Univ. Press, Baltimore, Md (1981).

(2) J. W. Howe and J. J. Tarrant, An Alternative Road to the Post-Petroleum Era: North-South Cooperation, Overseas Development Council, Washington, D. C. (1980).

(3) J. C. Shah, Towards an Institutional Synthesis for Energy Development, pp. 206-315 in International Energy Studies, by R. K. Pachauri (ed.), Wiley Eastern Ltd., New Dehli, India (1980).

(4) State Petroleum Enterprises in Developing Countries, UN Centre for Natural Resources, Energy and Transport, Pergamon Press, New York, N. Y. (1980).

(5) R. Turvey, D. Anderson, Electricity Economics. Essays and Case Studies, The Johns Hopkins Univ. Press, Baltimore and London (1977).

(6) R. J. De Lucia, Infrastructure for Rural Energy Systems, A Background Paper prepared for the Ad Hoc Expert Group in preparation for the conference UN Conf on New and Renewable Sources of Energy, Nairobi, Aug. 1981 (Jan. 1981).

(7) Mobilizing Renewable Energy Technology in Developing Countries; Strengthening Local Capabilities and Research, World Bank, Washington, D. C. (July 1981).

ACKNOWLEDGEMENT

This work has been made possible by a co-operation agreement concerned with energy policies in LDCs with Cesen (Centro Studi Energia, Genoa, Italy). The author particularly wants to express his gratitude to L. Baldo, O. Bernardini, G. Leoni and B. Musso for the long friendly days passed together in Genoa and elsewhere to discuss and try to understand how countries in the World can find better institutions, develop collaboration and achieve mutual understanding.

ANNEX I

THE ENERGY PLANNING AND MANAGEMENT PROGRAMME

Background

The seminar during which the contributions in this book were presented constituted one module in a management training programme held under the auspices of the Commission of the European Communities from 3rd to 26th May, 1982. The seminar occupied the first two weeks of the programme and was held in the Ispra Establishment of the Commission's Joint Research Centre. The other modules in the programme were:

- . parallel workshop sessions, conducted in tandem with the seminar (reported in Annex II);
- . a ten-day energy study tour, with visits to projects and installations in Italy, France, Scotland and the Netherlands;
- . a three-day final session at the headquarters of the Commission of the European Communities in Brussels.

The programme was conceived at the request of the European Commission's General Directorates for Development (DG VIII) and for Energy (DG XVII). Although open to participation from all those developing countries in which English in the main European language used, the programme was specifically directed to fulfilling commitments to energy training in the framework of the Communities' cooperation with the ACP, Maghreb, Mashreq and ASEAN groups of associated developing countries.

Actual organization of the programme was in the hands of the Joint Research Centre's Training and Education Service ("Ispra Courses") in collaboration with the JRC's Systems Analysis Division. Ispra Courses has ten years of experience in organizing short scientific courses, mainly at post-university level, while the Systems Analysis Division has substantial experience in energy modelling, techno-economic evaluation of energy projects and energy technology assessment.

The form and content of the programme were established following discussions with various European institutions, notably the Institut Economique et Juridique de l'Energie (IEJE) in Grenoble, the International Institute for Environment and Development in London, the University of Strathclyde's Energy Studies Unit in Glasgow, the Irish Management Institute and

H. Neu and D. Bain (eds.), National Energy Planning and Management in Developing Countries, 417–422.

the Industrial Development Authority in Dublin, and UNIDO in Vienna. The first round of discussions led to draft proposals which were debated and elaborated by a group of outside experts (François LeGuay of IEJE, John Twidell of Strathclyde ESU, György Dobos of UNIDO and independent consultant I.D. Stancescu) in collaboration with representatives of the General Directorate for Development.

Programme philosophy

In the light of perceived needs and training opportunities a number of guidelines were established:

i) It was decided to design the programme (the first of its kind) to fit the needs of senior decision-makers in developing countries. Not only was there felt to be a major task of management training to be accomplished at this level but it was also considered necessary to obtain guidance from such officials as to the nature of future training programmes.

ii) It was agreed that there would be a conscious effort to make a distinctively European contribution to energy management and planning, drawing on the different traditions and approaches of the various EC countries, rather than simply reproducing national training programmes. In terms of a programme for "anglophone" countries this entailed special attention to the (mainly French) "planning" approach in addition to the "managerial" approach more commonly used in English-speaking countries.

iii) It was agreed to avoid any attempt at "selling" particular technologies, techniques, approaches or ideologies. In particular it was decided to avoid giving any impression that the European experience was either uniformly successful or suitable for adoption in developing countries. Participants were to be left to judge for themselves what aspects of the management and planning tools available in Europe were relevant or transferable to their home countries.

Taken together with the various restraints under which the programme operated - no control over choice of participants, incomplete foreknowledge of participant's background or skill levels, no experience of similar programmes to draw on - these guidelines led to a natural evolution of the programme as not simply a didactic exercise but as a management training experience. Instead of being presented with a standard set of fixed solutions to the problems of energy planning and management, participants were faced with a large amount of information which they had then to evaluate, select, adapt, emphasize or discard as appropriate to their own particular needs. It was constantly emphasized to them that the most important resource-base available to them was their own combined experience and that it was for them to draw on both this and the seminar presentations to achieve an integrated appreciation of how best to handle their own countries' energy problems. By being forced to address situations where multiple choices of action were possible, where information overload existed rather than information scarcity, where differing and sometimes conflicting advice were an offer, participants were confronted with a situation resembling real-life decision-making.

Participants

The organizers had asked that participants be senior officials but the actual choice of personnel was in the hands of the developing countries themselves. In the event, as will be partly evident from the designation of participants (see Annex III), the general ability and experience of participants was very high indeed. This was an essential factor in the success of the programme.

Energy Study Tour

The study tour proper was preceded by some project visits during the course of the seminar section of the programme. At the Ispra Establishment participants visited various energy projects, particularly the European Solar Testing Installation (ESTI) and the ESSOR nuclear material test reactor. During the first week-end there was a trip to Fossano, in Piedmont, to visit an agricultural training establishment which uses the TOTEM (Total Energy Module) system to convert animal waste into electricity and heat.

The study tour itself was conceived as an opportunity for participants to see at first-hand some of the high-level research into energy supply technologies currently under way in Europe.

The first visit was to the EC-sponsored EURELIOS solar power station at Adrano, in Sicily. This 1 MW(e) station, the first of its kind in Europe, operates on the tower and field of mirrors principle to produce pressurised water vapour which is then turned into electricity. The 182 mirrors, with a total of 6,200 square metres surface area, are centrally adjusted via individual motors with microprocessor controls, to focus the sun's rays on the receiver/boiler unit on top of the tower. This receiver/boiler unit, of the advanced "Francia" design, has a control system which enables the water vapour to be kept at a constant 510°C and 64 atm., temporary fluctuations in sunshine levels (due to clouds passing) being offset by a storage facility, based on molten salt, which can keep the system functioning for up to thirty minutes.

The next visit was to the long-established geothermal power station complex at Lardarello, in Tuscany. The 14 power plants in the complex are of two types - natural condensation steam power plants (ranging in size up to 120 MW(e)) and atmospheric back-pressure direct steam power plants (in the range 0.9 MW(e) to 18 MW(e). With a total capacity of 400 MW(e) this "non-conventional" energy supply complex is larger than the total electricity supply capacity of many of the participants' countries.

Following this participants went to Paris, to the headquarters of the state electricity corporation, Electricité de France (EdF). There they had the opportunity to learn details of EdF's planning and management procedures. There was also a visit to one of EdF's many residential training colleges, at "Les Mureaux".

Scotland was the next destination in the study tour. Participants went first to Pitlochry to visit the Control Centre of the North of Scotland Hydro-Electric Board where they saw how a scattered network of mainly small generating stations, many of them in very remote areas, could be managed centrally using modern electronic control systems. There was also a visit to the neighbouring research laboratories to see various projects related to other alternative energy systems (wind and wave power), to energy conservation, environmental control systems and physical modelling of proposed and existing hydroelectric installations.

The next day the participants were split into two groups. The largest of these went first to the U.K. National Engineering Laboratory at East Kilbride to see the research being carried out on the oscillating water column variety of wave power device. This device, which resembles and can serve the function of a breakwater, is based on currently available and proven technology, With units ranging in size from 0.5 MW(e) upwards the technology is particularly interesting for small island and remote coastal communities.

The group then went to the University of Edinburgh's Wave Power Research Centre to see a very different type of wave power machine. Known as Salter's Duck (as a result of its bobbing motion and after its inventor, Stephen Salter) this is a device using very high technology to obtain energy conversion efficiencies in excess of 90%. While more futuristic than the NEL concept the Duck incorporates many intriguing features, such as a frictionless (and thus maintenance-free) power generating system and a 24-hour energy storage capacity in its gyroscopes.

The other group went to the University of Dundee to see the advanced work on improving the cost-effectiveness of photovoltaic cells which is being carried on there. The Dundee researchers, under the leadership of Professor William Spear, have developed vapour deposition techniques, using a radio frequency discharge, to produce low-cost photovoltaic cells based on amorphous silicon.

The two groups were reunited again in the Netherlands for the final series of visits. The first of these was to the Netherlands Energy Research Foundation (ECN) at Petten where participants were given descriptions of both the Dutch national wind energy research programme (for domestic applications of wind power) and of the work of SWD - the government-sponsored Steering Committee Wind Energy Developing Countries. This was followed by a tour of the wind turbine test sites at Petten, including a visit to the 25 m diameter experimental horizontal axis machine used as a testing facility for the eventual creation of wind parks.

The Eindhoven Wood Stove Project formed the subject of the next item on the study tour. For various logistic reasons it proved impossible to go to Eindhoven to see the work being done on producing detailed technical advice, based on fundamental research into the physics of combustion processes, for transfer to stove-manufacturing industries in developing countries. Participants were able, however, to participate in a seminar on "a systems approach to the dissemination of stove technologies" presented by Dr. Krishna Prasad, the Eindhoven project leader.

The final visits in the study tour were to two projects located in Hook of Holland. The first was a visit to the energy exhibition and information centre run by Stichting Energie Anders (the Alternative Energy Foundation). This is an initiative by private citizens (although now in receipt of funds from public bodies) designed to inform ordinary citizens about the energy options available and the scope for investment in conservation measures and in renewable energy devices (such as solar collectors).

The other (and final) visit was to the University of Delft's Hook of Holland test centre to see their research project on wind turbines using tipvanes at the ends of the rotor blades. Windtunnel tests on scale models indicate that such machines have the potential for a very substantial increase in energy output for a modest increase in capital costs. If full-scale tests prove successful this will greatly improve the economics of wind energy technology and will also make wind power an option to be considered even in those countries where wind speeds are too low to justify the use of existing windmill designs.

The tour provided possibilities for more project-oriented discussion of energy planning and management issues than had been possible during the seminar and workshop sessions. It also gave participants ample opportunity to demonstrate the extent to which their individual and collective skills in evaluating energy projects had been sharpened during the programme.

Final Session in Brussels

This "briefing and evaluation" session consisted of a series of descriptions of European Community institutions, programmes and policies, with special reference to cooperation with developing countries. These briefings were given, for the main part, by senior officials of the General Directorates for Development, for Energy and Information and of the European Investment Bank.

Topics covered included EC energy and development policy, energy projects in developing countries and training activities. Participants also had opportunities to establish contacts with officials from various European Community institutions, including the European Investment Bank, the ACP-Secretariat and the Centre for Industrial Development.

ANNEX II

REPORT ON WORKSHOPS

A. van Buren and G. Foley

Editors' Note

Given the importance which the programme organizers attached to the workshop sessions, it was decided to devote an important effort to ensuring their success. To this end we were extremely fortunate in obtaining the assistance, as workshop co-ordinators and reporters, of Gerald Foley and Ariane van Buren. Their reputation for successfully undertaking similar tasks for other meetings proved wholly justified and, beyond this, their enthusiasm, professionalism and appetite for work played an important part in the success of the programme.

Their report, reproduced here, attempts to synthesize the outcome of the workshops from the participants' point of view. It was endorsed by the participants themselves during the final meeting (in Brussels) of the programme.

H. Neu and D. Bain (eds.), National Energy Planning and Management in Developing Countries, 423–437.

1.0 ORGANISATION

The workshops were designed with two objectives. The first was to obtain feedback from the participants on identifying energy problems which might be resolved by external aid - hence feeding back into the policy-making apparatus of the Directorates for Energy, Development, and Research. The second aim was to obtain feedback which would help define the content and organisation of future Ispra seminars and training courses.

Due to the heterogeneity of the participants, no basis for prior allocation to working groups existed; neither was there any indication that a grouping around specific topics would necessarily be effective or creative. Accordingly, an initial purely random allocation of participants into three working groups of ten to twelve people was proposed. This was accepted by the participants, and was maintained throughout the seminar, despite the fact that a mid-way re-grouping had been planned.

A further decision was to exclude lecturers, Ispra staff and the workshop organisers from working group sessions, except by special invitation. Provision was made to enable available lecturers and staff to be brought into particular sessions for special advice, if required. This decision to exclude even the staff, which runs counter to the didactic instincts of many lecturers was, in fact, seen by the participants as an important element in the cohesion achieved in the working groups.

It should be noted, however, that these arrangements were based on the requirements and objectives of this particular group of participants. Time for workshop meetings was extremely limited, with just two formal sessions in the first week; and a further three in the second week. Moreover, the final session was devoted to a systematic evocation of formal feedback from the participants. Had there been more clearly identifiable nodes of interest, or more opportunity for meeting, particularly in the evenings, a series of subsidiary or possibly radically different groupings would almost certainly have begun to emerge.

1.1 Group Structure and Tasks

Each working group appointed its own chairman, reporter and liaison officer to ensure that communication with other groups was maintained. Thus a basic operating structure common to each was established from the outset.

It was suggested in the opening session with the organisers that the first week might, in parallel with the lecture series, devote itself to clarifying analytic issues, or defining problems and contextual constraints; and that the second week should study some of the institutional and other requirements for their solution.

In addition, the following possible tasks were outlined:

a) An evaluation of the lectures and other events in the seminar.
b) A discussion of the positions to be taken up by representatives from each working group in the round-table discussions and debates scheduled in the programme.
c) Collective views on the major issues or themes of the seminar.
d) The identification of 'solvable problems' -- that is, energy problems to which solutions could be envisaged, granted certain actions were taken.
e) Views on the research, development, and deployment of technology in response to the energy problems of different developing countries.

These tasks were prepared in advance of any certain knowledge of the full range of participants, and were necessarily tentative. Nevertheless, the importance of the workshop results had already been stressed and in the briefing notes for participants it was stated that the report of the seminar would:

> 'be used in the design of subsequent training seminars. It will undoubtedly carry weight also, if it is credible and well done, in helping aid donor agencies to define their priorities and areas of action.'

It was also necessary to make it clear that participants were not being asked to take official positions or represent their governments or organisations. The notes stated:

> 'The objective of the workshops is to exchange views in a friendly and candid fashion. It is fully appreciated that people in official positions may feel constrained in their expression of views. We can, however, guarantee that in the final documentation, and the records of the seminar full confidentiality will be respected.'

Without such a guarantee and the creation of a climate of trust between participants and all concerned with the organisation, discussion becomes unproductively constrained.

1.2 Operation: Week 1

The ways in which groups organised themselves and set about their tasks demonstrated considerable effectiveness. The initial responses were made by arranging brief report-back sessions between the organisers and officers of each group; these were quickly followed by hand-written, or typed minutes of the working groups' own meetings.

These early responses set up a mechanism for systematic communication between the participants and the seminar administration which carried considerably greater authority than merely random personal comments. It enabled alterations to be made to the programme within the limits of the already committed lecture time.

In the first week three extra sessions, or fringe events, were organised in response to requests from the working groups. Efforts were also made to inform incoming lecturers of the participants' requests and particular needs, and ask them to tailor their lectures accordingly. Some were able to react with quite remarkable responsiveness to this last-minute briefing; others found it more difficult, given the nature of the presentation they had prepared.

The working group reports prepared after the meetings in the first week were circulated to all participants as rapidly as possible. By the end of the first week, therefore, a substantial amount of written material was available from the workshops.

It was also clear that the original formulation of tasks was being adapted by the participants to match their interests more closely. A particularly sophisticated method of assessing the presentation, relevance, and completeness of lectures had been devised; and very useful guidelines were being given for the development of future training programmes.

1.3 Interim Report

Midway through the seminar, while the participants were on a weekend programme of visits, an interim report was prepared by the organisers. This was a 10-page informally written document which brought some of the emerging themes of the working group papers and the discussions together so that they could be examined more closely in the remaining workshop sessions.

A plenary session was held at the beginning of the second week together with the organisers. Full consideration was given to making changes in the composition, method of operation, and objectives of the working groups, but none were felt to be necessary. In particular, the desire to maintain the autonomy of the groups and avoid the engagement of course staff in their deliberations was reiterated.

1.4 Operation: Week 2

Three sessions of the working groups were held during the second week. In addition, three informal lunchtime meetings were organised in which participants described the energy position, or aspects of it, in their own countries.

Transparencies used in lectures and the informal presentations were copied and distributed. Two lecturers were together invited to remain for a full plenary working group session of further questioning on the material they had presented.

By now it was clear that had there been opportunities for small ad-hoc groupings in the evenings, or further free time in which people could put events together in an informal manner, a great deal more would have been possible.

On the last day of the seminar the afternoon plenary session was devoted to a final evocation of individual comments from the participants. This enabled everyone who had a desire to make a contribution to do so in a systematic way. This final session played a key role in establishing the feedback from the groups and is worth describing in some detail.

Before the session a presentation was organised incorporating the latest available synthesis of the working group material. The process to be followed was also carefully outlined to the participants. An outline structure of the content of this report was presented; and the remaining stages in its production were described. The areas in which feedback was not needed, or in which feedback was already adequate, were excluded from discussion. This left four major areas in which further guidance from the participants was required. These were:

a) major themes from the discussion and from the entire seminar;
b) solvable problems;
c) priorities in aid programmes;
d) future courses.

No speeches or further reflections on the issues were permitted by the chairman. For each separate item the microphone was passed from one participant to the next. This enabled each participant, if they wished, to prepare a sentence or two on the issue. This was recorded by the organisers.

It is vitally important that this final feedback is carried out systematically and patiently, and at the same time is prevented from turning into a rediscussion of issues. Experience shows that this final crystallisation of thoughts into as succinct a form as possible is extremely fruitful in enabling people to make their maximum contribution.

1.5 Preparation of the Final Report

A draft of this report was distributed to all the participants. Each page was reviewed in a last plenary session and comments on the accuracy, tone, and balance of the material were accepted. Whenever possible, these were given in the form of precise textual additions, deletions, or amendments. While editorial responsibility rests with its authors, the final version endeavours to incorporate as accurately and fully as possible the views expressed by the participants through the workshop reports and in the final plenary session.

Thus, while no individual participant would necessarily assent to everything here, everyone has had the opportunity to review and criticise the report right up to its final draft stage. It thus represents a reasonable consensus of the views of the participants in dialogue with each other, and with the workshop organisers.

Needless to emphasise, it does not commit the official position of any participant, organisation, or government; neither does it prejudice the position of the Ispra JRC, or any aspect of the Commission of the European Communities and its organisations.

2.0 RESULTS OF THE WORKSHOPS

A number of major concerns emerged from the workshop reports, debates, round-tables, and discussions as the Seminar progressed. They were widely shared among the participants. They are drawn together here and examined for their relevance in the formulation of future development assistance programmes with an energy component, whether in the EEC or elsewhere.

This is an area of considerable complexity, and in the time available only the broadest indicators could be obtained. Nevertheless, the message they convey is clear. In some cases, it is an important counter-balance to some of the views conventionally expressed on energy assistance for developing countries.

2.1 Widening the Context of Energy Assistance

The context within which the energy problems of developing countries are considered is frequently too narrow. Their basic and immediate problem is not that of finding a total transition away from oil, but of finding ways to pay for the oil essential for their economic progress.

The perspective on developing country energy problems must be wide enough to accommodate other measures beyond the mere substitution of oil. Factors such as the lack of physical and technical infrastructure, world market conditions, restricted access to capital, and low prices for export commodities are as important as the shortage of energy itself, and are often its root cause. Restricting assistance to the introduction of particular energy sources or energy technologies without taking this wider context into account runs the risk of gravely reducing the effectiveness of such energy programmes, and diverting resources from where they might be more usefully employed.

Energy assistance must be given in a sufficiently wide context to permit the deployment of the full range of technical, economic, and managerial measures required to increase the availability of conventional, as well as new energy sources, and the productivity with which energy is used. Insisting on the substitution of oil by inferior or technically immature technologies, for example in the realm of new and renewable energy sources, may, in fact, have seriously detrimental economic and technical consequences.

Neither should it be automatically assumed that oil consumption must be reduced. There are many instances where additional oil consumption is economically justifiable and, in fact, essential to development. One such application may well be the substitution of kerosene for depleting woodfuel resources.

2.2 Conservation and Demand Management

The productivity of existing uses of energy could generally be improved. Conservation techniques have scarcely been applied in many of the developing countries up to now and this should be remedied in the future. Much could be learned from successful programmes of energy conservation which have been implemented in Europe.

Training in the maintenance and improved running of all kinds of mechanical plant would reduce fuel consumption and the foreign exchange costs of spare parts and maintenance. Assistance should also be given for the creation and implementation of programmes of proven techniques in energy conservation. Demand management methods should be used to optimise the use of all available resources.

2.3 Coal and Nuclear Power

Coal is an important resource in some developing countries and it can frequently be imported much more cheaply than oil. It should be considered seriously as a possible choice for certain uses. Initially these could include cement manufacture and power generation.

While an apparent economic case can be made at present for using coal instead of oil, the technical problems of coal transport and use and the infrastructural requirements need to be taken into account. Due attention should be paid to possible adverse environmental effects and the cost of overcoming them. A judgement must also be made on the likely future movements in coal and oil prices. The question of whether a dependence on imported coal is necessarily an improvement over a dependence on imported oil needs to be weighed carefully in all its aspects.

It is, however, also recognised that coal is playing an increasingly important role in the industrial countries and that its use is likely to increase. It is essential that developing countries are not excluded from these evolving patterns in world coal trade, particularly where long-term joint ventures between exporting and importing countries are involved, and that they seize any opportunities to become actively engaged in them. Countries with their own coal resources should be enabled to prove them and develop them to an appropriate level for their own use and benefit.

For the immediate future the role of nuclear power in most developing countries is likely to be minimal. It should come low in the order of priorities for large-scale investments in energy.

2.4 The Traditional Fuel Sector

In many developing countries there is a large gap between the traditional non-commercial fuel sector and the commercial energy sector. Scant reliable data exist on non-commercial fuel consumption; in some instances the quoted figures may be based on such poor evidence that they are actually misleading and hence impossible to use as a basis for policy. Nevertheless the problems of energy supply in the rural areas are already grave in many countries and are likely to become very much worse in the future unless adequate means of dealing with them can be devised.

In approaching the traditional fuel sector it is necessary to recognise its special position and the fact that it is almost totally outside the scope of existing planning techniques. This is particularly the case when the fuel is not traded commercially. There were few indications that participants, from their own knowledge of their own countries, have any ready answers to the problem of ensuring energy supplies in the rural areas. Nevertheless, purely technical initiatives, such as the promotion of solar cookers, were regarded as almost totally irrelevant.

The gradual commercialisation of fuel consumption was, however, felt to offer some hope. As commercialisation occurs there is an associated encroachment of conventional economic methods of balancing supply and demand upon the traditional sector. The opportunities for fuel substitution, investment to improve the efficiency of consumption, demand management, and optimising resource-use are then more numerous, and techniques can be developed to plan and promote them. But the effects of commercialisation upon the rate of depletion of the resource and upon income distribution are poorly understood. New techniques for dealing with the non-commercial and commercialising traditional fuel sector will have to be devised.

The lack of reliable data is a serious handicap to any systematic approach to the problem of rural energy supply. But the expenses of data collection are high. The planning of surveys must take into account, and strike a balance between the kind of data required, the use to which they will be put, and the effort required to obtain them.

2.5 New and Renewable Sources of Energy

The potential for hydro-power, whether on a small-scale or in large power projects, was seen as extremely great. This is a mature and well-proven technology. For many countries, in locations where it is economically justifiable and otherwise acceptable, hydro-power would have the highest priority for energy investment if finance were available.

There was no desire to downgrade the possible ultimate significance of other new and renewable energy sources, nor to adopt a negative stance towards present efforts to develop them. There were, nevertheless, grave reservations about certain aspects of current approaches to new and renewables in some technical assistance programmes.

It was agreed that much of the optimism about new and renewables in the past had been unwarranted or, at least, premature. A considerable amount of development is still required for most of these technologies. This was particularly borne out in the discussions at the solar house during the tour of the Ispra Joint Research Centre, where a range of flat-plate solar collectors are being compared in use. It was made clear by the Ispra researchers that even in the case of this comparatively well-developed technology there are still many problems of reliability and long-term endurance to be resolved. Until these questions are answered, economic evaluation and, hence, rational planning for the use of solar collectors remain impossible.

It was felt that when funds are being made available for energy, undue emphasis should not be placed on new and renewable energy sources. There was a widely expressed disapproval of forcing these technologies on developing countries or not permitting an alternative use of funds committed in this way. This sentiment particularly applied to solar photovoltaic pumps which many felt are being prematurely promoted.

In general, a limited role was seen for new and renewables. They should not be seen simply as possible substitutes for oil. Efforts should instead be concentrated on bringing them to technical maturity and finding the right applications for them in the various energy sectors in developing countries.

2.6 Computer Modelling as an Aid to Energy Policy-Making

There was considerable discussion of this topic, and it formed a significant part of the technical presentation during the second week of the seminar. There was agreement that computer models can be used as planning tools, but that they must be handled with extreme caution. They cannot be used as a basis for policy, only as an aid in its formulation.

If there is a definite articulation of economic or social goals, models can render assistance in working out the energy demand which could occur if these goals are reached. They may also be helpful in working out a possible method of allocating scarce energy resources between different possible end-uses. But models by themselves cannot be used to predict energy demand or to formulate energy policies.

In the non-commercial energy sector, data are usually extremely unreliable for the present, and completely lacking for the past. The commercial and non-commercial sectors must be dealt with differently and due account must be taken of this when both are integrated into the same computer model.

Most available developing country energy models are still too highly aggregated to be useful as practical planning tools. Smaller models with applicability at a sectoral level, or even for individual installations of industrial or commercial plant, would probably be found more useful.

There was a universal objection, expressed in quite strong terms, to consultants coming for a short stay to a country, collecting data without seriously attempting to verify it, going away to process it abroad, and coming back with an energy policy. Such a process was felt to be thoroughly unacceptable. Energy policy can only be created in collaboration with the people concerned with its implementation, and with those who will bear the consequences of its success or failure.

2.7 Access to Aid Agencies

For many developing countries, particularly the smaller ones, the processes of obtaining funds are slow and cumbersome. Waiting to know the results of an application for project funding prevents other decisions being made and alternative paths of action being followed. It often appears as though the most effective access to information on what projects are fundable, and how, is best obtained through informal networks of contacts; and those who do not have access to such networks are at a consequent disadvantage.

The loss of time in preparing projects and waiting for responses can be very serious to hard-pressed developing country planners. There are many capital investment projects ready for funding but held up because of lack of experience in project preparation.

It would help considerably if clear aid-policy guidelines were prepared and made available. The setting-up of a short EEC course in project preparation would speed up the project cycle and avoid a great deal of wasted time on both sides. The establishment of focal financing, such as an Energy Development Fund, could further help clarify investment opportunities. Above all, it would improve matters in the future if recipient countries were consulted about the guidelines to be applied to them before these became rigidly established.

2.8 Use of Local Professional Expertise

If expert studies are financed in a particular country, it is essential that local people are not merely used to furnish data. Local knowledge, local experience, and local capabilities must be harnessed in the collection and evaluation of data, the analysis of options, and the formulation of policy. Professional consulting expertise within the country should be used wherever possible. Aid should be used principally for equipment and for training local people; it should not be spent virtually entirely on visiting experts.

There are often severe problems in creating an effective interaction between visiting experts and local professionals and administrators. But such problems must be overcome if indigenous capabilities are to be strengthened, and truly applicable policies are to be evolved. Unless this integration between indigenous and foreign expertise can occur, the policies which are devised will have little chance of truly reflecting local priorities and matching local capabilities and aspirations.

Another aspect of the same question is that of making developed country experience relevant to developing countries. Much expertise is available in the developed countries but it is matched to the context, needs, and infrastructure of a developed country. Consequently it is frequently not directly applicable to the needs of a weaker economy. There is a need to adapt the hardware and planning and management techniques to the real conditions in developing countries before using them in training programmes and courses. In this task of adaptation a genuine collaboration between local expertise and visiting professionals is essential if success is to be achieved.

2.9 Concluding Comments

Although circumstances vary greatly between the different countries represented at the Seminar, a substantial measure of agreement was recorded on all the above points. It was not the intention of the participants to convey detailed recommendations on particular aid programmes in energy; that is a matter for negotiation between governments. It was rather to give voice to the views of senior energy planners and administrators from developing countries in the worldwide debate on securing the energy required for survival and development.

ANNEX III

LIST OF PARTICIPANTS

Africa

Ghana: Mr. Joseph AGGREY, Secretary, Ghana Atomic Energy Commission, P.O. Box 80, Accra.

Kenya: Mr. Lincoln BAILEY, Planning Officer, Ministry of Energy, P.O. Box 30582, Nairobi.

Malawi: Mr. Rex MAGOMBO, Senior Economist, Office of the President and Cabinet Economic Planning Division, P.O. Box 30136, Lilongwe 3.

Sierra Leone: Mr. Ogunlade DAVIDSON, Member National Energy Council; Lecturer, Department of Mechanical Engineering, University of Sierra Leone, Freetown.

Somalia: Mr. Omar MOHAMED, Head of Energy Section, National Technical Committee of Energy, P.O.Box 606, Mogadishu.

Mediterranean

Egypt: Mr. RAMADAN ABDEL RAHMAN MOHAMED, Ministry of Electricity and Power, Qattara Hydro and Renewable Energy Projects Authority (Q.P.A.), 18 Hoda Sharawi street, Bab el Look, Cairo.

Mr. Mahmoud EL TANTAWI, Ministry of Electricity and Power, Qattara Hydro and Renewable Energy Projects Authority (Q.P.A.), 18 Hoda Sharawi street, Bab el Look, Cairo.

Syria: Mr. Rifaat IDRIS, General Director, Public Establishment for Electricity, Damascus.

Turkey: Mr. Muammer CETINCELIK, Chairman Alternative Energy Sources Projects, Ministry of Energy and Natural Resources, M.T.A. Institute for Research, P.O. Box 37, Bakanliklar, Ankara.

Yugoslavia: Mr. Zelimir CRNOGORAC, Senior Councillor, Federal Committee for Energy and Industry, Omladinski Brigada n° 1, 11070 Belgrade.

Mr. Petar NOVAK, Fakulteta za Strojnistvo, National Commission for New Sources of Energy, National Commission for Air Pollution, Murnikova 2, 61000 Ljubljana.

South America and Caribbean

Ecuador:	Mr. Carlos QUEVEDO-TERAN, Executive Director, Instituto Nacional de Energia, Italia 438, Casilla 007-C, Quito.
Guyana:	Mr. Abel FELIX, Deputy Chairman Guyana National Energy Authority, 295 Murray street, Georgetown. Ms. June BUDHOORAM, Regional Energy Specialist, Caribbean Community Secretariat, 3rd floor, Bank of Guyana Building, Georgetown.
Suriname:	Mr. André HAAKMAT, General Director, Energy Bureau of Suriname, Ministry of Natural Resources and Energy, Mr. Dr. J.C. de Miranda-street, Paramaribo. Mr. Oswald SPONG, Ministry of Natural Resources and Energy, P.O. Box 9290, Paramaribo.

South Asia

Pakistan:	Mr. Daud BEG, Member (Power), Pakistan Atomic Energy Commission, P.O. Box 1114, Islamabad.
Sri Lanka:	Mr. G.B. Aelred FERNANDO, Director for Energy Planning, Ministry of Power and Energy, P.O. Box 540, Colombo. Mr. B.P. SEPALAGE, Chief Engineer, Energy Unit, Ministry of Power and Energy, Ceylon Electricity Board, P.O. Box 540, Colombo.

People's Republic of China

Mr. Bao Hua WANG, Engineer, State Planning Commission, Bureau of Fuel and Power Planning, Beijing.

Mr. Yun-feng LI, Engineer, Beijing Coalmine Planning and Designing Institute, Ande Road, Beijing.

Mr. Bao He ZHU, Engineer, Ministry of Electrical Power, Beijing.

Ms. Zhung-Bei CHEN, Engineer, Shanxi Research Institute of Science and Technology Information, Taiyuan, Shanxi Province.

Mr. Ji Jun XU, Engineer, China Energy Research Association, State Commission of Science and Technology, Beijing and Power Engineering Department, Kuao-Tong University, Shanghai.

South East Asia

Indonesia: Mr. Suwarto MARTOSUDIRJO, Director, Lembaga Fisika Nasional LIPI (LFN-LIPI), JLN CISITU, KOMPLEKS LIPI (National Research Centre, Indonesian Institute of Sciences (LIPI), Bandung.

Malaysia: Mr. MOHAMED NAWAWI, Ministry of Energy, Telecommunications and Post, Kuala Lumpur.

Philippines: Ms. Lydia TANSININ, Chief, Planning and Programming Division, National Science and Technology Authority (NSTA), Bicutan, Tagig, Metromanila, P.O. Box 3596, Manila.

Singapore: Ms. Aminah KASMOH, Assistant Director, Ministry of Trade and Industry, 39th floor, CPF Building, Robinson Road, Singapore.

Thailand: Mr. Supot TADYU, Chief, Energy Planning Survey Unit, National Energy Administration, Rama 1 Road, Bangkok.
Mr. Piromsakdi LAPAROJKIT, Director, Energy Planning Sector, National Economic and Social Development Board, Krung Ka Road, Bangkok.

Pacific

Fiji: Mr. Peter JOHNSTON, Director, Energy Planning and Development, Ministry of Energy, Government of Fiji, P.O. Box 2256, Govt. Buildings, Suva.

Papua New Guinea: Mr. Noki MAKAP, Office of Environment and Conservation, Central Government Offices, P.O. Wardstrip, Waigani.

Tonga: Mr. Sione TONGILAVA, Permanent Secretary, Ministry of Lands, Surveys and Natural Resources, P.O. Box 5, Nukualofa.

ANNEX IV

LIST OF CONTRIBUTORS

Mr. Paul S. BASILE, International Energy Development Corporation, 18, rue Corbusier, 1208 Genève, Switzerland.

Dr. Oliviero BERNARDINI, Centro Studi Energia (CESEN), via Serra, 16122 Genova, Italy.

Prof. Ramesh BHATIA, Institute for Economic Growth, University of Delhi, 110007 Delhi, India.

M. Jean-Pierre CHARPENTIER, International Atomic Energy Agency, P.O.Box 200, 1400 Vienna, Austria.

Mr. Bernard CHATEAU, Institut Economique et Juridique de l'Energie, Université des Sciences Sociales de Grenoble, B.P. 47x, 38040 Grenoble, France.

Mr. Dominique FINON, Institut Economique et Juridique de l'Energie, Université des Sciences Sociales de Grenoble, B.P. 47x, 38040 Grenoble, France.

Mr. Gerald FOLEY, EARTHSCAN, 10 Percy Street, London WIP ODR, United Kingdom.

Dr. Markus FRITZ, Oklahoma State University, Institute for Energy Analysis, Stillwater, OK 74078, USA.

Prof. Sergio GARRIBBA, IEFE, Università Commerciale L. Bocconi, via R. Sarfatti 25, 20136 Milano, Italy.

Mr. Gerald LEACH, International Institute for Environment and Development, 10 Percy Street, London WIP ODR, United Kingdom.

Dr. Mohan MUNASINGHE, 10, de Fonseka Place, Colombo 5, Sri Lanka.

Dr. Helmut NEU, Joint Research Centre, Ispra Establishment, 21020 Ispra (Va), Italy.

Dr. Jyoti PARIKH, International Institute for Applied Systems Analysis, 2361 Laxenburg, Austria.

Ms. Ariane VAN BUREN, International Institute for Environment and Development, 10 Percy Street, London WIP ODR, United Kingdom.

Mr. Pierre VERNET, IREP Développement, B.P. 47x, 38040 Grenoble, France.

GPSR Compliance

The European Union's (EU) General Product Safety Regulation (GPSR) is a set of rules that requires consumer products to be safe and our obligations to ensure this.

If you have any concerns about our products, you can contact us on ProductSafety@springernature.com

In case Publisher is established outside the EU, the EU authorized representative is:

Springer Nature Customer Service Center GmbH
Europaplatz 3
69115 Heidelberg, Germany

Batch number: 10809439

Printed by Printforce, the Netherlands